Post-Construction Liability
and Insurance

Other Titles From E & FN Spon

Avoiding Claims
A practical guide for the construction industry
M. Coombes Davies

Building Regulations Explained
J. Stephenson

Construction Contracts
Law and management
J. Murdoch and W. Hughes

A Guide to Construction Quality Standards
G. Atkinson

An Introduction to Building Procurement Systems
J. W. E. Masterman

The Management of Quality in Construction
J. L. Ashford

Presentation and Settlement of Contractors' Claims
G. Trickey

Spon's European Construction Costs Handbook
Davis Langdon & Everest

Spon's Asian and Pacific Rim Construction Costs Handbook
Davis Langdon & Seah

For more information on these and other titles please contact:
The Promotion Department, E & FN Spon,
2–6 Boundary Row, London SE1 8HN.
Telephone 071-522 9966

Post-Construction Liability and Insurance

Edited by

J. Knocke

The National Swedish Institute for Building Research

E & FN SPON

An Imprint of Chapman & Hall

London · Glasgow · New York · Tokyo · Melbourne · Madras

Published by E & FN Spon, an imprint of Chapman & Hall,
2–6 Boundary Row, London SE1 8HN

Chapman & Hall, 2–6 Boundary Row, London SE1 8HN, UK

Blackie Academic & Professional, Wester Cleddens Road, Bishopbriggs, Glasgow G64 2NZ, UK

Chapman & Hall Inc., 29 West 35th Street, New York NY10001, USA

Chapman & Hall Japan, Thomson Publishing Japan, Hirakawacho Nemoto Building, 6F, 1-7-11 Hirakawa-cho, Chiyoda-ku, Tokyo 102, Japan

Chapman & Hall Australia, Thomas Nelson Australia, 102 Dodds Street, South Melbourne, Victoria 3205, Australia

Chapman & Hall India, R. Seshadri, 32 Second Main Road, CIT East, Madras 600 035, India

First edition 1993

© 1993 International Council for Building Research, Studies and Documentation/Conseil International du Bâtiment

Typeset in 10/12pt Plantin by Best-set Typesetter Ltd, Hong Kong
Printed in Great Britain by St Edmundsbury Press, Bury St Edmunds, Suffolk

ISBN 0 419 15350 0

A catalogue record for this book is available from the British Library

Library of Congress Cataloging-in-Publication data
Post-construction liability and insurance / edited by J. Knocke (The
   National Swedish Institute for Building Research). — 1st ed.
      p.   cm.
   Includes index.
   ISBN 0–419–15350–0
   1. Construction contracts.   2. Construction industry—Insurance—
Law and legislation.   I. Knocke, Jens.   II. Statens institut för
byggnadsforskning (Sweden)
K891.B8P67   1993
343′.07869—dc20
[342.37869]   92–41908
                                                              CIP

And it is worth consideration, whether the practical expectation of being thus called to account, has not a great deal to do with the internal feeling of being accountable; a feeling, assuredly, which is seldom found existing in any strength in the absence of that practical expectation.

John Stuart Mill (1806–73)

La notion de responsabilité est évidemment une pièce nécessaire de notre édifice moral. Qu'on la retire, il s'écroule.
(The concept of responsibility is clearly a necessary element in our moral structure. Remove it and the structure collapses.)

Lucien Lévy-Bruhl (1857–1939)

Contents

List of members

ACTIVE MEMBERS

BELGIUM

Professeur Maurice-André Flamme*
Avenue A.-J. Slegers 129
1200 Bruxelles

Telephone: +32 2 770 33 52

Monsieur Philippe Flamme
Belgian Building Research
Institute
'Green Corner'
Lozenberg, 7
1932 Zaventem

Avocat honoraire
Conseiller juridique
Telephone: +32 2 716 42 11
Fax: +32 2 725 32 12

Monsieur Ph. Fontaine
Les Assurances Fédérales
Rue de l'Étuve 12
1000 Bruxelles

Conseiller juridique
Telephone: +32 2 509 04 11
Fax: +32 2 509 04 00
Telex: 24921 afvbru

Maître Dr Vera Van Houtte*
Stibbe & Simont
Rue Henri Wafelaerts 47–51
(bte 1)
1060 Bruxelles

Avocat
Telephone: +32 2 533 52 11
Fax: +32 2 533 52 12
Telex: 24519

* Member of Steering Committee.

CANADA

Maître Josette Béliveau
  Régie des Entreprises de
  Construction du Québec
  577 Boulevard Henri-Bourassa est
  Montréal, Québec H2C 1ES

Directrice, Recherche et politiques
Telephone: +1 514 383 1010
Fax: +1 514 873 6750

Maître Daniel Alain Dagenais
  Lavery, de Billy, Avocats
  1 Place Ville Marie
  Montréal
  Québec H3B 4M4

Avocat
Telephone: +1 514 871 1522
Fax: +1 514 871 8977
Cables: barrister montreal

DENMARK

Hr Ratko Cikusa
  The Copenhagen Reinsurance Co
  Ltd
  Amaliegade 39
  1256 Köbenhavn K

MSc, Underwriter
Telephone: +45 33 14 30 63
Fax: +45 33 32 52 70
Telex: 19617 copre dk

Hr Flemming Lethan*
  National Building and Housing
  Agency
  Stormgade 10
  1470 Köbenhavn V

Telephone: +45 33 92 61 00
Fax: +45 33 92 61 64

FRANCE

Professeur Gérard Blachère*
  AUXIRBAT
  3 rue Copernic
  75116 Paris

Telephone: +33 1 47 55 01 16
Fax: +33 1 47 04 24 66

Professeur Jérôme Chapuisat
  Académie d'Amiens
  518, rue Saint-Fuscien
  Amiens

Recteur-Chancelier

Monsieur Jean Desmadryl*
  42, Avenue Junot
  75018 Paris

Expert près la Cour d'appel de Paris
Telephone: +33 1 46 06 36 11

* Member of Steering Committee.

Monsieur J. Heller
  SMAbtp
  114, Avenue Emile Zola
  75739 Paris Cédex 15

Relations extérieures
Telephone: +33 1 40 59 70 00
Fax: +33 1 45 78 87 40
Telex: 250 612

Monsieur Claude Mathurin
  Agence qualité construction
  30 Place de la Madeleine
  75008 Paris

Telephone: +33 1 34 62 50 67
Fax: +33 1 47 42 81 71

Dr Nicole Victor-Belin
  Fédération nationale du bâtiment
  33 Avenue Kléber
  75784 Paris Cedex 16

Telephone: +33 1 40 69 51 00
Fax: +33 1 45 53 58 77
Télex: 611975 fedebat f

ITALY

Prof.ing Nicola Sinopoli
  DAEST
  Ca'Tron, S. Croce 1957
  30125 Venezia

Telephone: +39 51 583 721
              +39 41 796 201
Fax: +39 51 331 312
              +39 41 5240403

JAPAN

Professor Kunio Kawagoe
  Japan Building Equipment
  Safety Center Foundation
  Dai-Ichi Tentoku Building
  1-13-5 Toranomon, Minato-Ku
  Tokyo 105

President
Telephone: +81 3 3591 2421
Fax: +81 3 3591 2656

Mr Hiroyuki Nakamura
  Institute of Technology
  4-17, Etchujima 3-chome, Koto-ku
  Tokyo 135

Telephone: +81 3 3643 4311
Fax: +81 3 3643 7260
Telex: 252 2373

THE NETHERLANDS

Ir Arie van den Beukel
  TNO Building and Construction
  Research
  PO Box 49
  2600 AA Delft

Telephone: +31 15 84 22 77
Fax: +31 15 84 39 90
Telex: 38270

Mr André Dorée
  University Twente
  Faculty of Public Administration
  Department of Civil Engineering
  PO Box 217
  7500 AE Enschede

Telephone: +31 53 89 40 04
Fax: +31 53 35 66 95

Meester Wilma S. M. Lloyd-Schut
  Van Anken Knüppe Damstra
  Mathenesserlaan 233
  3001 HA Rotterdam

LLM, Advokaat
Telephone: +31 10 224 9224
Fax: +31 10 414 3644

  Van Anken Knüppe Damstra
  100 Fetter Lane
  London EC4A 1DD

Telephone: +44 71 242 6006
Fax: +44 71 242 3003

Meester D. E. van Werven
  Instituut voor Bouwrecht
  Postbus 85851
  2508 CN 's-Gravenhage

Telephone: +31 70 324 55 44
Fax: +31 70 328 20 74

NEW ZEALAND

Mr John Hunt
  Building Industry Authority
  PO Box 11846
  Wellington

Executive Director
Telephone: +64 4 471 0794
Fax: +64 4 471 0798

Mr Bill Porteous
  Victoria University of Wellington
  School of Architecture
  PO Box 600
  Wellington

Senior Lecturer
Telephone: +64 4 471 5361
Fax: +64 4 495 5233

Mr Alan K. Purdie
  New Zealand Institute of
  Architects
  PO Box 438
  Wellington

General Secretary
Telephone: +64 4 473 5346
Fax: +64 4 472 0182
Telex: 31525

NORWAY

Hr Jan Einar Barbo
  Institute of Private Law
  University of Oslo
  Karl Johansgt. 47
  0162 Oslo

Assistant Professor
Telephone: +47 2 42 90 10
Fax: +47 2 42 69 57

Hr Knut Ekeberg, MRIF
Hjellnes COWI AS
PO Box 91 – Manglerud
0612 Oslo 6

Managing Director
Telephone: +47 2 57 48 00
Fax: +47 2 19 05 38
Telex: 76420 hjell n

Hr Per Jæger
Byggeindustriens Landsforening
Postboks 112 Blindern
0314 Oslo 3

Director
Telephone: +47 2 96 55 00
Fax: +47 2 46 55 23

Ms Elin Rønningen
Norwegian Council for Building
Standardization
Boks 129
0314 Oslo

Legal Adviser
Telephone: +47 2 96 59 50
Fax: +47 2 60 85 70

Hr Hans Jakob Urbye
Norwegian Council for Building
Standardization
Boks 129
0314 Oslo

Legal Adviser
Telephone: +47 2 96 59 50
Fax: +47 2 60 85 70

PORTUGAL

Professor J. Ferry Borges
LNEC
Avenida do Brasil 101
1799 Lisboa CODEX

Telephone: +351 1 848 21 31
Fax: +351 1 849 76 60
Telex: 16760 lnec p

Dr Artur Ravara
GAPROBA
Av. Eng. o Arantes e Oliveira, 3,
s/l A,B
1900 Lisboa

Chairman
Telephone: +351 1 80 87 97
             +351 1 80 89 21
Fax: +351 1 848 48 44
Telex: 64 548 gapres

SINGAPORE

Dr Quah Lee Kiang, FCIOB
School of Building & Estate
Management
Faculty of Architecture &
Building
10 Kent Ridge Crescent
Singapore 0511

Co-ordinator, CIB W70
Telephone: +65 7723562
Fax: +65 7755502
Telex: 33943 vmid sbem

SPAIN

Sr José Antonio Aparicio
INTEMAC
Monte Esquinza, 30
28010 Madrid

Engineer, Chief, Quality Assurance
Department
Telephone: +34 1 319 7202
Fax: +34 1 410 2580

Sr Alfredo Cámara Manso
Colegio Oficial de Aparejadores y
Arquitectos Técnicos
Colón, 42, 1°
46004 Valencia

Telephone: +34 6 351 97 10
+34 6 351 98 03
Fax: +34 6 394 00 47

Sr Antonio Castrillo Canda
Consejo General de Colegios
Oficiales de Aparejadores y
Arquitectos Técnicos
Paseo de la Castellana, 155-1°
28046 Madrid

Secretary General
Telephone: +34 1 570 15 35
+34 1 570 55 88
+34 1 571 05 16
+34 1 571 05 26
Fax: +34 1 571 28 42

Sr Manuel Olaya Adan
Instituto E. Torroja
P O Box 19002
28080 Madrid

Engineer, Lawyer
Telephone: +34 1 302 04 40
Fax: +34 1 302 07 00

SWEDEN

Hr Håkan Albrecht
HSB
Box 8310
104 20 Stockholm

Head, Real Estate Dept
Telephone: +46 8 785 33 87
Fax: +46 8 785 31 97
Telex: 11724 hsbcoop s

Hr Hans Ekman
The Swedish Association
of Local Authorities
Hornsgatan 15
116 47 Stockholm

Head of Sect. for Civil Law
Telephone: +46 8 772 4437
Fax: +46 8 772 4490

Hr Kåre Eriksson
AB Bostadsgaranti
Box 26029
100 41 Stockholm

LLB
Telephone: +46 8 679 90 20
Direct: +46 8 679 62 85
Fax: +46 8 611 37 83

Hr Christer Högbeck
The Swedish Federation for
Rental Property Owners
Box 1707
111 87 Stockholm

Legal Adviser
Telephone: +46 8 613 57 00
Fax: +46 8 21 06 24

Hr Kjell Jutehammar
AB Bostadsgaranti
Box 26029
100 41 Stockholm

Director, BA (Pol Sc)
Telephone: +46 8 679 90 20
Direct: +46 8 679 58 65
Fax: +46 8 611 37 83

Hr Anders G. Kleberg*
Advokatfirman Anders
G. Kleberg
10 Hill Street
London W1X 7FU

Lawyer
Telephone: +44 71 629 3634
Direct: +44 71 409 1843
Fax: +44 71 493 1106

Hr Jens Knocke*
The National Swedish Institute
for Building Research
Box 785
801 29 Gävle

Co-ordinator, MSc
Telephone: +46 26 14 77 00
Direct: +46 26 14 77 95
Fax: +46 26 11 81 54
Telex: 47396 byggfo s

Hr Thomas Kruuse
The Swedish Construction
Federation
Box 27308
102 54 Stockholm

Legal Adviser
Telephone: +46 8 665 35 00
Fax: +46 8 662 97 00
Telex: 10976 byggen s

Hr Bengt Leffler
Trygg Hansa SPP
Fleminggatan 18
106 26 Stockholm

MSc (Civ. eng.)
Telephone: +46 8 693 18 70
Fax: +46 8 650 12 63

Hr Bo Linander
The Swedish Construction
Federation
Box 27308
102 54 Stockholm

Legal Adviser
Telephone: +46 8 665 35 00
Fax: +46 8 662 97 00
Telex: 10976 byggen s

Hr Gunnar Lindgren
The Swedish Construction
Federation
Box 27308
102 54 Stockholm

Head of Legal Department
Telephone: +46 8 665 35 00
Fax: +46 8 662 97 00
Telex: 10976 byggen s

Hr Åke Lindsö
Byggjuridik Åke Lindsö
Hembyvägen 6
181 48 Lidingö

Lawyer
Telephone: +46 8 767 29 80

* Member of Steering Committee.

Hr Lars W. Lundenmark
  The Swedish Association of
  Consulting Engineers
  Box 22076
  104 22 Stockholm

Lawyer
Telephone: +46 8 654 08 60
Fax:　　　+46 8 650 29 72

Ms Monica Nilsson
  Djurgårdsvägen 49
  136 71 Haninge

LLB
Telephone: +46 8 776 41 49
Fax:　　　+46 8 702 01 15

Hr Kurt-Jörgen Olsson
  The National Board of Public
  Building
  Karlavägen 100
  106 43 Stockholm

Lawyer
Telephone: +46 8 783 10 00
Direct:　　+46 8 783 13 78
Fax:　　　+46 8 783 11 80

Hr Hans Pedersen
  Advokatfirman Hans Pedersen
  Box 3256
  103 65 Stockholm

Lawyer
Telephone: +46 8 723 01 90
Fax:　　　+46 8 796 79 06
Telex:　　15721 pedlaw s

Hr Arne Sembrant
  AB Max Matthiessen
  Insurance Brokers and
  Consultants
  Birger Jarlsgatan 64
  114 20 Stockholm

Director
Telephone: +46 8 613 02 07
Fax:　　　+46 8 613 02 80
　　　　　+46 8 613 02 90
Telex:　　19560 maximum s

Hr Christer Skagerberg
  Trygg Hansa SPP
  Fleminggatan 18
  106 26 Stockholm

Legal Adviser
Telephone: +46 8 693 23 28
Fax:　　　+46 8 654 57 41

Hr Hans Sundström
  AB Bostadsgaranti
  Box 26029
  100 41 Stockholm

Managing Director
Telephone: +46 8 679 90 20
Direct:　　+46 8 679 80 65
Fax:　　　+46 8 611 37 83

Hr Bo Wolwan
  SBC-The Swedish Federation of
  Housing Co-operatives
  Svartbäcksgatan 8
  753 20 Uppsala

Area Manager
Telephone: +46 18 69 66 60
Fax:　　　+46 18 12 34 93

TANZANIA

Dr J. A. K. Msina
  Makali Contractors Limited
  PO Box 20327
  Dar Es Salaam

Chairman, Tanzania Building
  Contractors Association
Telephone: +255 51 32963
           +255 51 63908
Telex: 41378 gemini tz

TURKEY

Dr Heyecan Giritli
  Yapi Uretimi ve Teknolojisi
  Birimi
  Mimarlik Fakultesi
  Teknik Universite
  Taskisla-Taksim 80191
  Istanbul

Associate Professor

UNITED KINGDOM

Mr Michael Ankers
  The Chartered Institute of
  Building
  Englemere,
  Kings Ride,
  Ascot
  Berkshire SL5 8BJ

Director of Professional Affairs
Telephone: +44 344 23355
Fax: +44 344 23467

Mr George Atkinson
  Romeland Cottage
  3 Romeland
  St Albans
  Herts AL3 4EZ

OBE BA(Arch)RIBA
Consultant architect
Telephone: +44 727 51246

Mr Ron Baden Hellard
  Polycon Consultants
  70 Greenwich High Road
  London SE10 8LF

Chief Executive, Dip Arch, FRIBA,
  FBIM, FCIArb, FAQMC
Telephone: +44 81 691 7425
Fax: +44 81 692 9453

Mr David Barclay
  RIBA
  66 Portland Place
  London W1N 4AD

Director, Practice
Telephone: +44 71 580 5533
Fax: +44 71 255 1541
Telex: 24224

Professor Donald Bishop★
8 Pondwicks Close
St Albans
Herts AL1 1DG

Telephone: +44 727 54167

Mr G. P. Cottrell
The Electrical Contractors'
Association
ESCA House
34 Palace Court
London W2 4HY

Head of Commercial Contracts and
 Legal Department
Telephone: +44 71 229 1266
Fax: +44 71 221 7344
Telex: 27929

Mr Ron Denny★
R. M. Denny Associates
Maplescombe Oasts
Farningham
Kent DA4 0JY

MA.DIP TS, MCIT, FIHT, FBIM
Telephone: +44 322 86 43 63

Mr J. L. Derry
Parratt, Reynolds & Young
Sea Containers House
20 Upper Ground
London SE1 9LZ

FRICS
Telephone: +44 71 633 9966
Fax: +44 71 928 2471

Dr Padraic Doyle
National House Building Council
NHBC
58 Portland Place
London W1N 4BU

Public Affairs
Telephone: +44 71 637 1248
Fax: +44 71 580 2874

Mr Peter Fenn
Dept of Building and Engineering
UMIST
PO Box 88
Manchester M60 1QD

Lecturer
Telephone: +44 61 200 4242
Fax: +44 61 200 4252
Telex: 666094

Mr Andrew Fryer★
Willis Faber and Dumas Limited
Casualty Professional Risk
Division
10 Trinity Square
London EC3P 3AX

Executive Director
Telephone: +44 71 488 8111
Direct: +44 71 975 2312
Fax: +44 71 488 8276
Telex: 882141

Professor David Jaggar
School of the Built Environment
Clarence Street
Liverpool L3 5UG

Director, Research and
 Development
Telephone: +44 51 231 3602/3
Fax: +44 51 709 4957

★ Member of Steering Committee.

Mr J. C. Kings
　Park Nelson Solicitors
　1 Bell Yard
　London WC2A 2JP

Telephone: +44 71 404 4191
Fax:　　　+44 71 405 4266
Telex:　　894760 pardev g

Dr Anthony P. Lavers*
　School of Estate Management
　Oxford Brookes University
　Headington
　Oxford OX3 0BP

LLB, MPhil, PhD, ACIArb
Telephone: +44 865 74 11 11
Direct:　　+44 865 81 94 89
Fax:　　　+44 865 81 99 27
Telex:　　83147 VIA

Miss Siobhán Leslie
　48 College Road
　Norwich
　Norfolk NR2 3JL

Solicitor
Telephone: +44 81 299 0960

Mr Peter McDermott
Building Research Establishment
　Garston
　Watford WD2 7JR

Head of Marketing and Information
　Division
Telephone: +44 923 89 40 40
Direct:　　+44 923 66 42 43
Fax:　　　+44 923 66 40 10

Mrs Frances A. Paterson
　CNA International
　5th Floor Fountain House
　130 Fenchurch Street
　London EC3M 5DJ

Solicitor
Telephone: +44 71 283 0044

Mr Neil T. Pepperell
　RIBA Indemnity Research Ltd
　New Loom House
　101 Back Church Lane
　London E1 1LU

Managing Director
Telephone: +44 71 283 2000
Fax:　　　+44 71 481 0513

The Royal Incorporation of
　Architects in Scotland
　15 Rutland Square
　Edinburgh EH1 2BE

Telephone: +44 31 229 7205
　　　　　+44 31 229 7545
Fax:　　　+44 31 228 2188

Mr Derek Tadiello
　Wilde Sapte
　Queensbridge House
　60 Upper Thames Street
　London EC4V 3BD

Civil Engineer, Solicitor
Telephone: +44 71 236 3050
Fax:　　　+44 71 236 9624
Telex:　　887793

---

*Member of Steering Committee.

Mr Roger Wakefield★
  Construction Department
  Nabarro Nathanson
  50 Stratton Street
  London W1X 5FL

Telephone: +44 71 493 9933
Direct: +44 71 491 6876
Fax: +44 71 629 7900
Telex: 8813144 nabaro g

Mr Alan Wightman
  Alan Wightman Associates
  6 Melville Terrace
  Stirling FK8 2ND

AADip FRIAS ARICS
Telephone: +44 786 62614
               +44 786 64616

Mr N. A. Wilson
  Mott Mac Donald
  20/26 Wellesley Road
  Croydon CR9 2UL

CEng FICE FIStructE MSocIS deF
Telephone: +44 81 686 5041
Fax: +44 81 681 5706, 1814
Telex: 917241 mottay g

Miss Rhona Wyles
  IBC Legal Studies and Services
  Ltd
  57-61 Mortimer Street
  London W1N 7TD

Managing Director
Telephone: +44 71 637 4383
Fax: +44 71 631 3214

UNITED STATES

Mr James V. Atkins, CPCU
  Design Professionals Insurance
  Company (DPIC)
  PO Box DPIC, Monterey
  CA 93942

Senior Vice President
Telephone: +1 408 649 5522
Fax: +1 408 649 5192
Telex: 171030

Mr Alastair G. Law
  MMP International Inc
  8260 Greensboro Drive, Suite 501
  McLean
  Virginia 22102

FRICS, Chairman
Telephone: +1 703 821 5980
Fax: +1 703 821 6657

Mr Paul M. Lurie, J. D.
  Schiff Hardin & Waite
  7200 Sears Tower
  Chicago
  Illinois 60606

Partner
Telephone: +1 312 258 5660
Fax: +1 312 258 5600
TWX: 910 221 2463

★ Member of Steering Committee.

Mr John B. Miller
  Massachusetts Institute of
  Technology
  c/o Gadsby & Hannah
  125 Summer Street
  Boston
  Massachusetts 02110

BS, MS, JD, LLM, Of Counsel
Telephone: +1 617 345 7000
Fax: +1 617 345 7050
Telex: 6817512 gadhan bsn

INTERNATIONAL ORGANIZATIONS

European Builders Confederation
  46 Avenue d'Ivry
  BP 353
  75625 Paris Cedex 13
  France

Telephone: +33 1 45 83 03 00
Fax: +33 1 45 82 49 10

Mr C. E. Pollington⋆
  International Council for Building
  Research, Studies and
  Documentation (CIB)
  Postbox 1837
  NL-3000 BV Rotterdam
  The Netherlands

Deputy Secretary General
Telephone: +31 10 411 0240
Fax: +31 10 433 4372
Telex: 22530 bouwc nl

⋆ Member of Steering Committee.

List of contributors

Håkan Albrecht, Head of Real Estate Department, HSB National Federation, Stockholm

Peter Anderson WS, Legal adviser to the RIAS and to RIAS Insurance Services, Edinburgh

Jan Einar Barbo, Department of Private Law, University of Oslo, Oslo

Jenny Baster, Solicitor, Ove Arup Partnership, London

Donald Bishop, CBE, St Albans, England

Gérard Blachère, AUXIRBAT, Paris

Robert Blair FCIOB, former Scottish director of NHBC, Edinburgh

George Burnet WS, Hon FRIAS, former Legal Adviser to the RIAS, Edinburgh

Alfredo Cámara Manso, MUSAAT, Madrid

Daniel Alain Dagenais, Lavery de Billy, Montréal

Jean Desmadryl, Ingénieur, Expert près la Cour d'appel de Paris, Paris

Padraic Doyle, NHBC, Amersham, England

Stephen Dunmore, Department of the Environment, London

Mark C. Fell, McKenna & Co, London

J. Ferry Borges, Laboratório Nacional de Engenharia Civil, Lisbon

Maurice-André Flamme, Brussels University, Brussels

Philippe Flamme, Belgian Building Research Institute

Philippe Fontaine, Les Assurances Fédérales, Brussels

Freehill, Hollingdale and Page, MLC Centre, Sydney

Andrew Fryer, Willis Faber and Dumas Limited, London

Kjell Jutehammar, Director, AB Bostadsgaranti, Stockholm

Kunio Kawagoe, Japan Building Equipment Safety Center Foundation, Tokyo

Anders G. Kleberg, Member of the Swedish Bar, Advokatfirman Anders G. Kleberg, London

Ian Lancaster, Citicorp, Singapore

Anthony Lavers, Fishburn Reader in Law, School of Estate Management, Oxford Brookes University, Oxford

Flemming Lethan, National Building and Housing Agency, Copenhagen

Bo Linander, Legal Adviser, The Swedish Construction Federation, Stockholm

Peter Madge, Peter Madge Risk Consultancies, London

John B. Miller, Massachusetts Institute of Technology and Gadsby & Hannah, Boston, Massachusetts

Hiroyuki Nakamura, Institute of Technology, SHIMIZU Corporation, Tokyo

Eric Purves, NHBC, Edinburgh

Nicola Sinopoli, Istituto universitario di architettura di Venezia

Christer Skagerberg, Legal Adviser, Trygg Hansa SPP, Stockholm

George Thomas, Heath, Hudig, Langeveldt and Aruno Salvi, Citicorp, Singapore

Sebastian Tombs, Deputy Secretary, Royal Incorporation of Architects in Scotland, Edinburgh

A. Torres Mascarenhas, Research Co-ordinator, Laboratório Nacional de Engenharia Civil, Lisbon

Hans Jakob Urbye, Norwegian Council for Building Standardization, Oslo

D. E. van Werven, Instituut voor Bouwrecht, Hague

Alan Wightman FRIAS, Alan Wightman Associates, Stirling

Rhona Wyles, non-practising barrister, Managing Director of IBC Legal Studies and Services Ltd, London

Acknowledgements

HISTORY

In 1983 the National Swedish Institute for Building Research suggested that CIB create a Working Commission on Post-Construction Liability and Insurance. The CIB Programme Committee concluded that there was, indeed, a vast interest for this topic and an inaugural meeting was held in Brussels in March 1985, under the auspices of the Belgian building research institute, the CSTC. After a feasibility study was produced by the Secretariat the CIB Board of Directors approved the project. The Commission was formed later the same year. Also in 1985 it held its first plenary meeting, in Montréal. The Commission has since had similar meetings in London, Copenhagen, Paris, Edinburgh, Lisbon and Las Palmas and London.

AUTHORS

Both the Key Concepts (Part One) and the National Overviews (Part Two) are the result of an international, bilingual co-operation between experts from two dozen industrialized countries under the auspices of CIB, the International Council for Building Research, Studies and Documentation.

The introductory chapters and the text describing Key Concepts are by Jens Knocke who has served as the Commission's Co-ordinator since its inception. He would like to give special thanks to Messrs John Derry, Roger Wakefield, C. E. Pollington and Sebastian Tombs of the UK whose detailed comments have improved the presentation. He would also like to thank Professors Donald Bishop (UK) and Gérard Blachère (France), Maître Daniel Alain Dagenais (Québec), Dr Alicia Montarzino (Sweden), Mr Neil Pepperell (UK) and Professor Nicola Sinopoli (Italy), for inspiration and for drawing his attention to some factual errors in the original manuscript.

Professor Gérard Blachère contributed the chapter on what can be done on the national level to reduce risks.

Maître Daniel Alain Dagenais wrote the chapter on Law and Contract.

The overviews, in some cases the result of a collective undertaking, are written by national experts.

Preface

In the industrialized world buildings are long-lived, expensive and complicated, and they consume land, a limited resource which, as Mark Twain is rumoured to have said, 'isn't made any more'. Buildings are also the object of investment and taxation and they form a very large part of any industrialized nation's wealth.

The production of buildings is a perplexing process, split into more or less distinct stages and, as often as not, involving actors – 'clients' and 'producers' – with dissimilar skills, diverse professions and trades, and differences in motivation. Compared with other artefacts needed for survival and enjoying life, buildings are, indeed, curious creatures.

This is why the production of buildings requires intricate administrative frameworks, partly public – such as construction legislation, technical building regulations and codes – and partly private – mainly based on contractual and commercial practice and agreed standards. The balance between public and private features in the framework varies from one country to the next and, as demonstrated by some of the national overviews in Part Two, even within one and the same country.

Thus a more or less transparent legal system attaches to any process of commissioning and producing buildings.

Jens Knocke

Introduction

Jens Knocke

THE ESCALATING RISK OF BUILDING FAILURES

There are at least three trends which point towards a rise in the occurrence of building failures.

Firstly, as the general standard of living increases, so do the public's expectations as to the quality of goods and services. This undoubtedly has an impact on people's demands on those who produce the goods and services which provide the basis for their living standard. Buildings are, of course, an essential part of that basis. The public – whether as 'consumers' of dwellings or as users of and investors in private or public building projects – feel increasingly entitled to complain about low quality in their 'built environment'. This trend is probably independent of whether the public's complaints about shortcomings are justified or not.[1]

Secondly, and further to this 'subjective' trend, another reason for the increase in risk is that the legal dividing line between justified and unjustified complaints is moving. Whether the line moves smoothly or in jumps, the general trend, reflected in nearly all the overviews, is towards giving the public increased legal rights in their dealings with those who produce their buildings. There may be momentary breaks, regressions and back-lashes, but the overall direction of jurisprudence is clear and obviously responds to the perceived increase in living standards. True, any building endeavour still entails a certain risk for client and investor, but the distribution of this risk between practitioners, on the one hand, and clients–investors on the other, is moving towards the practitioners. Also new insurance arrangements are put on the market, spontaneously or imposed by the legislator, to protect – at a cost – the client and the investor.

Thirdly, there are numerous reasons for foreseeing also an increased risk because of new building technologies. One reason is the unprecedented – and as yet not always mastered – possibilities for climate engineering, acoustic comfort, energy conservation, built-in 'smartness', even artificial

---

[1] Therefore most home warranty schemes have found it necessary to deter owners from filing 'frivolous' claims. The 'claims deterrent' techniques – described in the overviews in Part Two – range from quite substantial deductibles to a fee to be paid by the claimant before filing a claim and returned only if the complaint proves justified. Such disincentives are rarely, if ever, found in the non-residential fields of building.

'intelligence', to save on human intervention, and new techniques for transport and communication within the building and with the outside; clearly, all these advances entail risks of system breakdowns. Another technological cause is the development of new ways to save on total costs (erection plus running). The new technologies often introduce new risks, as do clients' and investors' requests for speedy conclusion of site works to reduce interest on capital outlay. Simultaneously, in many countries urban areas are becoming saturated, so urban clients become more adamant that the space which the planners have allocated them be exploited to the full: this leads to requests for such features as low-pitch roofs, deep cellars and reduced space set aside for the – ever more sophisticated – technical services. Finally, with shrinking space available for urban development, less amenable land is being built upon, which increases geotechnical and climatic risks.

So, despite the nascent quality awareness in the building industry, there is no reason to expect that the issue of how to cope, technically, administratively and legally, with building failures will become less important as time goes by.

'LIABILITY SYSTEMS'

This book is about liability and insurance 'systems': what are the owner's rights vis-à-vis the people who built the building if something in the completed works is not as it should be? Are there any insurance arrangements which will protect the original and subsequent owners? How is the liability for malfunction allocated among the parties who contributed to the production of the building and can they insure themselves against consequent claims? Most of these questions are answered differently in each of the legislative[2] and contractual settings or 'systems'. In fact, the reader will find that all national systems differ, in one respect or the other.

Such systems evolve with the zeitgeist: they reflect attitudes to architectural and technical innovation; they are also a product of each era's distribution of different actors' roles, and they are influenced by the legislator's views on the protection of the national wealth invested in each country's buildings. The systems, in their turn, create a part of the ambience in which the nation's building industry produces its output. They also bring new insurance arrangements to the market, primarily nationally, ultimately across frontiers.

The national systems presented in the overviews – in Part Two of this book – amount to a good sample. There are, of course, still more systems than those described; a complete, worldwide coverage would fill volumes. But from the countries represented some archetypes, as it were, emerge,

---

[2] The UK, for example, is often said to have no specific building legislation; nevertheless, there are some 300 Acts which can be said to refer to construction matters in the UK.

present in all countries, and probably in all times. There is, indeed, and despite all national peculiarities, a common pattern to all known liability and insurance systems. The system imposed by the rulers in Mesopotamia some four thousand years ago could be fitted into the pattern of this book (though the reader will find no overview from King Hammurabi). The same probably holds for systems not yet existing.

How to use this book

Jens Knocke

NATIONAL OVERVIEWS

The book consists of two Parts, with two Appendices.

The national overviews in Part Two describe a series of national systems in force in the country or part of the country as of September 1992. Each overview is written by experts in the particular national system, in either English or French.

Each overview deals with one national system as seen by experts from that country. The overviews therefore are not primarily intended for information on cross-frontier transactions. The reader must also keep in mind that in some of the countries represented the industry is ambivalent in its attachment to the national system.

Authors of the overviews were asked to comply with the same strict arrangement of liability and insurance issues. Appendix A shows this arrangement given *in extenso*: (1) practitioners' liability, (2) property insurance and (3) liability and professional indemnity insurance. The adherence to such a rigid format has two main advantages: first, to ensure that each overview is similarly comprehensive and, second, to facilitate comparison, at a glance as it were, between characteristic features of national systems.

The Commission has English and French as its working languages. Of course flaws necessarily taint any translation. Four of the authors wrote in French and the publisher has translated three of these contributions into English while the overview from Quebec is translated by its author; other authors used English as a lingua franca. But further to considering the linguistic variance the reader must remember that pitfalls lurk in hasty parallels between legal systems from different national cultures, as explained below.

This being said it is worth noting that the arrangement is likely to be applicable to any combination of existing, defunct and yet-to-be-conceived systems. Thus the format in Appendix A offers one of two keys to understanding any liability and insurance system.

VOCABULARY

If the arrangement of the terms is one key, then the other key is a firm grasp of what the words mean. Vocabulary and taxonomy – an orderly classification – are the subject of Part One, which treats the issues independently of any national context. Part One is theoretical, albeit often taking its illustrations from existing national systems.

The reader will find that inevitably certain concepts do not lend themselves to an unequivocal definition. In other words there is not always only one way to read a meaning into a term; in the same way there is not always just one word to portray a given concept.

USING THE KEYS

The two keys provided by this book – the annotated vocabulary and taxonomy in Part One, and the national overviews in Part Two – open the door to an understanding of liability and insurance systems in the post-construction period. Once through that door, two main avenues are open to the reader: one is to gain a better understanding of the system in the reader's own country; the other consists of a series of beacons highlighting pertinent questions which need to be asked about any system, be it a foreign one or a new system developed, for instance, in a particular contractual context.

PART ONE

Key Concepts

CHAPTER ONE

Building

Buildings constitute a class within the larger concept of 'any significant structural improvement to realty' (see *Australia**). There is no distinction between new construction and upgrading, refitting, modernization and conversion of existing buildings.

What is, then, a 'building'? Is a tennis court a 'building'? Is an outdoor swimming pool? A telecommunications mast? Platforms in a railway station? A further question raised in post-construction liability and insurance is which works or parts of works belong to the category 'building works': which, if any, of the internal installations and fittings? Are access roads 'buildings'? Parking bays? External lamp-posts? The delimitation is clearly important in national systems where law or custom imposes a set of rules for the liability and insurance arrangements pertaining to building works, as distinct from other construction and engineering works.

Japan presents a legal definition of 'building' (see *Japan*, 1.1.1) and adds 'Building equipment shall be considered part of a building'; the Japanese overview also lists what is meant by this latter term. In Spain the 'permanent or general fittings from which each . . . portion of [a] building derives its own supply are considered an integral part of each building' (see *Spain*, 1.1.1). In Denmark 'a legal definition is required . . . There is no specific definition but this must be assumed from natural usage and a series of individual decisions' (see *Denmark*, 1.1.1).

Another country where the liability and insurance system requires a legal definition is France with its binding legislation pertaining exclusively to building works. As described in the corresponding overview, French law imposes a penal sanction for non-compliance with the obligation to insure 'building works', but has no such obligation for other construction ventures (though the liability system applies to all kinds of construction works). The rule of thumb used by French insurers[1] to define building works is quoted in 1.1.1 in the French overview, but it is no more than a rule of thumb. Because of the importance of the distinction the authoritative Office for Valuation of Construction Insurance (Bureau central de tarifaction de

---

* References to countries in italic in Part One are to the national overviews in Part Two.
[1] A more detailed version of the rule is quoted by the Société mutuelle d'assurance du bâtiment et des travaux publics in their CAP 2000, *Notice d'information* (Paris, 1990) p. 11 ff.

l'assurance construction) in Paris regularly publishes decisions and court rulings.

The international standard[2] *Building and Civil Engineering – Vocabulary – General Terms* appears to have made no inroad into the legal field in any of the 16 countries described in the national overviews in Part Two. The EC suggests the following definition: 'A building is a construction work whose purpose is to shelter human activities, domestic animals, certain goods and when it is linked with the ground and with various networks' (sic).

Whatever legal definition various countries adopt clearly not every 'significant structural improvement' is a 'building'.

[2] ISO 6707:1.

CHAPTER TWO

Clients and successive owners

CLIENTS

The client is the natural or legal person for whom the building is built. Although the term is clear enough in French, German and the Nordic languages,[1] alternative terms are met in many British and American texts: 'developer', 'owner', sometimes, even for new projects, 'purchaser'. In some English texts the term 'employer' means the client, in others it means the contractor.

Several national systems apply different administrative systems to certain types of clients. It is therefore useful to distinguish between categories of clients.

CLIENTS FOR RESIDENTIAL BUILDINGS

IN GENERAL

In several countries the legislator has been especially concerned with construction law applicable to residential buildings, sometimes to the extent of constraining the basic principle of 'freedom of contract' otherwise taken for granted in these countries, cf. Chapter 11. Secondary dwellings, such as weekend chalets, may or may not be included in the term 'residential' in this context. There is seldom any doubt as to what is meant, though borderline cases do arise, such as schools with accommodation for staff. Of course, there is no country with total 'freedom of contract', not even for non-residential buildings.

---

[1] *Maître d'ouvrage*, *Bauherr* and *byggherre* respectively.

CLIENTS COMMISSIONING A BUILDING FOR THEIR OWN RESIDENTIAL OCCUPATION: 'LAY' CLIENTS

In some countries particular liability and insurance systems exist for clients commissioning a building intended exclusively for the client's own residential occupation.[2] The overviews in Part Two describe systems catering for clients commissioning dwellings for themselves, such as the NHBC scheme (National House Building Council) in the UK and Housing Guarantee Ltd (AB Bostadsgaranti) in Sweden.

'Lay' – or, as they are also sometimes called, 'incompetent' – clients may be subjected to commercial incentives to comply with particular systems intended for them. Thus their obtaining certain loans may be conditional upon submission to a 'scheme', as is the case in the UK and Sweden. The system may also be imposed on such clients and their producers (concerning this latter term, see Chapter 3). Other countries – for instance, Italy and Spain – do not treat the construction of dwellings as a distinct class of construction works, though in some countries, such as Norway and Sweden, particular sets of standard contracts are intended to protect 'lay' clients, cf. Chapter 11. The distinctive features of such contracts, in particular their impact on the liability system and, as the case may be, the insurance arrangements are described in each national overview in Part Two.

In France, because of the practical difficulties of checking compliance with the general insurance obligations (see *France*, 2 and 3), no legal sanction for non-compliance is instituted against clients who commission a dwelling for themselves or their near relatives.[3]

CLIENTS WHO ARE PUBLIC INSTITUTIONS

In France certain types of public bodies are exempted from the insurance obligation normally imposed on all parties to a building project. This differentiation does not occur in the other countries discussed in the book, though many public authorities have developed their own contractual arrangements, common to the whole public sector or parts of it.

---

[2] Such clients are, for instance in Scandinavia, often called 'consumers' while in other countries a 'consumer' may be anyone who commissions a building, whether for a residential or a non-residential project. In yet other countries, 'consumer' simply means the public at large which 'consumes' any part of the built environment. Further, in the UK the party buying a completed dwelling from a developer is also sometimes called a 'consumer'. In international contexts the word 'consumer' therefore should be used with care and its meaning made explicit.

[3] Cf. the French *Code des assurances*, art. L.243–2, para. 2. The insurance obligation nonetheless applies also to such clients. Furthermore, transfer of ownership of the building is commercially more difficult if the normal French post-construction insurance cover is lacking or incomplete, so there is, in practice, a strong commercial incentive also for this type of client to comply.

CLIENTS USING PUBLIC FUNDS

In some countries particular liability and insurance systems are applied to building projects where the client is totally or partially financed from public funds. This is not the case in Belgium, Italy or the UK, for instance, while in Denmark and Portugal the use of public funding triggers the application of a distinct liability system, with or without an insurance component. Also, a 'subvention criterion' may be linked with a destination criterion; thus in Sweden a special liability system applies exclusively to the construction of bungalows with an input of public funds.

CLIENTS BUILDING LARGE OR COMPLEX BUILDINGS OR BUILDINGS INTENDED FOR USE BY THE PUBLIC

In France, particular rules imposing qualified technical supervision and inspection (*contrôle technique*) of the design or the works or both apply to very large or very tall buildings and to buildings open to the public. The use of qualified technical supervision is, of course, reflected in the insurance conditions applied to the project and may also have an impact upon the allocation of liability.

SUCCESSIVE OWNER AND SUCCESSIVE OWNERS

In this book the terms 'successive owner' and 'successive owners' mean any party who has bought a building from a 'client', as defined above, or from a 'successive owner' as defined here. Sections 1.6.1, 2.6.1 and 3.6.1 in the national overviews deal with the relationship between a successive owner and the (client's) producers, as opposed to the parties to the sale.[4]

---

[4] Legal liability pertaining to transfer of property is outside post-construction systems since the actual construction of the building is not the crucial issue in these transactions. Nevertheless a few national overviews (e.g. *Quebec*) outline the liability of the vendor of an existing – new or old – building vis-à-vis the buyer, but liability and insurance arrangements between vendor and buyer pertaining to such transactions are in general outside the book's scope. However, if the sale creates a relationship in liability law between the buyer and any of the original client's producers (a term explained below), then the overviews in Part Two will mention the characteristics of that relationship.

CHAPTER THREE

Producer

The term 'builder' is often used loosely. In some cases it is taken to mean only the building contractor, in others it also covers electricians and installers of heating, ventilation and air-conditioning equipment and other trades. Similar terms in other languages may or may not include architects, consulting engineers and even suppliers. Another vague term is 'engineer', sometimes synonymous with architect, sometimes – as in the FIDIC series of contracts – meaning the client's representative on site. The muddle is by no means limited to the English language; only the French have a clear-cut definition of *le constructeur*, namely in the modern version of the Civil Code.

In this book, 'producer' means any party who is liable to the client for the quality of the building as delivered. Although this definition is, of course, akin to a tautology it is not possible to advance a universally acceptable alternative: there simply is no definition of 'producer' applicable in all legal contexts.

Producers can be divided into two main categories: those who have entered into a contractual relationship with the client, and those who have not but nevertheless owe the client a duty and are therefore 'producers'.

PARTIES WITH A CONTRACT WITH THE CLIENT

In most – but not all – countries any party commissioned by the client is, of course, liable to him during a specified period after delivery of the completed works. An explicit example of this general principle is found in the French Civil Code (see *France*). It states that any architect, contractor, consultant or other person contractually linked to the client by an agreement to execute work is 'deemed to be a producer'.

Technical inspectors and other supervisors and inspectors commissioned by the client are therefore – except in Italy, but see below – 'producers'. The inspector nominated by the client for issuing the Certificate of Final Completion is a producer, whereas an inspector employed by a third party, for instance by the insurer, is not unless, of course, he can be held liable to the client 'in tort'.[1]

---

[1] Concerning the term 'tort' see *England, Wales and Northern Ireland*: in Scotland the corresponding term is 'delict'.

One deviation from the principle of 'payment implies liability' is met in Italy: here the Civil Code mentions only the contractor's liability to the client. This is interpreted so that errors and omissions committed by other 'producers' – such as the client's architect – are presumed to be discovered and corrected by the contractor. This therefore annuls other parties' civil liability for mistakes vis-à-vis the client. It is worth noting, though, that this principle is currently being revised in the Emilia–Romagna region. Here a Regional Law (no. 33, 1990) compels certain[2] producers to sign technical documents pertaining to the building project. Their signature is presumed to imply a declaration that the building conforms to municipal building and planning codes as well as to pertinent performance requirements. If, at the final inspection, an official inspector finds that the declaration is not truthful, then these producers may be subjected to penal sanctions for deceitful declaration to a public authority.

Singapore uses a completely different system of distribution of liability with a particular method of identifying producers: the Singapore Building Control Act 1989 has introduced the notion of the 'qualified person', the 'registered accredited checker' or the 'site supervisor' (see *Singapore*, 1.1.0). A client seeking compensation for damage caused by an error or omission committed, for instance, by his contractor, may, of course, seek to obtain this from the contractor, but he is entitled to sue the 'qualified person' instead.

PARTIES WITH NO CONTRACT WITH THE CLIENT

The second category of producers are parties who have no contract with the client but who nevertheless owe him a duty which the client may call upon after delivery in case of 'damage' (qv). Such parties are, by the above definition, producers. There are several such producers.

PUBLIC AUTHORITIES

In some countries the local government or municipal authority is under an obligation to check some or all building projects within its jurisdiction. Even if the authority is to check only the project's conformity with building regulations the client may hold it liable for negligence and, a fortiori, if the authority is supervising works on site or is otherwise actively engaged in the project, for instance by providing information on local soil conditions.

The law governing public bodies' liability may limit the amount which the client can claim (see, e.g., the two UK overviews), and in some countries the public building authority's liability is explicitly nil (see e.g.

---

[2] The producers mentioned in the Regional Law are the principal designer (the 'author' of the project in Italian parlance, cf. *Italy*), the site manager, the contractor, the technical supervisor and the inspector in charge of final inspection.

Singapore, 1.1.0, concerning the 'qualified person's' liability vis-à-vis the client).

TESTING LABORATORIES: NORMATIVE BODIES

Other bodies which may be liable to the client, and hence should be considered as producers, are testing laboratories, public and private. For instance, a testing and approval institution issuing 'agréments'[3] or similar documents may commit 'positive information errors' (see Chapter 8). They may in this case be liable to the client as is, for instance, under certain conditions the situation in the UK. If so, they are therefore 'producers'. In other countries the civil liability of such institutions could be nil, in which case, then, they are not 'producers'. Also, the liability of bodies issuing norms and standards varies from one country to the next and, in some countries, their possible liability has never been established or clarified by judicial decisions.

SUB-CONSULTANTS, SUB-CONTRACTORS

Sub-consultants and sub-contractors – parties who are not contractually bound to the client but who have entered into a contract with the client's (main) consultant or (main) contractor – are not normally 'producers'. However, there are exceptions to this rule in countries where the grounds for pursuing claims 'in tort' (see e.g. *England, Wales and Northern Ireland*) are broadly defined as used to be the case, for instance, in the UK.

In such legal systems these parties must be regarded as producers since their liability vis-à-vis the client may be called upon even though there is no contract between client and the parties in question. But normally in, say, Belgium, France or Sweden the term 'producer' would not be applied to a sub-contractor or a sub-consultant. If, as is sometimes the case in the UK, the 'sub-consultant' or the 'sub-contractor' enters into a direct contractual agreement with the client, then the prefix 'sub' is, strictly speaking, no longer applicable.

SUPPLIERS

Although in some countries (see e.g. *England, Wales and Northern Ireland*, 1.1.0) the contractor's suppliers may be held liable to the client 'in tort', a supplier's contractual liability is to the contractor with whom he concluded a contract of sale.

---

[3] An agrément is an expert assessment, published in an appropriate form, of the aptitude for use of a material or a construction technique not yet covered by a standard or a building regulation. See also Chapter 4.

The only exception to this rule is France, where suppliers of a certain category of components are in a special legal situation vis-à-vis the client: such suppliers – but none others (see *France*) – are under a presumption of *in solidum* liability with the contractor and are therefore 'producers'. The components in question are therefore called 'EPERS' (*élément pouvant entraîner la responsabilité solidaire*), that is, 'component which may entail liability *in solidum* [with the contractor]'; prefabricated timber elements, for example, are considered to be EPERS. The French Civil Code states expressly the liability of suppliers of EPERS.

SUMMARY

In a number of instances the client may invoke the post-construction liability of participants whom he has neither appointed, engaged nor paid. But, on the other hand, the very fact that the client engages a party in a contractual relationship does not, in all legal systems, generate that party's legal liability towards the client during the post-construction period.

In other words, there is no universal definition of who is a 'producer', i.e. who is liable to the client for damage becoming manifest during the post-construction period.

The definitions used on a national basis are covered in the national overviews in Part Two.

CHAPTER FOUR

Building codes

The construction of buildings is governed not only by law but also by a complex framework of normative technical documents.

The legal concept of 'damage' (qv) is to a very large extent linked to these normative documents.

It is useful to divide normative documents according to the areas they mainly deal with:

- norms pertaining to zoning or physical planning: land use such as reservation of areas for distinct purposes, spaces to be left intact and the like;
- norms pertaining to town planning or urbanisation: distance between buildings, distribution of private and public areas, rate of services in different sectors, overall aesthetic aspects, permissible height above ground level, admissible depth of constructions; and
- technical regulations, pertaining to the building within its own site.

Of these three groups only the technical building regulations – not to be confused with possible administrative and financial rules and regulations – need special consideration in post-construction liability matters.

TECHNICAL BUILDING CODES

Technical regulations or 'norms' fall into two main groups: norms with overall validity and norms pertaining to buildings with a specific purpose.

Norms with overall validity deal with more or less basic requirements and are normally in the public domain.

Another set of norms with general validity are technical product standards and codes of practice: in France, for instance, the famous DTUs (*documents techniques unifiés*). In the words of a French expert these 'codify traditional technical know-how'.

This set of general norms, valid for all building projects, is crucial to the definition of 'defect' and 'damage' and hence also to a good many insurance matters.

The other set of norms it is useful to distinguish in liability and insurance issues is a series of 'purpose-specific' standards. These may pertain to particular kinds of buildings: schools, post and telecommunications centres, railway stations, civil aviation terminals, hospitals and military installations. Some countries also have 'purpose-specific' norms valid for buildings with premises to which the public has access, such as shops, hotels, theatres, churches etc.

Each nations's official building code lays down many of these norms. But official building regulations do not contain all the normative documents used in building. The agrément[1] is an example of a normative document outside the code. An agrément may or may not enjoy an official hall-mark, depending on the national frame of reference. Even without official recognition the agrément's role in post-construction liability matters may be decisive, both for analysing 'defects' (qv) and in the context of insurance.

A convenient collective term for public and private technical regulations and standards is 'technical codes'. Technical codes are, in any system, presumed to be known to and respected by producers.[2]

QUALITY REQUIREMENTS IN CODES

Putting up a building which does not conform to stipulations laid down in building codes is, with negligible exceptions, unlawful. Clients are therefore entitled to expect that any building satisfies at least the codes' requirements; and they need not instruct their producers to this end. In matters of post-construction liability this renders a particular importance to public codes: they define the minimum level of technical quality which any client, even without referring explicitly to a public building regulation, may count upon. Such regulations are, as it were, the outcome of ongoing political consensus decisions concerning the fundamental quality of buildings. The role of political decisions concerning the standard which the built environment should provide is perhaps most obvious in regulations concerning comfort – such as indoor climate, sound insulation and the like – but is felt also in the more basic requirements pertaining to the building's structural safety, its behaviour in case of fire etc. Public building regulations can be said to express a country's zeitgeist, though they may vary between

---

[1] Each agrément describes a particular building material or a technique not yet covered by a standard or a regulation and testifies to its aptitude, thus granting the material or the technique an 'approval'. When a standard covering the material or technique is established the agréments in question are annulled since specific approval is no longer needed. Agréments are issued by boards of experts and conform to a standardized format. The idea of using such documents pending formal standardization was established in France after the Second World War in order to facilitate a speedy reconstruction of the built environment. The idea of delivering agréments for new building materials and techniques has gained acceptance in several European countries, now working together under the auspices of the UEAtc (Union européenne de l'agrément technique dans la construction). In the UK the BBA (British Board of Agrément) is a member of the UEAtc. See also Chapter 12.

[2] A clear example is provided by France (see *France*, 3.5.6) concerning the consequences of non-respect of codes and what is conventionally called the principle of good workmanship.

locations in the same country; for instance, local regulations on fire escape routes may reflect the fire-fighting equipment in the location.

The EC Directive on building products provides a clear example of the political dimension of national building regulations. Member states may regulate what the EC calls the six 'essential requirements' (see below). But one proviso in the Directive's preamble declares that member states are free to choose the level of protection they want 'in order to take account of . . . different conditions prevailing in the member states'. For example, the requirement concerning safety in case of fire proclaims that structural elements must maintain their load-bearing capacity in case of fire during a specific period of time, but it is left to each state to specify the period, which is, of course, a decisive qualitative dimension; the same idea applies to 'protection against noise', which requires 'satisfactory conditions', again leaving the quantification to the appropriate national body. So non-compliance with a code in one country does not mean that there is damage according to another country's code.

EXPRESSING REQUIREMENTS

Codes express requirements in either of two ways, conventionally known as the 'performance' and the 'technique-specifying' approaches. The latter of these lays down approved building and installation technologies such as quality of materials, sizes, dimensions, standard of workmanship and the like. The former approach leaves technologies aside and concentrates on specifying what the building should actually 'do', not how its producers 'make it do' what it should.

An illustration of the principal difference is the following. Given that the public wants buildings to be free of rats, authors of the public code have two alternatives for prescribing this. They may – the technique-specifying approach – state the maximum size of openings liable to be used as passages by rats and if, for some reason, this cannot be assured, they may prescribe the placing and maintenance of rat traps. This approach of course precludes the introduction of 'rat avoidance' technologies. If the authors opt for the performance approach, then they will state the requirement as 'Premises should be so built that rats will not sojourn in them', or words to that effect. The producer now has a much larger choice of technologies for complying with this requirement: this may include limiting the size of openings and providing rat traps, but also, say, installing sonic pest repellent devices.

More concrete examples are the following:

• Water-pipes: dimensions, interior smoothness, shapes of bends and fittings, methods of joining and fixing etc. vs. the quantity of water to be transported safely within stated levels of pressure and temperature, and with a maximum level of noise, and perhaps other hydraulic parameters.

- Thermal comfort: dimensions, configuration of insulation, ventilating and heating elements etc. vs. a certain overall thermal comfort requirement expressed in quantitative terms (see below).

So in contrast to solving the problem of expressing requirements through an enumeration of approved technologies stands the performance code, historically the younger of the two. The performance approach is also sometimes called the 'functional' approach. It is, at present, gaining the upper hand in competition with its rival, the code stating approved building and installation techniques, quality of materials, sizes, dimensions, standard of workmanship and the like.

The EC has listed the six 'essential requirements' which all construction works must satisfy. They are:

- mechanical resistance and safety;
- safety in case of fire;
- hygiene, health and the environment;
- safety in use;
- protection against noise;
- energy economy and heat retention.

The six essential requirements are couched in performance – or 'functional' – terms. The terminology adopted by the EC for this is the following:

'A performance expresses in a quantitative way (value, grade . . .) the behaviour of a work, part of a work or product as regards an action [see below about this important term] to which it is subjected (for example, the thermal resistance of a wall) or which it generates (for example, the acoustic power of a boiler) under the intended service conditions (for works or parts of works) or intended use conditions (for products).'

The term 'action' is defined as follows:

'Actions which may concern the compliance of the works with the "Essential Requirements" are brought about by agents acting on the works or parts of the works. Such agents include mechanical, chemical, biological, thermal and electromagnetic agents.'

A performance-approach code lays down solely which functions the building – 'the works' – or its constituent elements – 'parts of the works' – must perform in order to satisfy the code's requirements and, as importantly, under which conditions – the 'agents acting on the work' – this performance must be assured. This considerably increases the number of technical options open to producers.

The approach obviously also has the clear and important advantage of clarifying not only the performance the client is entitled to expect from 'the works', but also under which circumstances the works should 'perform'. Incidentally, the spelling out of these circumstances contributes to solving the tricky liability problem of force majeure or 'Act of God' mentioned in Chapter 10.

In practice, the author of a code will sometimes find it convenient to use both approaches in the same text. In English literature on the theory of codes this approach is known as 'deemed-to-satisfy' codified advice. It consists both of listing the quality requirements and giving examples of how these requirements are 'deemed to be' satisfied by certain listed building techniques.

Today many national codes use both approaches simultaneously. Both have their merits but, as mentioned, the present-day trend is clearly towards the performance approach. This is so, as already mentioned, because codes prescribing technical specifications are considered to hamper innovation and the free circulation of products and services. But whatever the approach, the raison d'être of any technical code is to ensure that users of buildings enjoy a certain minimum quality. It is therefore – theoretically – always feasible to decide whether a given example of damage constitutes a valid ground for a claim on the basis of the code:

- if the damage[3] reduces the quality to a level below what the code – explicitly or implicitly – prescribes, then it is justified to claim that the building is of sub-standard quality.

Clearly, the modern performance-approach facilitates deciding whether a given occurrence of malfunction constitutes damage: such a code will lay down the minimum performance, i.e. both a functional criterion and the agents (see above) which are considered normal and which the producer therefore should have considered.

If, on the other hand, the regulation is 'technique-specifying', then the functional criterion may sometimes need to be deduced from the specification, and that is not always an easy task. Questionable sound insulation provides an example.

Assume that in an office building the occupant of one office can hear and understand a conversation conducted in a normal tone of voice in another office. Suppose that the client considers this to be an unsatisfactory state of affairs. Now, assume that the client has omitted to specify any particular 'client's requirements' concerning sound insulation or privacy for his office building, then the building code provides the sole yardstick. If the code is 'technique-specifying' it will indicate building materials approved for partitions between offices, the way each of these materials should be stored, handled and built in, how pipes and ducts from one office to the next should be placed and laid and how the perforation of the partition should be treated once the pipes are in place, and possibly other technical details. But it so happens that sound insulation between rooms depends on many

---

[3] Whether damage, thus proven, entails liability is a legal issue which depends on the 'system' in force for that particular building.

factors, some of which are linked to the partition itself, others to openings, including pipes, and others again to sound travelling via convoluted transmitters ('flanking effect'); also workmanship is often decisive for the sound insulation actually obtained. All this may make it difficult for the client to pinpoint the weak link – in post-construction terminology 'the defect' – in other words, to prove exactly what is wrong in relation to the technique-specifying code. Filing a claim is then difficult.

If, on the other hand, the regulation is 'functional', that is, it expresses an acoustic performance requirement for offices, then it will prescribe maximum noise levels in one office when sounds of certain amplitudes and frequencies are produced in any other office. This means that if the client's measurement of the noise level in the adjoining room shows that the prescribed level is exceeded, then the building code clearly has not been respected: there is a clear case of damage. In many systems (see the overviews in Part Two for details) the client need not pinpoint the underlying defect: the mere proof of damage – in this case provided by measurement of an inadmissible sound level – is sufficient to prove the presence of damage, whatever the defect causing it may be.

If the code is 'technique-specifying' and the client is unable to find the defect underlying the damage he wants rectified he may still have a recourse, namely to produce evidence of the performance the authors of the technique-specifying code had in mind when they codified certain technologies. Having done this the client can then proceed to proving that the implicitly required, and now established, level of performance, has not been achieved.

As mentioned, at present no country adheres exclusively to the performance approach in its public codes. However, the United Nations' *ECE Compendium of Model Provisions* adopts this approach throughout and offers examples of recent developments in the field, and so will the codes adopted by the EC. Once codes are based on the performance approach the issue of defining post-construction damage will be greatly facilitated.

Chapter 10 provides further discussion of public codes and their impact on legal liability.

Normative documents are largely based on building research which thus contributes to an important aspect of post-construction liability and insurance, including risk reduction. As Gérard Blachère has put it in an unpublished contribution: 'it is possible to say that through normative documents and regulations, research contributes to the reduction and prevention of malfunctioning'.

CHAPTER FIVE

Delivery

DEFINITION OF THE COMMENCEMENT OF THE POST-CONSTRUCTION PERIOD

The transfer of the premises from the contractor to the client is sometimes called 'hand-over' by which the client 'accepts' or 'receives' the works. In this book the term 'delivery' is used. In building contract terminology in the UK the term for delivery is 'Practical Completion'. The architect may issue a 'Certificate of Practical Completion' at this moment: 'When in the opinion of the Architect Practical Completion of the works is achieved, he shall forthwith issue a certificate to that effect', as stated in the standard JCT 80 Contract. In the US the term is 'Certificate of Substantial Completion'. Other national terms are given in Part Two.

Delivery marks the end of the construction period and therefore the commencement of the post-construction period: from this moment the client can exercise the rights and must respect the obligations which, in all countries, are attached to ownership of a building.

DELIVERY

The delivery instrument is used by the parties to settle some or all of the following[1] issues:

- patent differences between the works as executed and contractually agreed specifications;
- outstanding claims and counterclaims, if the parties have not already settled them. Failing that, they must at least clarify their positions. In many countries the time limit for the parties presenting claims against one another starts at this point. The limit may be contractually agreed or legally fixed; for instance, in Sweden it is normally contractually agreed that the client has 30 days from delivery in which to claim penalties for late completion;
- the schedule for the client's final payment to the contractor. So, for

---

[1] The list builds mainly on CREPAUC *Les marchés privés de travaux* (Editions du Moniteur, Paris, 1990) and Åke Lindsö *Entreprenadjuridik* (AB Svensk Byggtjänst, Stockholm, 1991).

instance, a French standard states that 'within 120 days from the date of delivery the contractor provides the client with the detailed final balance (*le mémoire définitif*) of the sums that [the contractor] considers to be outstanding' (unauthorized translation). In Sweden: 'Not later than 8 months after the termination of the Construction Time the Contractor shall submit the Employer an invoice . . .' and 'A later claim does not give the Contractor the right to compensation unless . . .' (AB 72, official translation).

Also delivery:

- defines the starting-point for many[2] guarantees and warranties;
- terminates the contractor's access to the site, and therefore to the premises. This means that he may no longer execute work without the client's consent;
- terminates the client's licence to order the contractor to complete work;
- starts the schedule for the writing down of the contractor's bond, if any.

At delivery:

- the client takes over from the contractor the responsibility of protecting the premises against harm (such as fire, vandalism, burglary). For example, French law states, *a contrario*, that 'loss resulting from destruction of or harm to the works caused by an Act of God [or similar] . . . is to be met by the contractor if the destruction or harm occurs prior to delivery' (unauthorised translation);
- in some countries the party who has requested the building permit (or 'planning permission') must report the closing-down of the site to the authority which gave the permission.

There are a few important differences between nations as to the legal impact deriving from the instrument of delivery, be it formalised or not. However, the general pattern is the same: the contractor, when he deems that he has completed the building project, invites the client to take over the premises. But, as noted by John Derry:[3]

'there can be pressure from . . . [client] or contractor for . . . premature [delivery], where the . . . [client] is keen to occupy the building, or where the contractor is anxious to demonstrate completion unreasonably early. There can be a reluctance by the . . . [client] to take possession – for example, housing units for a local authority just before the Christmas holiday, when tenants may be unwilling to move in, when frost may damage water systems and when there is a risk of squatting (illegal occupation by homeless people).'

---

[2] In France, however, the post-construction warranty concerning sound insulation starts running when the client occupies the premises.
[3] Private communication from John Derry, FRICS, London.

Within a stipulated time from the contractor's invitation the parties proceed to an inspection of the works on site. In some systems the client's consultant may assist or represent the client, while in others an independent inspector, nominated by the client, chairs the whole proceeding. In France it is often the architect who signs the certificate, but this normally binds the client only if the client has issued an express mandate to this effect.[4] In Sweden the standard document 'General Conditions of Contract' admits several variants: 'Inspection shall be carried out by a competent person appointed by the [client]'; or 'by a person appointed jointly by [the parties]'; or 'by a committee . . . [consisting] of three persons, of which each of the parties shall appoint one, and the two thus appointed shall jointly appoint a third, who shall act as chairman' (official translation).

In many systems, inspectors are liable to the client for negligence, for example an omission, in their inspection. If so, inspectors are 'producers' (cf. Chapter 3), often with high demands on their diligence.[5]

Delivery may be informal, namely through 'tacit acceptance'. However, the attitude to tacit delivery may not be universal.[6] There may be systems, though no such is described in the national overviews, where tacit delivery is not a proper and recognized form.

The apparently simple form of tacit acceptance is said to accout for 80 per cent of all deliveries in some countries. It consists of the client's occupation of the premises with little or no formality. Tacit acceptance may, however, jeopardise the client's right to claim reparation after this moment, i.e. 'in post-construction', if he has neglected some basic procedures.

At the other end of the scale delivery may be a heavily formalised procedure with important parts of the form laid down in law (France) or standard contracts (Sweden).

In formalised procedures a certificate is signed either by some or by all the parties to the contract(s), or by the inspector alone. In France, if the client has engaged a technical inspector – as required by law for some projects (see Chapter 1) – then this inspector must supply the client with a report prior to the implementation of the delivery procedure. Section 1.2.3 in the national overviews gives further details.

There are, of course, two possible outcomes of a delivery proceeding: either the client finds that the works are completely satisfactory or that they are not. This gives rise to four different situations:

---

[4] Slightly different rules apply if the client is a public body, see Jérôme Chapuisat *Droit des travaux* (Amiens, 1990), vol 2.

[5] Thus the commentary to the 'General Conditions of Contract' in Sweden states that the inspector must have made himself familiar not only with all contract documents, including those added to the contract after signature of the initial agreement, but also with what the parties may have agreed to but not confirmed in writing.

[6] See, for instance, the discussion in Jérôme Chapuisat *Droit des travaux* (Amiens, 1990), vol 2, about the rules in France when the client is a public body.

(a) If the client is perfectly satisfied with the works he will take over the premises, i.e. he accepts the delivery. This he does also – though perhaps unwittingly – by occupying the premises without reservation, that is, by the tacit acceptance mentioned above. The formalised procedure requires a certificate signed by the parties. In both cases delivery has taken place and the post-construction period commences; any post-construction insurance arrangements agreed to (or, as in some countries, imposed) start running.

The three formulae used in situations where the client is not completely satisfied with the works are the following:

(b) The client may find the works so unsatisfactory that he refuses to take over the premises. In this case there is no delivery.[7]

(c) The client accepts the works as they are, that is with their – qualitative or quantitative – imperfections, and the parties agree to a commensurate reduction in the monies paid (or to be paid) to the contractor. This arrangement is codified, for instance, in the Singapore Contract SIA 87. Strictly speaking, the building received is not the one agreed upon (see below) and the original contract is, in this way, modified. Nevertheless there is still a delivery – albeit for a different product – and the normal post-construction guarantees from this moment apply to the building which the client actually receives. The fact that this is not the building agreed upon may have an impact on the post-construction insurance arrangements established prior to delivery, i.e. for a different building. It may therefore be necessary to amend the insurance policies accordingly.

If the malfunction which the contractor has repaired or paid for is due to another party, then the contractor will normally attempt to recover his loss from that party. Possible exceptions to this rule may be found in Italy and Singapore (see overviews).

The formula may be used, for instance, when the contractor is late for one reason or another – including reasons beyond his control – and the client is pressed for time.

(d) Frequently the client makes his acceptance of the works conditional upon the contractor's pledge to make good the imperfections speedily, at no cost to the client. In the same vein, the contractor has the right to remedy the flaws himself, a right he may lose if he does not fulfil his engagement to repair speedily. One exception is found in Quebec where the client may claim that he has lost confidence in his contractor and therefore may have the repairs done by another, at the original contractor's expense.

---

[7] Compensation, for instance for delayed or imperfect fulfilment of a contract, is not dealt with since it does not belong to the post-construction period.

The certificate issued at delivery must – normally, but see below – list all flaws. So far as the UK is concerned, John Derry has stated that 'problems when building works are not in reality complete or where defects exist are often overcome by the issue of a qualified Practical Completion Certificate, or a certificate is issued in respect of part of the works only'. He also notes that where qualifications are made because defects exist it may be doubted whether the certificate is valid. If the delivery procedure is valid, then the post-construction period starts with the post-construction insurance arrangements running, as planned, from this moment.

The delivery concerns the building agreed upon at the agreed price. The 'building agreed upon' should not be confused with the building originally agreed upon: few building projects remain unaltered from brief to inauguration; however, as works progress the parties approve each modification of the project. Hence the final result in all situations except (c) is agreed upon prior to delivery.

DEVIATIONS

Many Anglo-Saxon countries and Belgium offer examples of departures from the general pattern outlined above concerning the commencement of the post-construction liability period.

Save for the 'long negative prescription period' or 'long stop', countries influenced by Anglo-Saxon tort law can be said to have no general starting-point of the period during which producers are liable (see *Australia, England, Wales and Northern Ireland, Singapore* and *Scotland*, 1.5.1, for details): here the starting-point does not necessarily coincide with the moment of delivery. The US follows the European pattern.

The Belgian variant occurs in situations akin to formula (c) above, though with no price reduction, and to formula (d) but with no making-good claim. In other words, the client has noted the flaws at delivery but has accepted the building, without claiming either making good of the imperfection or price reduction. Still, according to an authoritative Belgian court ruling the client may, if an imperfection later develops into damage, claim compensation within the normal Belgian (Napoleonic) post-construction liability period (ten years).

Qualified certificates – certificates with at least one reservation – are now probably used in all countries. Although none of the national overviews provides an example, this has not always been so and it is by no means self-evident that a certificate with reservations is valid, cf. above. It is, in fact, perfectly feasible to imagine a system where a qualified certificate is not regarded as a valid final certificate. Thus in France,[8] when the client is a

---

[8] See Jérôme Chapuisat *Droit des travaux* (Amiens, 1990), vol 2. Likewise, the system in Sweden was modified only in 1972, though discussions of the acceptability of a certificate

public body, a qualified certificate has been regarded as a valid final certificate only from 1986, and the parts of works certified with a reservation are not included in the producers' decennial guarantee until the contractor has remedied the flaw.

containing reservations are reported from the 1940s; the word 'approval' in the now applicable General Conditions (AB 92, 7:14, para. 2) reveals this. It reads: 'At the final inspection the Inspector shall approve the Contract Work [an especially defined term] if no deficiencies or defects [especially defined terms] are found.' However, in the same paragraph, and clearly an addendum, it states: 'Minor deficiencies or defects of limited extent may not, however, prevent approval.'

CHAPTER SIX

Post-delivery

The post-delivery period starts at delivery or 'hand-over' (see Chapter 5). Obviously the duration of the warranties and guarantees which the client – and, in some systems, a successive owner – may call upon is a matter of import to clients, producers and insurers alike. The reader will find a great many variants in the national overviews, ranging from very short periods, such as in Portugal, to 30 years, as in Quebec, Canada.

Most Anglo-Saxon countries, as mentioned in Chapter 5, require separate consideration because of the uncertainty produced by the complex legal system concerning the period's commencement. It should also be kept in mind that the duration of guarantees and warranties is far from being the only important parameter in post-construction liability.

FLAWS NOTED AT DELIVERY

In most countries, during the first period following delivery the contractor must remedy any flaw recorded in the document issued at delivery and the client must give him a reasonable opportunity to do so. In Quebec, however, the client may, at any moment after delivery, claim that he has lost confidence in his contractor and appoint another contractor to carry out repairs at the initial contractor's expense.

In some countries the parties state the duration of this 'remedying period' in the certificate itself, the rule being that the repair work should cause the client a minimum of nuisance. In other countries the duration is legally imposed. Should the contractor fail to fulfil his obligations during the period, then the client may have another party remedy the flaws and charge the expense to the contractor's account.

FLAWS BECOMING PATENT DURING AN INITIAL PERIOD AFTER DELIVERY

Concerning flaws which could have been noticed but were neglected, see Chapter 5 ('qualified certificate').

In most countries the contractor's obligation to 'make good' extends also to flaws which the client notices after delivery but before the expiry of the initial period.

Again, the contractor – as the only producer during this period – has both the duty and the right to remedy any flaw in the works of which the client notifies him. The fact that the client will call exclusively on the contractor during this 'guarantee period' of course does not prevent the contractor, in his turn, from claiming compensation from any other party he may consider as having caused the damage.

NAME AND DURATION OF THE PERIOD

The names given to this initial period vary: the French have their 'period for formal completion' (*période de parfait achèvement*), enacted in the Civil Code with the reform of 1978; its duration is defined in law (see *France*). Belgians and Swedes talk about a 'period of warranty' (*période de garantie* and *garantitiden* respectively), but in Belgium this *période de garantie* follows delivery only if the contract so specifies (see *Belgium*, 1.2 and 1.5). The British JCT Contract calls this the 'Defects Liability Period'.[1]

The duration of the period varies from system to system, ranging from six months (for instance in Australia, Singapore and often in the UK) to two years (Sweden) or more (Portugal, for certain clients); see 1.2.3 in the national overviews.

At the expiry of this first period – an event which in some systems is marked by a second certificate – the liability system changes.

In Belgium a formal final delivery (*réception définitive*) may follow at the end of the guarantee or warranty period (*la période de garantie*). In France the end of the time allowed for formal completion (*période de parfait achèvement*) is not marked.[2] In Italy, there is a certificate of reception after the one referred to in Chapter 5. In Denmark, Norway, Singapore and Sweden a 'guarantee inspection', often accompanied by a certificate, marks the end of the period. In the UK sometimes a 'Final Certificate' is issued at the end of the 'Defects Liability Period' or whenever the flaws have been made good (generally a 'certificate of making good defects'), whatever is the later. In the US the second formality seems to be rarely used. For further details, see 1.2.3 in the national overviews.[3]

---

[1] Somewhat confusingly since there is, in Britain, a liability, if not for defects then certainly for damage, long after the expiry of the Defects Liability Period. Also in Singapore the period is quaintly named: though the contractor has no obligation to 'maintain' the building delivered to and taken over by the client the term used is 'maintenance period'.
[2] It used to be, prior to the Spinetta reform of 1978.
[3] Since 1986 Denmark has had a unique system with an inspection some five years after delivery.

AFTER EXPIRY OF THE INITIAL PERIOD

At the expiry of the Defects Liability Period (UK) or 'warranty' or 'guarantee' period (in most other countries) another type of producers' liability, extending only to damage (qv), takes effect. In some countries this liability is nil (Portugal provides an example of this if the client is the state). In most other systems (but see *Italy*, which is an exception) the client may now take action against any producer to claim compensation for damage.

The rules and the duration of the period – whether decided in law or in contract – governing liability during this period are explained, country by country, in the national overviews in Part Two.

It is worth noting that in several countries the rules change as time passes. The overviews from the Netherlands and from Quebec provide examples of this and highlight the difference between the first years after delivery and the following period.

CHAPTER SEVEN

'Failure' and 'damage'

HOW THE TERM 'FAILURE' IS USED IN THIS BOOK

In the mechanical industry 'failure' is often defined as 'the cessation of vital or important functions of a system'. Thus 'failure' refers to expected performance. This reference to performance requirements is explicitly fundamental in legal systems based on a doctrine of 'fitness for purpose', and constantly underlies the triggering of claims also in those systems which are based on producers' duty of care.

In this book building failure is defined as follows: failure exists when the building, at any time after delivery, fails to fulfil the most exacting of the following three categories of performance requirements:

(a) requirements in building regulations;
(b) requirements stated in or implied by the client's brief;
(c) requirements taken for granted in the particular culture where the building is situated, 'implicit general requirements'.

HIERARCHY

Whether requirements 'taken for granted' (c) are more or less stringent than those laid down in the national building regulations (a) depends on national contexts; the overviews only indirectly provide information on this hierarchy.

Clients' requirements (b) may often imply that the regulations' requirements (a) must be satisfied. There may be exceptions to this, however. In some countries, for instance, agricultural buildings may deviate from national building regulations and the client may wish that a lower standard be adopted.

The borderline between clients' implied requirements (b) and the 'national', as it were, cultural requirements (c) may often be hazy. It will be sharp if, for instance, the building is expected to meet very special requirements presumed to be known to the producers and therefore not expressly stated by the client (a soup kitchen for the Salvation Army and the main dining room in the Ritz in London, say).

DURABILITY REQUIREMENTS

Several building regulations, among them the English and the Swedish, expressly prescribe a certain durability for particular elements constituting a building. British courts have stated that the client is entitled to expect a certain useful life span, for instance, from heating equipment.[1]

Durability is also an issue in the EC's Construction Products Directive which states[2] that 'requirements must, subject to normal maintenance, be satisfied for an economically reasonable working life'. The document leaves each member state free to define 'economically reasonable'.

A further stressing of the link between durability and quality has been provided by W. A. Porteous.[3] In his analyses of the literature on the concept of 'failure' several authors quoted characterize failure as an increased demand on maintenance, i.e. low-quality buildings require more frequent intervention than high-quality buildings. Further, as pointed out by Porteous (personal communication), 'all that good design, workmanship, quality materials etc. does is to delay the inevitable victory of the natural forces'.

'FAILURE', 'DAMAGE' AND 'DEFECT'

'Failure' is a larger concept than 'damage'. Whether 'failure' constitutes 'damage', as the term is used in this book, is discussed below and in more detail in the following chapters.

The 'cessation of vital or important functions of a system' constitutes a necessary and sufficient condition for 'failure'. The same 'cessation' is necessary, but not sufficient, for 'damage': damage is a legal term.

'DAMAGE'

The authors of the overviews have used the following definition: 'Damage is the material manifestation of a defect'.

The term 'defect' is defined in Chapter 9: a defect is a flaw which is innate to the building; it does not develop with time but either is or is not present – albeit possibly unnoticed – at the moment of delivery. Damage, on the other hand, develops with time.

---

[1] See, for instance, Tony Ventrella 'Liabilities and Guarantees' in *Refrigeration, Air-conditioning and Heat Recovery*, August 1983 (London).
[2] EC Directive 89/106/EEC, Annex I, para. 1.
[3] W. A. Porteous *Perceived Characteristics of Building Failures – A Summary of Recent Literature* (Sydney, Australia, 1985) and *Proposal for the Classification and Minimisation of Building Failures* (CIB, Paris, 1989).

CAUSES OF FAILURE

It is useful to distinguish between the following two groups of causes of failure:

- Group 1: Alien causes – unforeseeable human action; force majeure.
- Group 2: Producers' 'errors and omissions'.

The first member of Group 1, unforeseeable human action, may be action by a third party, i.e. neither a producer nor a client; the action may be intended to harm the building (e.g. sabotage and vandalism) or it may not (e.g. a traffic accident damaging the building). Another example of 'unforeseeable human action', though in this instance not by a 'third party', is wrongful use or maintenance by the client or a successive owner.

Of course, the definition of the terms 'unforeseeable' and 'force majeure' (or Act of God) is very much a legal matter. It is therefore often met in insurance contracts. In post-construction liability matters guidance can be sought in building codes since an adequate code will, at least to a certain extent, define which destructive agents – human or natural – producers ought to foresee, cf. Chapter 4.

The definition of the 'producers' of the second group is, again, a matter of legal doctrine. Chapter 3, 'Producers', outlines the participants; the concept of the 'errors and omissions' committed by these participants is discussed in the following chapters.

Failure originating in a source from Group 1 does not constitute damage as the term is used in this book. There may or may not be insurance arrangements covering some or all such events, but in no legal system reported in the national overviews is a producer liable for this kind of failure.

This means that there is damage only if the failure has its origin in Group 2, in other words if the failure is due to a defect.

This does not necesarily imply that in each and every post-construction liability system there is automatically a producer who may be held liable for the defect, and hence for the damage.

This is so, first, because assigning the cause to either Group 1 or Group 2 is a legal matter. For instance, what in one legal system is an Act of God may in another be an action which the producer ought to foresee. Second, the liability issue depends on the law applied in the particular circumstance. Third, liability is always extinguished with the passing of time.

AETIOLOGY

The analysis of failures or malfunctions in buildings is sometimes called 'building pathology'. For the unravelling of the chain of factors or events leading to damage – defined above – another term borrowed from medicine, 'aetiology',[4] is useful.

Building failure aetiology is often complicated. Kouhei Matsumoto[5] enumerates the following characteristics of buildings as making the task particularly difficult:

(a) individuality;
(b) unstable conditions of production;
(c) ambiguous functions;
(d) difficulty of measuring functions;
(e) unique role of drawings and specifications;
(f) long period of use; and
(g) high price.

He also states[6] that 'characteristics (c) and (d) are major causes of long disputes', and characteristic (g) normally means that the owner–client has little experience – compared to the buyer of a machine, say – of this kind of transaction.

It is now and then argued that the 'assignment of a cause' – aetiology – for the occurrence of damage to buildings is costly. Costly or not, there are several good arguments for undertaking aetiological analyses of damage to buildings. This is so because, as mentioned in the Preface, the risk of building failures – including damage – is in all probability increasing all over the industrialized world.

First, an important means to forestall the risk materializing is what is commonly called feedback, that is, the channelling back of knowledge gained through aetiological analyses to the building industry's many participants, either during their initial training or in the course of their continued professional development (CPD). Examples of this are provided by studies and information drives undertaken by RIBA (Royal Institute of British Architects) Indemnity Research, the National House Building Council in Britain and, eminently, by the Agency for the Improvement of Quality in Construction (Agence qualité construction) in France.[7]

Another means of risk reduction, as outlined in Chapter 12, by Gérard Blachère, is research resulting in adequate building regulations, standards, agréments (see Chapter 4) and other testing and approval systems. A good part of the input to this research originates in aetiology.

A further reason for emphasising the usefulness of aetiology relates to the

---

[4] Aetiology is '[the] assignment of a cause' (*Oxford Dictionary*); 'etiology or aetiology, also aitiology: . . . all the factors that contribute to the occurrence of . . . [an] abnormal condition' (*Webster's Dictionary*).
[5] Kouhei Matsumoto 'The New Home Warranty System in Japan' *Habitat International* (Pergamon Journals Ltd), vol 10, no. 4 (1986), pp. 71–92.
[6] Fn. 5, above.
[7] See, for instance, Knocke *Preventing Building Damage – The French Way* (Gävle, Sweden, 1992).

attitude of the industry's many producers. They are often, and in practically all countries, rightfully or not, regarded as morally lax. As pointed out by several moral philosophers, putting the blame where it belongs surely is one way of increasing producers' actual quality awareness. This, of course, requires aetiology. Also the public interest and common equity require that negligent or incompetent producers, and none other, be identified and made to pay for failures. It has been said in more than one country that there is an evident need for 'la moralisation du secteur',[8] raising the moral standards of the building industry.

Finally, insurance arrangements – with the possible and exceptional case of those providing a 'no fault' cover – require identification of the cause or causes of the 'claims event'. This they do in order to decide whether the damage is covered by the policy. This obviously goes for liability and professional indemnity insurance, but applies also to most property insurance schemes. The important French property insurance known as *dommage-ouvrage* insurance, for example, covers damage precisely as the term is used in this book, and not just any occurrence of failure.

So, difficult or not, aetiological analyses are important.

They are, of course, facilitated by rigorous definition. The rest of this chapter therefore defines and comments upon three key concepts, namely:

(a) Errors and Omissions, which may or may not lead to
(b) Defects, which, again, may or may not lead to
(c) Damage.

Each of these three concepts is dealt with also in specific chapters (Chapters 8, 9 and 10); here the emphasis is on the link between them.

ERROR AND OMISSION

Any producer may commit errors and omissions at any stage of the process ending in delivery. The definition adopted by the authors of the overviews is the following:

(a) any departure from correct construction (including checking and supervision) technical inspection;
(b) absence of adequate instructions for maintenance and operation of the building.

In practice, an important part of the rules defining 'correct construction' is to be found in compilations of technical texts. In the UK the Codes of Practice, in Sweden the AMA (general descriptions of materials and work) are examples, but by no means the only ones. In France the *documents*

---

[8] This was expressly stated in the French Government's terms of reference for the study which led to the current liability and insurance system, the so-called Spinetta reform.

techniques unifiés (DTU)[9] are mandatory only for public contracts, but a producer's non-respect of these rules may jeopardise his liability insurance cover.

The definition assumes that producers know what constitutes correct building practices and adequate instructions. This, of course, may not be the case, and therefore errors and omissions must first be divided into two main groups: those committed knowingly and the rest. A systematic classification – a 'taxonomy' – is given in Chapter 8. Suffice it to say here that errors and omissions are in the realm of intangibles: only human action can be erroneous, and action, as well as inaction, are abstract notions.

The consequence of an error committed prior to delivery is a defect in the building as delivered.

DEFECT

The authors of the overviews use the following definition: a building suffers from a defect if:

(a) the building, as actually built, does not conform to recognised rules (*secundum artem* or *lege artis* in Latin, *selon les règles de l'art* in French) or to specifications in contract or in normative documents;
(b) instructions for maintenance and operation are inadequate or lacking.

By definition then, a defect, as the term is used in this book, is not abstract; though not necessarily easy to discover, it is a tangible occurrence that can be established. For example:

- the vapour barrier in an exterior wall in a cold climate is of an ephemeral material (tests can establish this) and may deteriorate, leading to moisture damage;
- instructions for use of a warehouse given to a client overstate the permissible loads (the document with the correct text was not made available) and the building will deform excessively when loaded as intended by the client;
- the ventilation system is of inadequate dimensions (inspection of the fans or ducts will show this) and it cannot achieve the level of expected comfort;
- lifts in an apartment building are too small for requirements (inspection will reveal this). Their being frequently loaded beyond their capacity requires excessive repairs.

The definition also makes it clear that a defect is necessarily the result of an error or an omission. For instance, initially the risk of rot and intrusive odour inherent in using floor screedings rich in casein was not known, nor was the

[9] Reportedly, French DTUs have also been used by courts in Italy, just as English liability rulings have been quoted in Swedish courts.

structural degradation caused by high alumina cement. However, this only means that the risk is a 'development risk' (see Chapter 8): the underlying error falls within the class of errors caused by an objective lack of knowledge. The defect is still tangible (the casein with its biological degradation potential exists; the water-sensitive cement exists).

Any error leads to a defect, but not any defect leads to damage. This is further discussed in Chapter 9. W. A. Porteous[10] provides the following example: 'a flat roof without the proper fall may be defective [the error: omitting to see to it that the fall is proper, which constitutes a departure from correct building practice] because water ponds on the surface, but if . . . no leaks occur no building failure may be perceived'. In our terminology: no damage occurs.

But if the defect causes damage we are confronted with the final link in our aetiological chain: damage. This concept is further discussed in Chapter 10.

SUMMARY

An error or an omission – including errors due to an objective lack of knowledge – always leads to a defect.

There is no defect without an underlying error or omission.

A defect may or may not lead to damage, but there is no damage without an underlying defect.

Errors and omissions may be committed at any stage in the building process, by any party to the building process. Errors and omissions are in the realm of the abstract.

Defects are the physical manifestation of an error or an omission. They are neither abstract nor latent: the building as delivered tangibly suffers from a lack of conformity with correct construction methods or, in the case of instructions for use and maintenance, from a lack of provision of adequate instructions.

Damage is the physical manifestation of a defect. It flows, via a defect, from an error or an omission. Failure is, as already stated, a larger concept than damage; in other words, without an underlying error or omission there may be failure, but no damage.

---

[10] W. A. Porteous *Insurance Data – A Potential Guide to Quality Assurance* (Wellington, CIB Seminar in Copenhagen, 1989).

CHAPTER EIGHT

Errors and omissions

DEFINITION

When producers, voluntarily or not, do their job erroneously or omit to do something which they ought to have done, they commit errors or omissions. Such actions, or omission of actions, will normally lead to a defect in the building as transferred to the client at delivery, and a defect may lead to damage. For a more formal definition and a description of the chain of events leading from an error or an omission to damage, see the previous chapter.

Defects result from an error or an omission committed anywhere en route to delivery. As pointed out in the previous chapter a defect may reside in either of two 'places': either in any of the elements physically constituting the works, or in the instructions for maintenance and use of the building.

A first classification of errors and omissions derives from this distinction of the 'place' of the defect. A translation of 'place' into the different stages constituting the production of a building leads to the following demarcations:

(a) designing the building, during initial conception and in the course of technical inspection during the works;
(b) erecting the building through the physical site work; and
(c) providing the client with adequate instructions for maintenance and use of the building.

This, then, is one axis along which to classify errors and omissions.

Errors are committed for a variety of reasons which will be discussed in the second part of this chapter and which will form the second axis. The obvious distinction between errors and omissions which producers commit knowingly, and those they do not, is one such division, but it is far too gross to serve the present purpose, which is to offer a taxonomy – an orderly classification – and an exhaustive matrix in which to place errors and omissions.

A taxonomy serves several purposes:

(a) as an introduction to different legal doctrines: it facilitates the understanding of producers' 'duties' vis-à-vis the client, cf. Chapter 11 by Daniel Alain Dagenais;

(b) understanding insurance policies: the liability and professional indemnity insurance industries *nolens volens* use some 'error and omission' taxonomy – though not necessarily the one presented here – in their assessment of the actuarial levels of risks, i.e. the probability of claimable damage ensuing from each particular type of error or omission. These types therefore enter, explicitly or not, any such insurance arrangement, and have an impact on premiums, franchises and the wording of the policy;

(c) in planning quality assurance endeavours: analysing and grouping possible or confirmed errors and omissions in the chain of production facilitates the optimum allocation of resources to different kinds of quality control. It is therefore part of industry's efforts to reduce exposure to liability for 'failure', a term explained in the previous chapter. In quality control, as in the error and omission taxonomy presented here, it is often effective to distinguish between 'design quality' and 'production quality', the former referring to design, the latter to the implementation of the design. This distinction is surely also useful in the building industry. The most often quoted example, however, is from the automobile industry: the Czech Skoda automobile of the 1950s was, according to experts, a car with outstanding 'design quality' and poor 'production quality'.

The first categorisation is based, as mentioned, on the different stages at which an error or omission may be committed.

The following classification owes a great deal to Professor Børge Dahl.[1] His taxonomy is general in the sense that it applies to the production of any 'goods and services on the market'. Here Dahl's arrangement has been modified to suit the particular needs of the output of the building industry, where 'goods and services on the market' would correspond to the finished and delivered works, i.e. the building during its post-construction period.

There are two sets of distinctions: the stage at which the error is committed and the cause of the error.

DISTINCTION ACCORDING TO STAGE

As already mentioned, we shall distinguish between three stages:

(a) initial design and subsequent design, the latter being influenced by technical inspection, often on the site;
(b) the physical putting together of the building following the design, i.e. site work; and
(c) delivery of instructions for use.

---

[1] Børge Dahl *Produktansvar* (Copenhagen, 1973 (in Danish)).

'DESIGN ERRORS AND OMISSIONS'

'Design' in Danish is *Konstruktion*.[2] However, in the context of building it is preferable to use 'design' rather than 'construction' since 'construction' normally refers to site work, the realization of the design.

A design error which affects the initial conception of a building makes the drawings and specifications the 'carrier' of the defect. If not rectified the error will affect all buildings built according to that design. The manufacture of a building kit in Sweden provides an example: some time after the introduction on the market of a certain kit it was discovered that the ventilation under the roof was inadequate. The design – dimensions and distribution – of air vents was erroneous. The factory recalled the model and offered free repair of any house already built with the defective kit, thus assuming responsibility for its design error.

On many building sites a great deal of design is in fact done while work progresses, for instance during site supervision (or 'observation', the preferred term in the US). If the inspection leads to erroneous design the works will suffer from a design error. Whether the inspection is carried out by an independent technical inspector, by a local authority or by the designer who produced the initial design is, of course, immaterial in the present context.

However, the term should not be taken to mean that an error in the initial design or introduced during supervision or inspection is necessarily to be laid at the door of the initial designer or of the supervisor.

This is so because in most construction projects – except perhaps in the Third World – the formal designer is not required to draw or describe each and every detail of the works to be executed on site, nor is the technical inspector expected to check each and every detail. Both are, indeed, entitled to assume that the contractor is able to build 'in a workmanlike manner', to use a term often met in the UK.[3] In other words, the architect's design and the inspector's checking are, at least in industrialized countries – and see below about Singapore – presumed complemented by the contractor on site while work proceeds, or beforehand. This checking by the contractor may take place during his own design work or at his own or a hired testing facility.

Therefore the fact that a defect is due to a design error does not, per se, imply an error on the architect's or the inspector's part – the contractor also normally carries out both design and inspection. The Italian liability system

---

[2] 'Design error' in Danish is *Konstruktionsfejl*, cf. *Webster's Dictionary*: 'Construction: . . . The form and manner in which something has been put together.' An example from the automobile industry: after a particular model of car has been put on the market it is discoved that the model suffers from deficient brakes. In publicised cases the manufacturer then invites all owners of the model to have their cars repaired at no cost to them.

[3] Used verbatim in the UK Defective Premises Act 1972 and in many English language standard building contracts.

goes so far as placing the liability for errors and omissions in design and inspection on the contractor alone (see *Italy*, 1.5.6). The Italian doctrine is an extreme, as is, at the other end of the scale, the one found in Singapore, where the 'responsible person' may be held liable for any 'design error' (see *Singapore*, 1.1.0).

In other countries a balance is struck on the basis of 'good practice' within that nation's building industry in general and the particular contractors' or suppliers' competence in particular. This then decides the apportionment of the liability for a design error between the formal designers (the architect etc.) or the inspecting party and the contractor or contractors, sub-contractors and suppliers. In post-construction liability the issue of competence is often important. It will be met again under the causes of errors below.

In countries such as the UK and Scandinavia, where the local building authority also plays a technical inspection role, the designer sometimes leaves a particular part of the design to the local authority. It is thus, rightly or wrongly, assumed that the authority is responsible for parts of the necessary technical inspection.[4]

SITE[5] WORK ERRORS AND OMISSIONS

Site work errors and omissions occur on the building site. They happen despite impeccable design and are due to a mishap – of some kind, see 'Distinction according to cause', below – on that particular site.

A milestone verdict concerning the consequences of a 'site' omission – inadequate checking – in manufacturing industry was rendered in the British House of Lords in 1932:[6] when a rotting snail was found in a bottle of ginger ale, making the contents unpalatable, the House of Lords held the producer liable. This famous case has had a considerable impact upon the UK building industry. This kind of error has led Dahl[7] to suggest what he calls 'errors in checking' (*kontrolfejl* in Danish, cf. the French *contrôle*), a category surely of practical import to the building industry.

A common type of site work error is careless treatment on the site of otherwise satisfactory materials. Clearly, even if the architect does not mention it in his design, he is entitled – at least in industrialized countries, cf. above – to assume that sensitive materials be protected against rough

---

[4] For instance, architects' plans bearing the statement 'foundations accordng to local building authority' are encountered in some countries, thus leaving at least part of a design error to be attributed to the authority. Although this may not be considered good architectural practice it is sometimes the only way to solve a problem in the production of 'catalogue houses', where the same plans above ground are used for several buildings, on different sites with soil conditions which are, by definition, unknown to the designer.

[5] Site and site work correspond to the Danish *Fabrikation*, and 'site error' is *Fabrikationsfejl*.

[6] *Donoghue v Stevenson* [1932] AC 562, HL.

[7] Fn. 1, above.

treatment on the building site, that rain-soaked components not be built in while still wet etc. Likewise both the consulting engineer and the technical inspector may assume that the contractor will remove loose rust from reinforcement bars, cf. the standard assumption mentioned under the section on 'Design errors', above.

In principle, the contractor's building in of a defective component or his using a defective material is a site work error or omission, since normally he ought to check the quality of materials and components. There is, however, one important exception to this principle, namely in France concerning the EPERS[8] components. For these, the contractor is relieved of his obligation to check or test, meaning that the contractor may presume that an EPERS component is of the quality required (see *France*). Also, the supplier of the EPERS component is liable jointly with the contractor who installs the component. This doctrine is not stated so explicitly elsewhere, though it is assumed in a good many liability cases outside France.

As mentioned above, errors and omissions in technical inspection may be design errors committed by the technical inspector, but the contractor can also commit design errors. In the same way, not every error committed on site is necessarily a contractor's site work error.

INSTRUCTION ERRORS AND OMISSIONS

Instruction[9] errors or omissions exist if the instructions delivered to the client for use and maintenance are not adequate. Such errors may create a defect which may cause damage when the premises are used.

A French court case illustrating an instruction error causing severe damage – and a producer being held liable for the error – is reported in Chapter 10 (caretaker inadvertently flooding several floors with fuel because of wrong information, namely inadequate instructions for filling a tank).

Just as with design errors, the same instruction error may affect a whole series of buildings. As buildings and installations become ever more sophisticated it is to be expected that instruction errors and omissions become more frequent and costly to remedy.

A sub-division of instruction or information errors is sometimes used by the legal profession: a 'positive information error' exists when the information is not correct, for example a paint is wrongly claimed to have rust-proofing qualities. In the building industry, for instance, an agrément (see Chapter 4) could contain a 'positive information error'. The legal liability for positive information errors in an approval system such as an agrément is

---

[8] *Elément pouvant engager la responsabilité solidaire.*
[9] In Danish *Informationsfejl*, 'information error'.

treated differently in different systems. A few of the national overviews describe this.

DISTINCTION ACCORDING TO CAUSE

Another useful distinction consists in dividing errors and omissions leading to defects – and hence possibly causing damage to the building, see the previous chapter – into the following four categories: (1) Development, (2) System, (3) Jerry-building, and (4) Incompetence.

DEVELOPMENT RISKS

'Development errors and omissions' is the name given to errors and omissions caused by an absolute, though sometimes timebound, lack of knowledge about a technology. The term therefore implies that at the time the development error was committed not even the most advanced expertise could foresee the ensuing defect. The corresponding risk of 'development damage' is a 'development risk'.[10]

Development risks exist because the sum total of human knowledge is constantly developing and therefore limited at any given point in time. This limitation causes an 'error', but committing it is 'nobody's fault'. Of course, once the causal link between occurrence of damage and erroneous act is established the error moves out of the 'development category' and into one of the three other categories.

Examples of development risks are, of course, provided by many new technologies (materials and design) but, as Dahl[11] notes, 'However, development damage [that is, loss due to a development error] may also occur with well-known products, because much scientific progress consists in discovering new causal links.'[12]

In the building industry, the use of asbestos offers an example: it was not known, when asbestos-based sheets were first used in ventilation ducts, that the fibres released by some sheets are oncogenic. The 'aetiology' (see the previous chapter) is the following:

- the 'error' is that this material was installed;
- the 'defect' is the existence of a material which releases asbestos into the air;

[10] Development risks afflicting the pharmaceutical industry are often publicised, for example a drug prescribed to pregnant women and only later, after use, found to be harmful to the foetus.
[11] Fn. 1, above.
[12] Dahl cites the links between cigarette smoking and different diseases, links clearly discovered only well after production of cigarettes began and which scientific knowledge at its then prevailing stage could not reveal.

• the 'damage' is the inadmissible quantity of oncogenic fibres in indoor air.

Another example is the use of materials with a much shorter span of useful life than was assumed on the basis of knowledge available at the time of production, such as certain types of concrete currently requiring extensive repair works and some plastics.

National doctrines concerning liability for development risks vary widely, as do different insurance policies' cover of these risks. Nor is there unanimity within the EC member states.[13] In some legal systems compensation for 'development damage' is to be paid by the producer – or his insurer – despite the producer's innocence, namely in systems where the doctrine of duty of result (see Chapter 11) governs the liability system; the overviews from France and Quebec provide examples of such systems. In other systems errors and omissions which are 'no one's fault' will fall under the blanket heading 'bad luck' meaning, in practice, that the client or his insurer, if the insurance provides a 'no-fault' cover, will have to meet the loss.

'SYSTEMATIC ERRORS'

'Systematic errors' are errors which cause defects which are known, and knowingly accepted, by all parties.[14] The term is sometimes also used for defects which the producer could not possibly forestall. For instance, a contractor who uses a load-bearing component delivered on site and ready for installation cannot, without destroying it, check its quality (but see above regarding the EPERS components in French liability law).

Systematic risks are probably rare in the fields of instruction errors, site work errors and conception of building projects, but they are habitually encountered when it comes to technical inspection (see section on 'Taxonomy', below).

---

[13] The EC's Defective Products Directive (85/374/EEC) leaves each member state free to adopt its own doctrine.

[14] In medicine, for instance, it may be known that a certain vaccine suffers from the defect of being potentially harmful, meaning that it is known that it will harm a certain proportion of the population inoculated; the probability of damage, though not zero is, however, generally accepted because the vaccine is considered as having an overall positive effect on the population. Another example from the medical field is tobacco, also flawed by a defect, namely that it may cause, for example, lung cancer. Smokers are (now) presumed to be aware of this, and knowingly accept this systematic risk. Smoking, therefore, is an example of a risk which, with the development of medical science, has moved from being a development risk to a systematic risk.

JERRY-BUILDING

A builder who builds flimsily and cheaply to reduce his outlay is a jerry-builder.[15] The term jerry-building may mostly be used about contractors but it is, in fact, applicable to any party who knowingly delivers sub-quality work in order to maximize his immediate profit. For instance, an architect may cut costs by omitting to execute a geotechnical survey, a technical inspector by skipping a site visit, and so on, thereby increasing their (immediate) gain and, conversely, increasing the risk of damage to the finished building.

The same goes for deliberate carelessness: both carelessness and jerry-building are expressions of willingly and knowingly adopted attitudes. In both cases the producer chooses to replace care and diligence with jerry-building or carelessness in order to minimize his (immediate) outlay in money or in effort or both.

From a purist point of view it is, of course, a moot question whether the term 'error' is relevant when applied to jerry-builders or careless designers knowingly cutting corners: they probably do not consider that they commit any 'error' by maximizing their instant profit in money or effort.

However, a producer may wish, or be driven, to economise by using low quality materials or inadequate dimensions. This could be construed as intentional malpractice, *dolus* in legal terminology.

There is no noticeable difference in national doctrines in relation to liability for jerry-building.

Insurance[16] cover, on the other hand, may vary from country to country. The producer's liability insurer may refuse to meet the ensuing claim if the error is due to a deliberate deviation from correct procedures. So, for instance, in France (see *France*, 3.5.6): 'the insured party loses his cover in case of voluntary or inexcusable non-observation of the rules of the art . . .'; but in Quebec only intentional errors and omissions are uninsurable.

INCOMPETENCE

Lack of competence, on the other hand, is the result of a lacuna in the producer's knowledge or skills, and not a knowingly adopted attitude. The issue of competence is also present, albeit indirectly, in the definition of 'development risks', which is, of course, the proper term only if the knowledge needed to forestall the defect is lacking in the absolute sense. This means that if the necessary knowledge is available somewhere – even if

---

[15] *Longman's Dictionary of the English Language.* In American English the term is the same: *Webster's Dictionary* – 'to build cheaply and with inferior materials'. The Swedish term is *fuskbygge*, literally 'cheat-building'.

[16] The risk of fraudulent behaviour is one – of several – reasons why a good many national systems impose property insurance on lay clients. It is, however, by no means the only reason.

only at some specialised laboratory in another country – and therefore not 'absolutely' lacking, then the question of the individual producer's competence becomes relevant: in such cases there is no longer a development risk.

An example would be where a designer's lack of knowledge may be more or less normal within that particular class of producers: an average, reasonably diligent and skilful designer might not be aware of the risk, but there are, nevertheless, specialists, somewhere, who would not have committed the error of incompetence in question. If, however, the liability doctrine governing the system is that producers are under a duty of result (see Chapter 11), then the producer will nevertheless be liable, but if his duty is one of care, then a court or an arbitrator must decide which errors and omissions are excusable and which are not in each particular circumstance.

Clearly, the quality of the client's brief becomes significant in this context. For instance, laboratories for developing X-ray film should be free from even moderately radioactive building materials since background radiation – for instance from building materials – may render the building unfit for its intended purpose. An architect who is not aware of this requirement commits a 'design incompetence error' by prescribing, say, certain kinds of concrete which more competent designers know to be radioactive. But a court may find that the client should have briefed the architect on this particular matter and take this into consideration when apportioning the blame.

Another example of the client's possible contribution to an error is met when a producer, once he has taken on a job, finds that it is so badly paid that doing the job properly would mean a loss. This leads to an ongoing discussion of the client's role in risk allocation: does the client, by choosing, say, the cheapest producer, knowingly contribute to the risk of errors and omissions affecting the outcome? And if so, should the client then bear at least part of the loss resulting from the incompetent or insufficiently motivated producer's errors and omissions? Or, on the other hand, is a client always entitled to trust that any producer who holds himself out as being capable of doing a job implicitly undertakes to do it well? Legal attitudes vary.

National overviews describing legal systems built on a doctrine of 'duty of care' deal with the issue of the degree of competence or care which the client may 'reasonably' expect from his producers. The issue is, of course, without relevance in liability systems built upon a doctrine of a duty of result (see Chapter 11). Nonetheless, even in countries with contracts imposing a duty of result, the degree of competence is relevant to insurers' assessment of risks. So is the case when the task is to arrive at a fair apportioning of a claim for compensation among several parties if more than one of them contributed to the same occurrence of damage.

A particular aspect of the question of expected competence, mentioned under the section on 'Design errors', above, is how much of the design the architect 'reasonably' may presume that the contractor will do correctly on his own. Here again, the client's competence may be a matter of importance in these deliberations.

TAXONOMY

The complete matrix for errors and omissions looks as follows:

	Cause of risk			
Stage	*Development*	*System*	*Jerry-building*	*Incompetence*
Design	D1	D2	D3	D4
Site	S1	S2	S3	S4
Instructions	I1	I2	I3	I4

Examples are as follows:

D1: DESIGN AND DEVELOPMENT RISK

Two examples are given above: asbestos and shorter life-span than expected. Also damage in the shape of radon, and perhaps other emanations, belong to this group.

D2: DESIGN AND SYSTEMATIC RISK

It is (probably) a systematic error to reduce the rate of air change in a building in order to save on heating costs; the known risk (reduced air quality) is, nevertheless, imposed by some national building regulations, under the assumption that (a) the defect – reduced aeration – may not develop into damage, or (b) that the risk of ensuing damage is worth taking.

Building regulations are a design issue and so is technical inspection, but no technical inspector can check each and every detail. He must allocate his resources – make his choice of what to check – on the basis of his knowledge or intuition about the risks created by incomplete inspection. This implies assuming a systematic risk.

D3: DESIGN AND RISKS CAUSED BY JERRY-BUILDING

Examples, from contracting and design, were given in the section on 'Jerry-building', above. Some national overviews also briefly mention fraud and the consequences of attempts at deliberate concealment, but in principle acts which fall under penal law are outside the scope of the overviews.

D4: DESIGN AND INCOMPETENCE RISK

An architect not familiar with the climate on the future site of the building may omit to take the necessary precautions, as illustrated by a French court case:[17] damage to a roof was caused by 'admittedly unusual though not unforeseeable' rain and the architect should have taken the weather in that particular spot into consideration.

S1: SITE WORK AND DEVELOPMENT RISK

Until a few years ago it was not known that casein-based floor screeding materials, offering considerable savings in contractors' labour costs, constituted a risk of rot and offensive odours.

Before Luigi Galvano (who, after all, died as long ago as 1798) it was (probably) not known that one of two adjacent metals, when their joint is exposed to water, will corrode faster than if alone. So two hundred years ago it was (presumably) a development error to put copper nails in zinc.

S2: SITE WORK AND SYSTEMATIC RISK

Some paints should be applied with a brush or a roller and not sprayed on. However, it is nevertheless 'often' all right to spray them on, though 'sometimes' this leads to damage (rot).

The words 'often' and 'sometimes' indicate that the contractor who uses a spray takes a systematic risk. The same goes for the laying of floor coverings which prevent further drying out of an underlying concrete slab: 'often' one week is sufficient for allowing the water to evaporate from the concrete, but 'sometimes' it is not enough.

S3: SITE WORK AND JERRY-BUILDING

The infill around porous drainage pipes should have a certain coarseness in order to let the pipes absorb water in the vicinity. It is cheaper for the contractor – at least at the time of building – to neglect this and use any handy infill, covering it up before the next site inspection.

S4: SITE WORK AND IMCOMPETENCE RISK

An incompetently supervised workforce may inadvertently perforate the vapour barrier, thus causing deterioration of the thermal insulation.

---

[17] *Le bâtiment-bâtir*, June 1982, p. 22.

I1: INSTRUCTIONS AND DEVELOPMENT RISK

When linoleum flooring was first introduced some hundred years ago it was not known that linoleum floors require special care in cleaning. When, for instance in hospitals, they were scrubbed with detergents perfectly adequate for stone and wood flooring, the linseed oil was washed out of the linoleum with (extensive) damage to the floors as a result of the defective instructions.

It is (still) not known how best to maintain some plastics and instructions are (probably) imperfect.

I2: INSTRUCTIONS AND SYSTEMATIC RISK

In modern buildings there are no longer instructions for the use of water closet bowls. Formerly it was considered safer to put up signs saying 'No Rubbish In The Bowl'. Correct use is presumed to be known by users of buildings (though not by airline passengers) and the damage to drains caused by debris in water closets is (presumably) regarded as an acceptable risk.

I3: INSTRUCTIONS AND JERRY-BUILDING

The installer saves on translation costs by not bothering to have the instructions for maintenance of an imported boiler translated into a language understood by the client. Disaster may follow.

I4: INSTRUCTIONS AND INCOMPETENCE RISK

The contractor does not know that a special floor-covering which he has laid needs special maintenance and omits to call the client's attention to this.

The contractor knows that a water-borne heating system will need topping up but is not aware of the fact that tap water in the area is too hard and should not be used.

In both examples the defect resulting from the error, caused by incompetence, is 'no instructions', with ensuing damage to the floor and the central heating installation.

CHAPTER NINE

Defect

A defect is the result of an error or omission (see previous chapter) in one of the stages preceding delivery. The aetiology – the chain of events – is explained in Chapter 7, which also gives the definition of 'defect' used by the authors of the national overviews in Part Two.

As the definition shows, defects, as the term is used here, may be hidden but not 'latent', while damage, on the other hand, may very well be 'latent'.[1] As stated, a defect is a tangible thing and therefore 'patent'. Although it may be difficult to find – 'hidden' – it is not latent.

Not all defects lead to damage. Presumably many defects go unnoticed for ever or, as also happens, it is only when a building is being demolished or refurbished that a defect is discovered and may become a matter calling for remedies or, if it has no consequences, simply noted as a curiosity in the history of technology.

Post-construction attitudes to defects which do not cause damage are given under 1.2.5 and 1.5.5 in the national overviews. The systems described in Part Two differ depending, inter alia, upon whether the defect causes a monetary loss or not, but also upon the legal consequences of the delivery procedure. If the client discovers a defect which has not developed into damage prior to the building delivery – i.e. not in the post-construction situation – he may claim reparation or a reduction in the monies to be paid by him to the producer responsible for the error or omission underlying the defect (cf. Chapter 5). Such claims, however, are not post-construction matters and the overviews do not deal with them. For defects which have not developed into damage and are discovered during the immediate post-delivery period, see Chapter 6.

In other words, damage as distinct from defect always entails a monetary loss. Damage therefore, in principle though perhaps not in actual legal fact, is always claimable. However, the legal definition of damage may preclude compensation for a defect which develops into damage which is legally regarded as being 'minor'. Thus in Spain: 'There may be a hidden defect in a construction which, if it does not result in ruin, does not give rise to compensation' (see *Spain*, 1.2.5). But whatever the legal definition of

---

[1] 'Latent' means 'capable of becoming visible or active though not now so' (*Longman Dictionary of the English Language*, 1984).

damage, the discovery of a mere defect – i.e. with no ensuing damage at all – will normally not constitute a 'claims event'; cf. Chapter 6.

France, with its clear-cut law on post-construction liability offers a series of court decisions which throw light upon what is and what is not considered a defect in that jurisdiction.

Thus, referring to the law on decennial liability with its requirement for the building to be fit for its intended purpose (see *France*), a court[2] found that:

> '[a] mediocre heat transmission coefficient does not in itself constitute a defect in the legal sense . . . However, in view of the importance of thermal insulation for dwellings, and taking into consideration the current concept of minimum thermal comfort and the technical and financial constraints on heating systems, residential buildings whose thermal insulation characteristics are clearly deficient may, under certain assumptions, be regarded as unfit for their purpose.'

The court goes on to specify the damage – mould, humidity and condensation on the walls – caused by the defect.

Also in France, a defect remains a defect even if the producer takes other measures to mitigate the damage caused by the defect:[3] 'The dwelling's ventilation system was never intended to palliate the observed malfunction of . . . [the building's] thermal insulation . . .'. This means that the producers' liability is called upon to remedy the defect in the insulation, even though increased ventilation does away with the visible symptoms of its being defective. The defect, in this case, of course affects the thermal insulation, not the ventilation system.

If works to remedy the defect increase the building's value, then the client must normally pay for the added value. However, this is not necessarily so in a system based on a duty of result, such as the French. Thus the French supreme appeals court[4] has stated:

> 'Producers are held to deliver a building which conforms to its intended use and is free of any phenomenon which renders it unfit for its purpose. It is of no consequence that because of the supplementary works [to remedy the defect] . . . the building's value is increased in relation to what it originally was, insofar as it is not clearly proven . . . that the client contributed to the damage or aggravated its consequences by any act of his.'

This question has been treated in the same sense by other French courts[5] also:

[2] Unauthorized translation from the Tribunal administratif de Strasbourg, 15 December 1983, quoted in SYCODÉS Informations 7:1990, p. 64.
[3] Unauthorized translation from the Tribunal de grande instance, Tours, 30 April 1985.
[4] Unauthorized translation, Cour de cassation, 18 December 1984.
[5] Thus Cour d'Appel de Riom, 22 October 1981. Unauthorized translation, verdict quoted in SYCODÉS Informations 7:1990.

'It is not a valid argument that because the building was originally designed to be cheap, the clients cannot now [after delivery] insist on supplementary works which will make them richer without paying for them [the supplementary works]. The interested parties . . . are in their right to insist on having . . . [external] walls which achieve what is expected of them.'

In this case the issue was walls which ensure healthy indoor conditions and allow the building to be normally heated.

If the defect develops into damage only under certain circumstances and if these circumstances never occur then, of course, the category is that of a defect not manifesting itself in damage, an issue dealt with above.

Examples of defects which do develop into damage only under severe strain on the building are given by a series of French court cases in the field of condensation: the defect, in all the cases referred to, is a ventilation system which proved inadequate only when the occupants' use of the premises caused a load on the system which was heavy, though not, in the opinion of the court, totally abnormal. But the fact that the occupants put an extremely heavy strain on the installation may lead the court to find that they ought to share some of the responsibility.

In one such case an analysis of the owners' heating bills had shown that they had not heated their villa adequately. The court[6] therefore charged the owners with part of the cost for remedying: 'The owners of the villa, by heating the premises in a decidedly insufficient way . . . have themselves contributed to the occurrence of the damage.' The error underlying the defect is, then, that the producer had not considered that the building could be used in a particular way. Since the producer could not foresee this he is partially excused and the owners must meet a part of the costs to remedy the defect.

---

[6] Unauthorized translation from the Tribunal de grande instance, Bobigny, 2 February 1988.

CHAPTER TEN

Damage

As pointed out in Chapter 7 on failure and damage, these are not synonymous terms: 'failure' is a general term containing 'damage': 'damage' is 'failure' caused by a 'defect'. If there is failure to the building without an underlying 'defect' we are confronted with another type of 'harm' than 'damage'.

The character of the duty which producers owe their client constitutes a general dividing line between liability systems.

Producers are either under a duty of care or under a duty of result, cf. Chapter 11. Still, on both sides of the dividing line there are a good many other factors to be considered; each of the national overviews deals with them, particularly under 1.2.5–1.2.6 and 1.5.5–1.5.6. And in all systems the time elapsed between delivery (see Chapters 5 and 6) and the occurrence – or sometimes the discovery – of the failure is critical.

MINOR AND MAJOR DAMAGE

Several systems – see, for instance, overviews from Italy, Quebec, Spain, Sweden and the UK – distinguish between 'minor' and 'major' damage, with legal liability only for 'major' damage. There are no criteria valid for all such systems and legal practice is, to say the least, inconsistent.[1]

Whether the distinction between minor and major damage is important or not, and whether producers' duty is one 'of result' or one 'of care' there are, as mentioned in Chapter 7, three criteria for deciding whether a building 'fails'. They are:

(a) requirements stated in building regulations;
(b) requirements expressed in or implied by the client's brief; and
(c) requirements taken for granted in the particular culture where the building is situated, 'implicit requirements'.

---

[1] In the UK the Building Research Establishment has published guidelines classifying cracks in brickwork according to their structural significance.

REQUIREMENTS IN BUILDING REGULATIONS

As stated in Chapter 4, building codes are the outcome of ongoing decisions concerning the fundamental quality of buildings. For buildings requiring a building permit (or 'planning permission') the client normally[2] cannot demand that his producers build short of the public building regulations requirements.

In all liability systems, therefore, a code is a platform from which to decide whether a certain kind of observed failure in a building constitutes damage in the legal sense, i.e. whether there is a shortfall in relation to a codified requirement.

In other words, codes may serve to decide whether a certain type of failure is such that the producers' liability can be invoked on the basis of his breach of the code. Of course, there may be damage even if the code is observed, but the code offers a minimum quality level beneath which there is unequivocally failure. If this failure is caused by a defect (as defined in Chapter 9) there is, then, damage.[3]

The role of regulations for the delimitation of 'extraordinary events' is also often decisive. It will be recalled (Chapter 7) that Acts of God or force majeure belong to 'Group 1 – Alien causes', which may cause havoc to a building, but not damage. Of course, the definition of what is extraordinary is crucial. Again, just as for 'requirements', building codes may provide guidance: most building regulations indicate at least some loads and destructive agents which are presumed ordinary, such as suction and pressure from wind, load from snow, temperature ranges, sometimes even floods, certain earth tremors, attacks by rodents, limited impact from vehicles ramming the building and maybe others. If this is the case then only agents and forces more destructive than those stated in the code will be force majeure, and there is damage – i.e. no Act of God – if failure is caused by an event foreseen in the code.

But the code does not necessarily explicitly foresee any destructive agent to which the building may be subjected. If, for instance, local conditions are such that the producer ought reasonably to have anticipated exceptional – but not 'extraordinary' – destructive natural agents, then this could be construed as an 'implicit' code requirement. Thus, as mentioned in Chapter 8, a French court[4] has ruled that damage to a roof was due to 'admittedly unusual though not unforeseeable' rain. Although the code was silent on this particular climatic destructive agent the designer ought to have made himself familiar with the weather in that particular spot where heavy rain was by no means 'extraordinary'.

---

[2] There may, in some countries, be exceptions to this rule if the building is temporary. Also, at least in theory, so-called Crown buildings in the UK are not subject to building regulations and in Sweden some buildings for agriculture do not require a building permit.
[3] Whether this entails liability in any particular system is, of course, a legal matter.
[4] *Le bâtiment-bâtir*, June 1982, p. 22.

The definition of what is extraordinary may also depend on cultural patterns: as shown by the example in the section on 'implicit requirements', below, producers should anticipate that in France children swing on doors; a court in another country might rule differently.

Delivery of adequate instructions for use and maintenance may be prescribed in the code. If it is not, then it is often taken as an implicit requirement (see below).

As a general rule, if the client has stated no specific requirements, the answer to the question whether there is failure will depend only on public regulations and performance related to implicit requirements. Again, whether failure constitutes damage depends on whether the failure is caused by a defect or by an 'alien cause': see Chapter 7. And, again, whether damage entails compensation is a legal matter dealt with, system by system, in each of the overviews in Part Two.

CLIENTS' REQUIREMENTS

As mentioned in Chapter 7 on the concept of failure, a client may state specific requirements, for instance by notifying the producers of his intended use of specified chemicals within the building or of particular demands on the quality of air, maximum admissible vibrations and so on. Also classes of clients – educational, penitential, medical, military etc. – very often issue specific requirements (or 'standards', cf. Chapter 4).

Clients' requirements thus can be explicitly stated in the client's brief, but they may also be implicit, taken for granted by the client. The issue of clients' requirements therefore is sometimes complicated by the fact that the client feels entitled to assume that his producers – his consultants, say – are aware of his particular requirements, without the client having to state them in so many words.

An English court case illustrates this point.[5] A chemical manufacturer, who had used the same architect for several years, albeit not for purely industrial design, asked the architect to design a building which included a sizable storage tank. The architect, after consulting others, specified a stainless steel tank which, however, proved costly. In his attempt to save initial costs, the client made his own enquiries and ordered a replacement of the steel with a cheaper material. This tank failed by leaking. The client held the architect responsible for this failure. The court, although recognizing that the architect was not an expert in chemistry, found that the client was entitled to assume that the architect was aware of his client's particular requirements. The architect was therefore found to owe the client a duty to warn him that the substituted materials were unlikely to perform as required. Without that knowledge the architect could have claimed that he

[5] *Richard Robert Holdings Ltd v Douglas Smith Stimson Partnership* (1989) 47 BLR (QBD).

exercised all reasonable skill and care and although the failure undoubtedly constituted damage – being due to a defect – the architect would not have been liable for the damage. As it was, the court found that the architect had failed in this duty and held him legally liable.

What was said above concerning the distinction between 'failure' and 'damage' is of course valid also for damage defined as non-fulfilment of clients' – explicit and implicit – requirements. And, as always, the fact that some occurrences of failure constitute damage does not necessarily mean that the client is legally entitled to compensation.

'IMPLICIT' OR 'CULTURAL' REQUIREMENTS

Implicit requirements are those taken for granted, those assumed to be known by all parties to a building project. They are based on common sense, laid down in codes of practice or textbooks used by producers for reference, or can be assumed to be part of skills acquired during producers' initial or continued professional education and training or through their gaining professional experience.

Although it is difficult to quantify these requirements they may be crucial in post-construction liability matters. In penal and tax law, the letter of the law is decisive; this is different in civil law issues such as post-construction liability.

For instance, a French court took it for granted that a refuse chute, when a hatch is opened in the normal process of refuse disposal, must not spread dust from the duct through the open hatch.[6] When a chimney effect in the refuse disposal system in a very tall building blew out dust through an opened hatch, the court found the producers liable for the damage. The requirement 'no dust out of a hatch' is not stated in any French regulation but was taken as an implicit quality requirement in refuse disposal arrangements. Likewise, as mentioned above, there may be no regulation requiring that doors in a dwelling should carry the load of a child swinging on the open door but, again, this was to be taken for granted (French court case again. But only children under a certain weight!) Another French court[7] took it for granted that building elements in a concert hall should not emit noise during the performance. A client claimed, and obtained, compensation when the ceiling in such a hall vibrated audibly.

Another example of an implicit requirement follows: few building codes identify acceptable levels of vibrations, but some courts may find that a building 'fails' if the floors shake perceptively when the lift stops or the boiler starts. Building materials should not smell and glass panels in partitions should not shatter when a person falls against them, thus maiming the

---

[6] Quoted in *Le bâtiment-bâtir*, October 1982, p. 76.
[7] Cour de cassation, 3 chambre civile, 14 October 1992, Les Mutuelles de Mans, no. 1363 P.

person. These are all examples of requirements taken for granted in the particular culture (the UK) where the project is situated.

Delivery of adequate instructions for maintenance is often assumed to be a reasonable implicit requirement, whether or not expressed in the code. As stated in the section on 'Aetiology' in Chapter 7, failure due to wrong use or incorrect maintenance may constitute damage, namely if the producer provided the client with inadequate instructions at delivery, because this omission constitutes a defect (see Chapter 9).

A French court case already mentioned (see Chapter 8, p. 41) illustrated liability arising from non-delivery of adequate instructions: the caretaker of a clinic inadvertently flooded several floors with fuel when he intended to fill up the fuel tank of an emergency power plant on one of the upper floors of the building. The court found that the instructions given to the client by the producer were inadequate, and the producer was therefore held liable for the damage. Had the instructions been adequate and had the caretaker deviated from them, then the mishap would not constitute damage in the sense in which we use the word, since there would be no underlying defect (and hence no liable producer).

To avoid any misunderstanding: the fact that an event of failure constitutes damage, in other words, that it is due to a defect caused by an error or omission committed by a producer, does not automatically make it 'actionable'.

Whether the client is entitled to compensation will always, in the end, depend on the relevant legal system.

CHAPTER ELEVEN

Law and contract*

Daniel Alain Dagenais

INTRODUCTION

Compiling a comparative analysis of legislation originating from different sources requires the utmost care. In fact, laws establish or define rules chosen according to the particular social, historic, economic and judicial context in which they are to be applied and according to common usage.

Construction industries are themselves subject to each country's particular requirements: the climatic, geographic, geotechnic and economic conditions, customs and tastes of users, industrial needs and production means, professional bodies, means of governmental and other types of control.

Moreover, within each judicial system, the approach may vary according to the type of intervening party, the nature of the contract, the severity of the damage or the time elapsed after construction.

A proper understanding of each country's laws requires one to bear in mind all the background applicable to it and needs an analysis with a much greater scope than this brief account.

It should be noted, however, that a significant effort towards harmonization of the relevant international law has been made by the United Nations Commission on International Trade Law (UNCITRAL), which in 1988 made public the *Legal Guide on Drawing Up International Contracts for the Construction of Industrial Work*.

In fact, some of the major topics have a similarity, in particular, the intervening parties, the types of problem and the sort of risk encountered; therefore, the principles underlying judicial solutions are, without committing oneself, comparable. In order to administer the obligations and responsibilities of the various parties involved in a construction, each judicial system must make choices concerning these major topics, some of which will be discussed here.

---

* Original text in French.

DEGREE OF INTERVENTION BY LEGISLATOR

To start with, it must be stressed that certain legislations intervene more than others. Generally, codified systems of the Napoleonic type seem to prefer the establishment of stricter parameters which are obligatory: duration and extent of the guarantee, solidarity of the intervening parties, presumption of fault and responsibility, are established by legislation, that is to say, duties and responsibilities are defined by law, whatever wishes to the contrary are expressed in the contract. In this system, therefore, even non-liability or mitigation of liability clauses, freely negotiated between the parties, are without effect – unless perhaps the proprietor is himself an expert in construction.[1] This is explained by the priority given to public order, required here by the protection of property and public security – a collective heritage, the safeguarding of which has precedence over individual wishes.

Elsewhere, the principle of contractual freedom is preferred. From the start, construction is seen more as a private affair, the framework of which is formed by the wishes of the parties expressed in the contract which is their law, the strict rules of competition having finally to suffice in order to regulate the market and ensure good quality construction.

This liberal system allows one to limit the extent of liability, to restrict guarantees and to share risks according to the various wishes expressed in each particular contract. However, it is rare to find countries which allow total freedom. Norms, guidelines and controls are set up which are imposed on parties. Because the financial and human resources invested in building, and the consequences of an overly casual attitude, are inevitably sooner or later assumed by society as a whole, the latter imposes standards in order to protect itself and those co-contracting parties who are weaker or less aware of the risks involved. Nowhere is it accepted that the producer – deemed competent – has complete freedom of movement and should profit from his broader experience in order to negotiate with a less capable client conditions and terms which are dangerous and greatly unbalanced.

Moreover, in such a context it has been noted that due, no doubt, to social and governmental pressures, agreements are negotiated between groups (professionals – contractors – consumers) in order to make stipulations in contracts which will either be obligatory for the members of these groups or offered to whoever wishes to make use of them.

Everywhere, these different types of intervention seek first to promote quality, together with efficiency, and thereafter to establish each individual's responsibility in case of damage.

---

[1] T. Rousseau-Houle *Les contrats de construction en droit public et privé* (Wilson & Lafleur, Montreal, 1982) p. 317.

EXTENT OF DUTY OF PRODUCER

The producer will be held to a duty of care or a duty of result.

The producer who is tied by a simple duty of care will not be held responsible for the defects he has brought about or allowed unless it is established that he has not taken the reasonable measures that a normally competent and informed professional would have taken during the execution of his work. It should be noted that these measures, the rules of art, are not necessarily the best, but simply those in successful current usage under prevailing circumstances.[2]

However, the producer who is held to a duty of result will in no way be excused if the desired result – a sound building – is not attained. Therefore, on him rests the burden of responsibility associated with the limits of technical and scientific knowledge (the development risks). Held to a duty of result, a producer who uses to his best possible knowledge an innovative process about which it is believed everything is known but which turns out to generate damage, cannot claim ignorance. He is responsible because the result is not attained. In the majority of these cases, even the clear acceptance of risk, freely given by the client, does not constitute a valid defence.

Therefore, where there is a duty of result, is more restraint exercised with regard to the application of innovative techniques than where the mere respecting of the duty of care disclaims responsibility?[3]

Probably not. Certainly the producer on whom a duty of result rests takes all the risks. However, amongst the rules of art – the duty of care to which the other is held – is the one that is the most adequate for the technology in question; the producer would neglect his duty of care by using a too risky or insufficiently tested technology if it proves deficient. It must then be established if the producer has failed to recognize a risk or if he had a generally shared confidence in the safety of the technology.

The position of the client, in that he is less competent, less capable of evaluating the procedure than the producer, should also be considered. Has he been consulted and, if so, was his acceptance clarified with regard to the danger, known or unknown, to which he was being exposed? Is that even a pertinent factor?

At this point, duties of different natures meet and it is not so much these duties as professional conscience and the mastering of the rules of art which are the best guarantees for the quality of the building. It is, in other respects, the rules of competition and the advantages obtained by innovations which impose the innovations.

---

[2] *The London and Lancashire Guarantee and Accident Company v Compagnie F. X. Drolet* (1944) SCR 82 (Canadian Supreme Court).
[3] V. van Houtte *Risks in Construction: Damages which are not Caused by Negligence*, unpublished contribution to CIB Working Commission on Post-Construction Liability and Insurance, 1987.

In conclusion, with regard to the scope of duties of producers, it should be noted that, in some systems, after a certain period the error or omission which has caused the defect must be substantial, that is, either a gross mistake or showing a degree of negligence not to be expected of a professional.

DAMAGE WHICH TRIGGERS LIABILITY

Small day-to-day difficulties, resulting from the inevitable behaviour of even the best materials – for example, the normal drying out of structural timber or slight shrinkage of concrete – do not involve liability (unless the contract specifies it) since they do not, strictly speaking, constitute damage.

Likewise, minor damage resulting from simple bad workmanship, which was visible at the time of agreed completion of the building by the client, is in nearly all jurisdictions presumed accepted by him.

The courts seek above all to compensate the client for damage which was hidden at the time of completion and which only manifests itself subsequently, or the initial manifestation of which does not allow the client to foresee the major subsequent deterioration. In fact, delivery usually constitutes release from apparent defects.

Major damage – such as large cracks threatening solidity, substantial deterioration of insulation, dangerous chimneys etc.[4] – usually requires reparation. It is not necessary, in order to call for recourse, that the building is already in ruins and the damage has manifested itself to its full extent. The threat of damage already justifies corrective intervention and therefore a claim.

In the same way, the fact that the works are unsuitable for the use they were destined for opens the door to a request for compensation. The safety requirements of a police building are certainly greater than those of an administrative building; a hospital must be easier to maintain than a mechanical workshop; a music room requires special acoustic insulation. All these are requirements which a normally competent producer must take into account.

SOME MEANS OF EXONERATION

Whatever the nature of the duty of the producer, there are certain means of exoneration which are fairly usual.

One of these is 'Acts of God'. If it is established that the damage is

---

[4] For a comparison of major and minor faults under Quebec law – Napoleonic system – see D. A. Dagenais *Classification des désordres (mineurs, majeurs)*, unpublished contribution to CIB Working Commission on Post-Construction Liability and Insurance, 1987. Under Belgian law, see M. A. Flamme and P. Flamme *Le droit des constructeurs, L'entreprise et le droit*, 1984 (Brussels) p. 218 ff.

caused by an external, irresistible, unforeseeable event which has made correct execution of the works impossible,[5] then a producer cannot be held responsible. But what are these events in the context of building? The criterion of externality precludes consideration of whatever is within the producer's capacity to palliate; the unforeseeable character bars consideration of most major natural phenomena (earth tremors, tempests etc.), of which science, with the help of statistics and exact records, has learnt to recognize at least forebodings and patterns, if not to forestall them. Therefore, construction must be carried out in accordance with them. Irresistibility does not exist unless the obstacle is insurmountable, so that the circumstances which merely make the execution of a contract more difficult or onerous, are not means of avoiding liability.

One can conceive of certain extremely rare natural phenomena occurring without warning and not reasonably foreseeable as Acts of God, but it is more difficult to think of other unforeseeable events which might result from human intervention. Labour disputes might seem to constitute such an unforeseen event, but although they may result in acceptable delays in completion or perhaps increased costs, they could not excuse poor execution of works.

Other generally accepted means of exoneration from liability, even if the duty is a duty of result, are the acts of a third party, a breach of duty by the client or user and passage of time.

The first calls forth truisms: when a lorry which is out of control hits a building, when an arsonist or a vandal gets to work, when a neighbour working in his own home causes excessive vibration which leads to cracks in an adjoining building, one does not usually seek to lay the responsibility on the producer. It should be noted, however, that in Sweden the technical standards require façade structures to be designed with a view to possible traffic accidents. The factor of intervention by a third party is therefore no longer a means of full exoneration from liability.

The client does not usually intervene, either at the time of initial design or in the period of construction of the building. He establishes his requirements, approves certain architectural options which are presented to him and fulfils his payment obligations. If, however, he does more than that, if he intervenes with the job of the producers or imposes concepts, materials, specific techniques or methods of work, does he take part of the responsibility if a defect occurs due to his intervention? It might be thought natural to reply in the affirmative. But this is not always the case. In fact, the duty of the producer comes into play at this point, especially that of the professional, to advise his client correctly, a duty which might go as far as taking responsibility for the choices the client has made when agreed to by the professional, or even, for choices made by the client which the

---

[5] D. A. Dagenais *Cas fortuits*, unpublished contribution to CIB Working Commission on Post-Construction Liability and Insurance, 1987.

professional has neglected to contest. It seems logical, however, to take into consideration the client's degree of expertise. If his knowledge of what he proposes is greater than that of his adviser, it would seem natural that he assumes the risk since he is better able to judge it.[6]

In other respects, the owner must protect the building by adequate maintenance bearing in mind, however, that it is the producer's duty to explain particular requirements to him. He must only use the building for the purpose it was destined for. If he damages the roof whilst removing snow, if he changes a building destined as a hotel into a centre for violent delinquents, or if he omits to have the chimneys swept, he can only hold himself responsible for the consequences of his actions or omissions.

Finally, the passage of time is always a means of extinguishing liability. There is, first, mitigation of damages which can go as far as full proscription of the claim if, from the moment he discovers the damage and the time he institutes his claim, the client allows too great a period of time to elapse. The length of time varies according to circumstances but the principle remains the same: because rapid intervention usually makes repairs easier and less onerous, there is a requirement that the party one wishes to hold responsible for a defect is advised in time to intervene early on. In other respects, it is also generally accepted that works of a human nature are perishable. Mitigation may therefore go as far as complete suppression of the claim, if the damage occurs after a certain time, either fixed arbitrarily or varying according to the normal life expectancy of the works at issue.

TRANSFER OF THE BUILDING

Let us underline, finally, that in principle the transfer of a building after its construction, by sale or otherwise, does not usually suppress claims arising from the law. In many countries the rights follow the building; the right to claim belongs primarily to whoever is the owner at the time the damage occurs. This concept is understandable by the fact that, in such countries, it is not only the first owner of the building whom legislation seeks to protect, but also the building itself being, as it is, part of the common assets; in such a context, the owner's identity matters little.

However, when the claim is of a contractual nature, the identity of the co-contracting parties is important and subsequent owners, not being party to this contract, may not necessarily find the basis for a claim in it, unless this right can be transferred.

---

[6] *Davie Shipbuilding v Cargill Grain* (1978) 1 RCS 576 (Canadian Supreme Court).

SUMMARY

This brief account of some major questions being raised all over the world seeks merely to outline some of the legislative and statutory choices imposed in each jurisdiction. Other subjects could have been dealt with, such as the effect of the type of contract (estimate and contract, turnkey contract, sales contract), the different responsibilities for each producer (service contract, execution contract, sub-contract, advisory, checking and supervisory duties) or even the extent of indemnisation due to the victim, either by virtue of the contract or outside this framework. The aim of this chapter has been to contribute to a discussion of the issues and awaken curiosity.

CHAPTER TWELVE

The role of national governments in risk reduction

Gérard Blachère

National interest requires that building malfunction be controlled: malfunctions can result in personal injuries and economic losses. What is required is a balance between the costs of prevention on the one hand and the global, direct and indirect, real and social cost of malfunctions on the other.

The optimum balance between expense to avoid malfunctions and the cost of remedying malfunctions will not be discussed here. This presentation focuses on the nature of the measures taken by governments.

Primarily, the origin of malfunction is human error or human ignorance. The answer to error and ignorance is training and education based on research. This research takes several forms.

RESEARCH

THE NATURAL ENVIRONMENT

There are a variety of ways in which the natural environment acts on the building:

- Research on the soil: Governments organize the drawing-up and publication of geological atlases. Government agencies or universities study soil investigation techniques and soil mechanics; they also do research on earthquakes and deliver seismic charts, while governments issue seismic regulations improved through research.
- Research on the atmosphere: This involves better knowledge of the velocity and turbulence of winds; the flow and direction of rain, including driving rain; and of frost, temperature variations and natural radiation.

All this research on atmospheric agents is done by national meteorological agencies, building research units and, to a lesser degree, by universities.

BUILDING MATERIALS

This second branch of research in relation to building malfunction is chiefly directed on their durability, durability being the time span during which the material retains its original properties, or at least those which are taken into account in the design of the building.

In various countries this research is done at universities or in building research centres. Such centres may be linked to a trade or a material, e.g. wood, cement or clay, and are sometimes instigated by the government (e.g. the French law of 1948 on Professional Technical Centres) or they may deal with the whole area of building.

The latter – among them the BRE[1] in the UK, the CSTB[2] in France, the LNEC[3] in Portugal – constitute in most countries the kingpin for governmental strategy in the matter of building research.

Although building research centres were originally developed to improve labour productivity and reduce costs in building they often yield a spin-off effect in the field of malfunction prevention.

BUILDING WORKS AND TECHNIQUES

A third kind of research deals with building works and techniques of execution. This kind of research was born with the wave of innovation that followed the Second World War. Up till that time evolution and innovation had been so slow that acceptance could take place through use and adoption or rejection. However, in the fifties, when innovations became fairly abundant, it was necessary to evaluate innovative techniques at short notice to keep clients, architects and contractors informed.

The big national research centres have been essentially devoted to the *evaluation of innovations*, which was necessary to permit the use of innovations, the aim being to introduce new building techniques for enhancing productivity without reducing the performance of the building including, first and foremost, its durability.

Research centres devoted to building materials have the mission not only to improve the quality of the products but equally the productivity of their manufacture and to develop new uses of the materials.

Some countries have institutions specially geared to the prevention of malfunction. Such is the case in France with the Agence qualité construction. Financed by a levy on insurance contracts it carries out research and studies malfunction statistics by a permanent and intensive inquiry into all

[1] Building Research Establishment.
[2] Centre scientifique et technique du bâtiment.
[3] Laboratório Nacional de Engenharia Civil.

malfunctions observed, thus making it possible rapidly to detect 'epidemics' of malfunction.

In the UK, the NHBC[4] has had an impact on quality by making contractors respect its rules.

DISSEMINATION OF RESEARCH

The results of research are disseminated among professionals by publications such as the *BRE News*[5] and similar publications, or by the *Cahiers du CSTB*,[6] but it is generally asserted that building professionals and tradesmen are poor readers . . .

A better means of dissemination is training and education, including CPD.[7]

However, the best way to make the results of research known is by transcribing them into normative documents: products standards, codes of practice, technical regulations and, first and foremost, through agréments.

Normative documents channelled through different means are commonly respected in every country. In countries like Germany and the UK adherence is imposed by regulations and checked by local authorities; in other countries, like Belgium and France, adherence is established by insurance companies which cover the building team's ten-year liability.

Normative documents are largely based on the results of research:

- design criteria concerning, e.g., solidity, fire resistance and watertightness;
- product standards and, even more, performance standards; the definition of testing and the level of performance prescribed are essentially based on research.

But the type of normative document which more than any other relies on research is the agrément (or technical assessment or, today in France, *avis technique*): the agrément is an assessment of the aptitude for use of innovative ways of building with new materials or new technologies. By definition it is not possible to evaluate these innovations along the lines of descriptive standards or codes of practice which are so important for traditional building, and it is difficult for an isolated practitioner to carry out such an evaluation.

It is therefore necessary to carry out research on a collective level to evaluate how far these innovations satisfy the requirements of the users, be they defined by the brief or by regulations.

The agrément procedure requires performance tests to be established and calibrated by research.

[4] National House Building Council.
[5] A bulletin from the Building Research Establishment.
[6] A bulletin from the Centre scientifique et technique du bâtiment.
[7] Continuing Professional Development.

PUBLIC BUILDING REGULATIONS

Finally, public building regulations are very frequently based on research. For example, regulations about protection against noise close to highways or airfields is based on recent research as are regulations aimed at energy saving. It is noticeable that countries where research on thermal losses from buildings was better developed have been able to adopt more modern regulations than others. The same can be said also about regulations concerning fire resistance of load-bearing structures: research made it possible to base the regulations on calculation as well as on direct fire resistance testing.

Summing up, it is possible to say that through normative documents and regulations research contributes to the reduction and prevention of malfunction, this research being largely supported by governments or, if not by governments, then by national scientific and professional organisations.

ENFORCEMENT OF BUILDING LAWS AND REGULATIONS

Authorities issuing building laws and regulations have various concerns: occupants' hygiene and safety, but also, at various levels, their comfort – and also the durability of the building – or again, mainly in periods of crises, good use of materials and energy saving. Some nations' legislation lists the law's aims and the EC has, in its Directive 89/106 on the free circulation of construction products, produced a list of the essential requirements which are to be met by buildings:

- Mechanical resistance and stability
- Safety in case of fire
- Hygiene, health and environment
- Safety in use
- Protection against noise
- Energy economy and heat retention.

A stipulation about durability is included in these requirements.

There are two fundamental ways of ensuring that buildings satisfy the requirements. The first relies on public administrative supervision and checking of design, execution and qualification of participating parties: architects, surveyors, engineers, contractors, materials producers etc. This is the traditional approach to the problem; there is little trust in competence and professional conscientiousness. The other way, more modern even if it is two centuries old, is based on the idea that competent people may be made liable for respecting regulations and meeting users' requirements. This system prefigures modern quality assurance.

The first system is in use in Germany, Scandinavia and the UK. The second appears in the Civil Code of Latin countries, following the example of the French revolutionary civil code, commonly called the Code Napoléon.

However, in practice, even in countries which adopt the principle of responsibility, supervision is brought in – not an administrative one but private supervision linked to the necessity for the members of the building team (*les constructeurs*) to contract insurance to cover their liability. Even if the insurance companies do not require total and systematic supervision of design and work at the site, they exercise a certain amount of supervision to normalize the risks. This supervision is based on the same kind of normative documents as those used by administrative supervisors in other countries.

On this second point we may note, as a conclusion, that nations are not indifferent to the risks of building malfunctions and they impose, in one way or another, basic requirements concerning safety and durability and, often indirectly, a broader obligation to respect standards, codes of practice and agréments.

NORMATIVE DOCUMENTS

These normative documents are generally prepared and issued by national bodies enjoying some level of governmental recognition. Thus a nation devotes a part of its resources to these endeavours and, implicitly, recognizes that the work is worth the effort. For obvious reasons an exact balance between efforts and results can rarely be achieved. But the system could not survive if a sufficient agreement did not exist, at least among building professionals and politicians.

The national effort to establish normative documents is first the task of standardization institutions. Each of them contains a building branch issuing standards of various kinds: vocabulary, identification, classification, descriptive standards, performance standards, codes of practice and design criteria. All these texts, particularly the last, aim to ensure safety, even if the primary goal of standardization is an economic one.

Codes of practice constitute a particular kind of normative document. Fundamentally, codes of practice are normative documents: they describe a recommended method of technical planning and erection of buildings. They arise from the field of engineering and architecture, where standards originated among building materials producers. In some countries they are produced by national organizations of architects and engineers, in others they are the fruit of multiparty collaboration between all parties participating in the building team.

The latest national organizations contributing to the issuing of normative documents are the agrément institutions: agréments (i.e. technical assessments) exist in every country of Western Europe – Switzerland and Greece excluded – and exist nowhere else. Created in the immediate post-war period in France, the Netherlands and Portugal they spread throughout Europe. As said earlier, they are the answer to the need of introducing innovation in building, for the sake of labour productivity, materials and energy saving, with limited risks of malfunction. In that way, they constitute a centrepiece in national efforts to reduce the occurrence of building failure.

SUMMARY

As a conclusion, let us remember what was said at the beginning: it is in the national interest to control malfunction in buildings and the consequence is undertaking a national effort in building research, building standards, agréments and building regulations.

PART TWO

National Overviews

Australia

Liability

Freehill, Hollingdale and Page

Insurance

Ian Lancaster

Commissioned by Anthony Lavers

1 LIABILITY

1.1 General

Australia has a federal system of government. The federal body, the Commonwealth of Australia, has been granted, under the Australian constitution, powers to make laws in certain specified areas.

The Commonwealth has power over, inter alia, interstate trade and commerce, taxation, defence, posts and telegraphs, banking, trading and financial corporations and international affairs. The Commonwealth Government has powers not contained in the US and Canadian constitutions over a number of legal and commercial matters (e.g. bankruptcy and insolvency and the conciliation and arbitration of industrial disputes extending beyond any one state).

Laws in other areas are made by the six Governments and by the Legislative Assemblies of the two territories, the Northern Territory and the Australian Capital Territory.

Each State has its own Supreme Court with several jurisdictions (such as the equity, probate, admiralty, common law, commercial, criminal, family law and protective jurisdictions) and a lower court system for minor criminal matters and civil claims of lower value.

There is a Federal Court which exercises jurisdiction under Commonwealth law but which does not have general jurisdiction. Rather its powers are conferred as taxation, judicial review of Commonwealth administrative decisions, intellectual property, bankruptcy, appeals from the Industrial

Relations Commission, matters covered by the Trade Practices Act 1974 (Commonwealth), customs and excise, immigration and more.

The final court of appeal is the High Court of Australia which possesses an original jurisdiction over disputes between the States and for federal matters. It also possesses an appellate jurisdiction from the Supreme Courts of each State and Territory.

Before the passing of the Australia Act 1986, it was possible to appeal from a decision of a State Supreme Court to the Judicial Committee of Privy Council. It was necessary to elect whether to appeal to the High Court or the Privy Council (appeals from the High Court to the Privy Council were abolished in 1974). The standing in the various Australian jurisdictions of Privy Council decisions is not clear. However, it is at least true to say that Supreme Courts must treat decisions of the Privy Council and the House of Lords made prior to 1986 as persuasive. Decisions of superior English courts are of continuing persuasive effect. Since 1986, appeals to the Privy Council are no longer possible.

The Cross-Vesting Act 1987 (Commonwealth) establishes a system of cross-vesting of jurisdiction in Federal, State and Territory courts.

The Federal Act and the parallel State legislation are part of a general scheme which provides for cross-vesting of civil jurisdiction between the two Federal Courts and each State and Territory Supreme Court, and between each of the State and Territory Courts.

The scheme vests in the State and Territory Supreme Courts all the civil jurisdiction (other than certain industrial and trade practices legislation) of the Federal Court and the Family Court, and vests in the Federal Court and the Family Court jurisdiction of the State and Territory Supreme Courts. It also vests in State and Territory Supreme Courts civil jurisdiction of other State and Territory Supreme Courts.

1.1.0 LAW OR CONTRACT OR BOTH

The post-construction liability of a provider of materials and services to a recipient is primarily governed by the contract between the parties. Beyond the contract, liability may be imposed by the common law of negligence. The contract may also be regulated by State and Commonwealth legislation such as the Trade Practices Act, Fair Trading Act, Limitation Act, Contracts Review Act (NSW), Sale of Goods Act and, in Queensland, the Professional Engineers Act.

The contract may incorporate one of the Australian standard form General Conditions of Contract (for example, AS2124) or may be individually drafted for a particular project.

1.1.1 TERMS AND CONCEPTS

The terminology used in this chapter has been standardised for ease of understanding. Irrespective of the language utilised in standard form contracts, for the purpose of this chapter 'owner' will be used for the person or company for whom the building is being built and 'builder' will be used to denote the person or company carrying out the construction. 'Contract administrator' will be used for the architect, engineer or superintendent who is supervising construction and administering the construction contract.

The terms 'building', 'producer', 'building authority' and 'tenant' have no special meaning under Australian law. The meaning may vary depending upon whether these words are set out in a contract, the subject of decisions by the courts or defined by legislation. Different meanings may apply to different documents in the same project for different purposes.

A 'producer' (i.e. a builder or professional engineer) can in certain instances under the law of negligence and under legislation be held liable to tenants and third parties, i.e. persons who are not party to the construction contract.

(a) A 'building' would be any significant structural improvement to realty.
(b) The word 'producer' is not usually used in Australia in the construction context. The appropriate descriptions for persons doing construction work would be 'contractor', 'builder', 'prime contractor', 'head contractor' or 'supplier'. 'Builder' is used in this chapter.

An 'architect' is a person who designs and/or supervises construction. An architect is required to be registered as an architect under the Architects Act in each State.

A 'professional engineer' also designs or supervises construction but, unlike an architect, an engineer is not required to have any set qualifications in NSW. However, in Queensland, an engineer must be registered under the Professional Engineers Act.

'Contract administration' is used in this chapter to denote an architect's or engineer's construction supervisory role.
(c) 'Building authority' is not a word used frequently in Australia but it could be a town, city or shire council which receives building applications, gives approval for building work to commence and inspects completed work, or it could be a governmental authority or procurement committee that enters into construction contracts as an 'owner'. There is little authority in Australia for the proposition that a building authority (which term is used in the former sense) is liable to owners and tenants for defective construction.
(d) A 'tenant' is a person to whom property is leased. It is defined in property legislation but this definition does not necessarily apply to construction claims. If a tenant suffers injury or physical damage as a

result of the negligence of the producer, then if the preconditions for recovery in negligence are met (e.g. breach of standard of care, duty of care, causation, proximity, loss) the tenant may sue the producer in negligence.

It is difficult to succeed in recovering for pure economic loss.

1.2 During the Guarantee (Maintenance, Defects Liability) Period

1.2.1 DETERMINATION OF THE COMMENCEMENT OF THE PERIOD

In Australia, the period specified in a contract, during which the builder is contractually required to rectify defects is commonly called the Defects Liability Period. The word 'guarantee' is not generally used in this context. Defects existing at Practical Completion or becoming apparent during the Defects Liability Period may be the subject of a direction to rectify by the Superintendent.

According to Australia's three most common standard form contracts, the Defects Liability Period commences:

JCCA/B, clause 9.11

On the date on which the works reached Practical Completion (term defined in clause 1.06.09).

JCCA/B, clause 11.12

The Final Certificate shall be evidence in any proceedings that the works have been completed in accordance with the terms of the agreement to the reasonable satisfaction of the architect. The benefit of this evidentiary principle does not apply in the case of:

- fraud, dishonesty or fraudulent concealment relating to the works;
- any defect in the works or any part thereof which was not apparent at the end of the Defects Liability Period; and
- any accidental or erroneous inclusion or exclusion of any work, materials, goods or figure in any computation or any arithmetical error in any computation.

AS2124 – 1986, clause 37

On the date of Practical Completion (term defined in clause 2).

AS2124 – 1986, clause 42.8

No evidentiary presumption flows from the issue of a Final Payment Certificate.

NPWC3, clause 37.1

On the date of Practical Completion (term defined in clause 2).

NPWC3, clause 42.5

The issue of a Final Certificate shall constitute conclusive evidence that all work under contract has been finally and satisfactorily executed by the contractor.

This presumption does not apply where it is proved that the Final Certificate is erroneous by reason of:

(a) fraud, dishonesty or deliberate concealment, on the part of the contractor or any of his sub-contractors or of any of the employees or agents of the contractor or of any of his sub-contractors, relating to the works or any part thereof or to any matter dealt with in the said Final Certificate; or

(b) any defect, including any omission, in the works or any part thereof which reasonable inspection at the time of the issue of the said Final Certificate would not have disclosed; or

(c) any accidental or erroneous inclusion or exclusion of any work, materials, goods or figure in any computation, or any arithmetical error in any computation.

If an owner uses or enters into occupation of the project or part, the Defects Liability Period may commence earlier (e.g. under clause 9.10.04 of JCCA/B on the date of occupation or use).

1.2.2 SIGNATORIES TO AND NAME OF THE DOCUMENT STATING THE COMMENCEMENT OF THE PERIOD

JCCA/B, clause 9.09.02

The contract administrator signs the Certificate of Practical Completion. Practical completion is deemed to have occurred earlier in certain circumstances.

AS2124 – 1986, clause 42.5

The contract administrator signs the Certificate of Practical Completion.

NPWC3, clause 42.2

The contract administrator signs the Certificate of Practical Completion.

1.2.3 THE PERIOD'S DURATION AND TERMINATION

This is a matter for agreement between the parties. The usual duration for construction of commercial buildings in Australia is 12 months unless legislation (such as the builders' registration legislation) imposes a longer period. It is often shorter for residential buildings (e.g. 90 days).

There may be a separate Defects Liability Period for rectification work (a period not exceeding the original Defects Liability Period).

Is the Defects Liability Period voluntary or mandatory? Generally, the period is a matter for negotiation and agreement. In some Australian States and in particular in relation to residential developments, the duration of the Defects Liability Period is governed by legislation.

There are no formalities associated with the termination of the period. However:

JCCA/B, clause 11.02.02

The builder cannot make his final claim for payment until he has completed rectifying the defects. The contract administrator may not issue his Final Certificate before the builder has rectified defects.

AS2124 – 1986, clause 42.7

A final payment claim cannot be made until expiration of the Defects Liability Period.

NPWC3, clause 42.7

A Final Certificate is not issued until all remedial work is completed. The contract administrator may order testing or remedial work.

A second Defects Liability Period on remedial work may occur automatically in some contracts or if so ordered by the contract administrator. It could also be altered by agreement between the contract administrator and builder or between the owner and builder.

1.2.4 TIME ALLOWED FOR DIFFERENT CATEGORIES OF PRODUCERS FOR MAKING GOOD, REPAIRING OR PAYING DAMAGES

JCCA/B, clause 9.12

The builder is required to rectify the defects within a reasonable time but if not done within a reasonable time, the owner may engage others to rectify the defects and charge the builder for the cost of rectification.

AS2124 – 1986, clause 37

The builder must rectify any defects as soon as possible after Practical Completion.

NPWC3, clause 37.2

Minor omissions and minor defects which existed at the commencement of a Defects Liability Period must be rectified by the builder as soon as possible.

1.2.5 KINDS OF DEFECTS AND DAMAGE COVERED BY THE GUARANTEE

There is no guarantee in the strict legal sense. The builder is required as a primary obligation to make good defects existing at Practical Completion or becoming apparent during the Defects Liability Period.

JCCA/B, clause 6.11

'. . . faults, omissions, shrinkages or other defects in the Works' which are apparent.

AS2124 – 1986, clause 37

'. . . omission or defect'.

[1.2.6: *Not applicable*]

1.2.7 CONSEQUENCES OF PRODUCER'S NON-COMPLIANCE WITH UNEQUIVOCAL DUTY TO MAKE GOOD, REPAIR OR PAY DAMAGES

(a) If he is able but unwilling: Whether or not the builder is incorporated as most builders of commercial projects are, a builder who fails to repair or make good or to pay for the cost of another builder making repairs may be proceeded against in the usual way for damages for breach of contract. The builder's property and assets may be seized and sold to satisfy any judgement rendered against him.

   A complaint may also be lodged against the builder, in certain circumstances, with the State builders' licensing authority. This could result in the deregistration of the builder.

(b) If he is unable due to bankruptcy or liquidation: It is open to the trustee in bankruptcy or the liquidator, as the case may be, to disclaim the contract as an onerous contract. In the absence of such a disclaimer, proceedings could be initiated against the trustee in bankruptcy or liquidator, however, recovery would be controlled by available assets and priority rules for distribution.

(c) If he cannot be traced: If the builder cannot be traced then, subject to the rules as to the service of proceedings, action could be taken against him in his absence.

1.5   From expiry of the Guarantee (Maintenance, Defects Liability) Period

Under the law of the various Australian States and Territories, time does not necessarily begin to run from the date of expiration of the Defects Liability Period.

For example, under the Limitation Act 1969 (NSW) in the case of actions founded on contract the limitation period is six years running from the date from which the cause of action first accrued to the plaintiff. The corresponding period is 12 years in the case of actions founded on a deed.

When does the limitation period begin to run? Under the law of the various Australian States and Territories the key issue is identifying when the cause of action 'accrues'. For breach of contract, the cause of action accrues when the contract is breached. For negligence the cause of action accrues when the damage becomes 'manifest'. There is a problem when the contractual obligation breached is of a continuing character (e.g. a duty to keep in good repair) as to when breach in a construction contract occurs. Does it occur at the time of defective construction or at the time of completion of construction?

Concerning variations upon the basic period of limitation for personal injuries the position appears to be:

PRE 1 SEPTEMBER 1990 CAUSE OF ACTION

(a) There is a six-year limitation period.

(b) Plaintiffs who were unaware at the relevant time of the fact, nature, extent or cause of the injury, disease, or impairment can bring extension of time applications at any time (provided this is within three years of becoming aware of all of these matters).

(c) Plaintiffs who were unaware of the fact, nature, extent or cause of the injury, disease or impairment at the relevant time can still bring applications outside that three-year period but only if they secured the relevant extension of time order before 1 September 1992 (the amnesty).

(d) The old extension of time regime (sections 57–60 of the Limitation Act 1969) still applies.

CAUSES OF ACTION ACCRUING ON OR AFTER 1 SEPTEMBER 1990

(a) There is a three-year limitation period.
(b) There is a discretionary extension of time for up to five years.
(c) Plaintiffs who were unaware of the fact, nature, extent or cause of the injury, disease or impairment at the relevant time, provided they bring an application within three years of becoming aware of all these things, can obtain an unlimited extension of time to bring proceedings.

The above regime relates to New South Wales.

1.6 Transfer of ownership and producers' liability

Under the law of the various Australian States and Territories the benefit of a contract is inherently assignable. However, each of the Australian Standard General Conditions of Construction Contract provide that neither party may assign the contract without the prior written approval of the other.

If it is proposed that the burden of a contract be transferred the appropriate vehicle is the novation agreement which is a tripartite agreement under which there is a complete ab initio substitution of one party by another.

2 PROPERTY (OR MATERIAL DAMAGE) INSURANCE

2.1 General

2.1.0 LAW OR CONTRACT OR BOTH

Property insurance is not required by law. It is always required by banks and mortgagees and may be required for a property management corporation. Owners sometimes take it out. Mortgage default loss may also be covered.

2.1.1 TERMS AND CONCEPTS

The 'policy holder' will be the owner, i.e. the client.

2.1.2 TECHNOLOGY AND INSURANCE TERMS AND CONDITIONS

Do terms and conditions depend upon:

(a) Insurers collecting technical information about the project?
Yes.
(b) Geographical location of the building, height etc.?
Yes, there are extensions of cover notably in North Australia for cyclones, floods etc. Also the height of buildings is relevant, particularly above six floors.
(c) Actual technology used?
Materials make a difference. Technology would not.
(d) The levels of competence of the producers?
In general no, but this could change for individual contractors or architects.
(e) Whether the insurer uses his own control organization?
Could do – insurers do have their own surveyors.

2.1.3 INSURER'S TECHNICAL CONTROL

Does the insurer:

● Have his own organization for control?
Yes.
● Consult the client's control bureau?
Yes.
● Consult the public control bureau?
Yes.

It is usually a combination of all three. This could influence the terms and conditions.

2.1.4 INSURERS' POSSIBLE INFLUENCE UPON BUILDING TECHNOLOGY

Do the insurers:

● Pool technological experiences?
Yes. The Insurance Corporation of Australia has its own specialist consulting service.
● Convey technological experience to producers?
Yes, insurers consult with bodies such as the Standards Association.
● Convey technological experiences to clients?

Clients are generally not well informed and therefore not so keen on such knowledge.

Is the insurers' experience taken into account:

- By authorities issuing codes and norms?
  Yes.
- By standardization organizations?
  Yes.
- By agrément organizations?
  No.
- By authors of textbooks/handbooks?
  To some extent.

2.1.5 NEW TECHNOLOGY

Assuming access to all relevant information would it be

(a) theoretically possible
(b) practically possible

to insure an innovative building against defects?

(a) Yes.
(b) Yes.

2.1.6 THE INSURANCE MARKET, GENERAL IMPRESSION

(a) Is there effective price competition?
   Yes, most definitely. It is severe.
(b) Is there a tendency to monopoly or oligopoly?
   About ten underwriters dominate the market as it concerns major buildings.
(c) Is competition national or international?
   Many of the Australian firms are part of international groups, although there are significant insurers without international connections.

2.2 During the Guarantee (Maintenance, Defects Liability) Period

2.2.0 EXISTENCE OF THIS KIND OF INSURANCE

Does property insurance cover losses due to damage during the 'guarantee' period?
No.

[2.2.7: *Not applicable*]

2.5   After the Guarantee (Maintenance, Defects Liability) Period

2.5.1   NORMAL DURATION OF COVER; CANCELLATION

(a) Is the cover always/normally annual?
    Normally.
(b) If so will premium and excess be the same for all consecutive years?
    No, it varies according to market conditions and specific claims.
(c) Under what circumstances can the policy be cancelled:
    - By the insurer?
      For non-payment of premiums; for non-disclosure of material facts
      under the Insurance Contracts Act.
    - By the insured?
      At any time.

2.5.2   LEVELS OF PREMIUMS; EXCESS; FRANCHISE

Information not currently available.

2.5.4   TIME ALLOWED FOR NOTICE OF CLAIM AND FOR PAYMENT OF DAMAGES

These are covered by the Insurance Contracts Act.

2.5.5   KINDS OF DEFECTS AND DAMAGE WHICH CONSTITUTE CLAIMS EVENTS

(a) Could a defect which has not caused damage constitute a 'claims event'
    for property insurance purposes?
    Normally not, but it could do if preventive work was necessary to avoid
    damage.
(b) Are there any rules as to what types of damage can constitute claims
    events?
    It depends upon the drafting of the policy. Structural problems are
    often excluded from property insurance. However, Industrial Special
    Risks and Industrial All Risks cover everything which is not specifically
    excluded. For example, subsidence is typically excluded.

2.5.6   EXCLUSIONS

(a) Defects which should have been discovered and included in Certificate
    of Practical Completion?
    Yes.

(b) Defects not developing into damage?
Yes. See question 2.5.5 above.
(c) Loss due to action of third party?
Normally covered, with right of subrogation for the insurance company.
(d) Loss due to Act of God/natural causes?
Would be covered, although flood or earthquake might be specifically excluded.
(e) Other force majeure?
Would be covered, but some exclusions such as nuclear attack.
(f) Loss caused by building user's failure to follow instructions or maintain?
Normally excluded, by requirement of due care and maintenance.

2.6   Transfer of ownership and insurance cover

Does the insurance cover follow the building or the owner?
The owner: the new owner must take out a new policy.

2.6.1   RULES

Are there any procedures for transfer of cover, e.g. inspections?
No, provided the same company is prepared to take on the risk.

3   LIABILITY AND PROFESSIONAL INDEMNITY INSURANCE

3.1   General

3.1.0   LAW OR CONTRACT OR BOTH

(a) Is contractors/professional indemnity insurance required by law?
No.
(b) Is it always/often/sometimes required by mortgagees/finance institutions?
Sometimes.

[3.1.1: *Not applicable*]

3.1.2   TECHNOLOGY AND INSURANCE TERMS AND CONDITIONS

Do terms and conditions depend upon:

(a) Insurers collecting technical information about the project?
Yes.

(b) The geographical location and height of the building?
    Yes.
(c) Actual technology/materials used?
    Yes.
(d) Levels of competence of producers?
    Yes.
(e) Whether insurer has own control organization?
    Yes.
(f) Claims record of insured party?
    Yes, especially this.

3.1.3 INSURER'S TECHNICAL CONTROL

Does this:

(a) Follow design and site work through its own control organization?
    If the insurance is on a project basis, yes. If it is annual, no, since this is turnover based.
(b) Consult the client's control organization?
    Probably not.
(c) Consult the public control bureau?
    No.

Does the organization of control affect terms and conditions of this kind of insurance?
The terms and conditions would vary little in this respect.

3.1.4 INSURERS' POSSIBLE INFLUENCE UPON BUILDING TECHNOLOGY

(a) Do insurers pool their experiences?
    No, only on large cases requiring re-insurance.
(b) Do they convey this information to clients?
    No, definitely not.
(c) Is their experience taken into account by public authorities?
    Yes, especially following major failures.
(d) Do insurers categorise, classify and rate producers?
    Not formally.
(e) Is this information published?
    No, it is sub judice during claims. Some information passes by word of mouth.

3.1.5 NEW TECHNOLOGIES

Assuming the insurer has access to all relevant information is it

(a) theoretically
(b) practically

possible to obtain this type of insurance on a project involving new techniques or materials?
Theoretically yes, but the key here is how much cover is needed. Reinsurance on large insurance often has to go out of the country, often to London.

3.1.6 THE INSURANCE MARKET, GENERAL IMPRESSION

There is limited price competition depending on the size of the cover. Large amounts may not be very competitive, or rather competition would be more London based than Australian. There are five main firms in professional indemnity, two Australian, the rest international. The market is hardening as the drift of rates is fuelled by litigation.

It is possible to get an extension under decennial insurance to cover architects and engineers. This is offered by the French company UAP, but it is believed that not much has been sold.

3.1.7 COLLECTIVE INSURANCE FOR PROFESSIONAL OR OCCUPATIONAL GROUPINGS. GRADATION OF INSURANCE CONDITIONS

There are such arrangements, notably for lawyers. Since reputation/status would include continuing education and development it could be said that this indirectly influences terms.

3.2 During the Guarantee (Maintenance, Defects Liability) Period

3.2.0 EXISTENCE OF LIABILITY AND PROFESSIONAL INDEMNITY INSURANCE

It is a claims made policy and is, therefore, retroactive in effect.

[3.2.7: *Not applicable*]

3.5   After the Guarantee (Maintenance, Defects Liability) Period

3.5.1   NORMAL DURATION OF COVER; CANCELLATION

(a) Is cover annual?
    Yes.
(b) Does failure to pay premium result in cancellation of cover?
    Yes. 90 days is normally given to pay premium.
(c) Can insurer and insured cancel the contract? Yes, insurer for non-
    payment of premium or non-disclosure of a material fact. The insured
    can cancel at any time.

3.5.2   LEVEL OF PREMIUMS, EXCESS AND FRANCHISE

Precise data are not available, but generally indemnity insurance is hardening and this has caused professionals to set up their own firms as it has
become expensive to buy indemnity cover.

3.5.3   TIME LAPSES BETWEEN OBSERVATION OF DAMAGE AND INVESTIGATION

See Insurance Contracts Act.

3.5.4   TIME ALLOWED BETWEEN INSURER'S RECOGNITION OF A CLAIMS EVENT AND HIS PAYING OUT

See Insurance Contracts Act.

3.5.5   KINDS OF DEFECTS AND DAMAGE WHICH CONSTITUTE CLAIMS EVENTS; ONUS OF PROOF

There are no rules determining what can constitute a claims event.
    The plaintiff bears the burden of proving fault by the relevant producer.

3.5.6   EXCLUSIONS

(a) Libel and slander are excluded.
(b) Dishonesty and fraud are excluded.
(c) If a professional joins a construction firm his indemnity cover may not
    operate.

Belgium*

Liability and Appendix II
Maurice-André Flamme and Philippe Flamme

Insurance and Appendix I
Philippe Fontaine

1 LIABILITY

1.1 General[1]

1.1.0 LAW OR CONTRACT OR BOTH

The contractual liability of professional producers (contractors, property developers, architects, engineers, technical offices, technical inspection or control offices) who have been involved in the construction of a building as the result of a contract for services (Latin *locatio operis*), concluded with the client, usually ends with the handover of the works to the client.

However, articles 1792 and 2270 of the Napoleonic Civil Code[2] – in Belgium[3] – impose a ten-year liability period on construction professionals after handover,[4] for defects (or damage) which simultaneously meet two conditions:

---

* Original text in French.

[1] The most recent academic work is *Le droit des constructeurs* (special issue 1984 of the Review *L'entreprise et le droit*, 34 Rue du Lombard, Brussels) by M. A. Flamme and Philippe Flamme.

[2] Article 1792: 'If a building constructed on a lump sum price basis (i.e. fixed price) suffers in whole or in part from construction defects, or from defects in the soil, the architect and the contractor shall be liable for ten years.' Article 2270: 'After ten years, the architect and contractors shall be released from the guarantee for major works they have carried out or managed.'

[3] These two articles have not been amended in Belgium, as has been done in France.

[4] Belgian practice is to effect handover in two phases (a 'provisional' handover and another 'final handover', separated by a 'guarantee' period, usually of one year's duration. In the absence of a specific clause in the contract, the courts lean towards the view that ten-year liability starts from the time of final handover, but the parties are free to specify commencement at the time of provisional handover (see M. A. Flamme and Philippe Flamme *Le droit des constructeurs* (1984), nos. 122 and 256).

(a) that they affect a building or major works (1.1.1); and
(b) that they are sufficiently serious, i.e. they affect the stability of the works or one of its main elements (1.1.1).

It should be noted that outside the conditions of application of articles 1792 and 2270 of the Civil Code, producers frequently[5] grant a 'post-handover' guarantee to the client, of variable duration (2, 5, 10 or even 20 years!), by means of a specific clause in the contract.

1.1.1 TERMS AND CONCEPTS

DEFINITION OF 'BUILDING' OR 'WORKS'

Contrary to the French Law of 4 January 1978, which uses the terms 'works' (*ouvrages*) and units of equipment of a 'building' (*bâtiment*), Belgian law has retained the original text of articles 1792 and 2270 of the Napoleonic Code.

These use the terms 'building' (*édifice*) and 'major works' (*gros ouvrages*).

Without making the least distinction between new construction and rebuilding works (nor between residential and non-residential buildings), it is case law and the ordinary courts which decide, case by case, if the damage for which reparation is requested by the victim actually affects a 'building' or 'major works'.

One gains the impression that the courts are often hesitant (for example, regarding heating installations, flooring, piping systems etc.) and that they are often motivated by their concern for increased protection of the consumer against building producers.

Therefore the following are usually considered 'a building' or 'major works':

- buildings or their main elements;
- the fabric of a building;
- a television aerial;
- roofing;
- a tennis court;
- the framework of a building;
- large sections of glass;
- a lift;
- a central heating installation;
- plastering which is not a simple finish but a true wall lining, intended as protection against the elements;
- civil engineering works (such as motorway bridges or locks);
- external works, sewage, drainage etc.

[5] For example, for waterproofing works, electrical or heating installations.

On the other hand, the following are usually considered 'minor works' which are not likely, therefore, to involve the ten-year liability of building producers:

- flooring (tiled);
- ceilings;
- partition walls between apartments;
- chimney flues;
- piping laid along walls and ceilings which is not flush.

It should also be noted that, according to Belgian case law, it is not sufficient that the damage cited affects 'a building' or 'major works', it must actually affect, i.e. endanger the stability of, the building or major works; in brief, it must be unquestionably serious. For example:

- defective waterproofing of a roof;
- poor mixing of reinforced concrete;
- persistent damp due to porousness of bricks;
- insufficient acoustic insulation.

On the other hand, the following are considered to be insufficiently serious to endanger the stability of a building:

- slight cracks in partitions or plastering, the result of normal settlement in every reinforced concrete building;
- poor quality of stones in a façade;
- defects affecting only so-called finishing works.

Recent case law points out, however, that although such a restrictive criterion (namely, the endangering of a building or stability of major works) could previously have been envisaged when the construction, stonework, stability and strength of a building were essential and not always obvious, current technical progress would render this kind of guarantee pointless.

In addition, construction today is a great deal more than the necessity for four walls and a roof. It is the attraction of a complete unit, furnished with the equipment essential for comfort and enjoyment that, because of advancing technology, is now a normal expectation from a building, whether it is intended as a dwelling or workplace.

In short, the court in Verviers[6] held that case law was correct in finding that the faults likely to activate the ten-year guarantee are not only those which, properly speaking, affect the stability and strength of the building itself, but also those which affect essential equipment, thereby rendering the works 'unfit for their intended use'.

The court also rightly added:

---

[6] 1st District, 2 December 1985, *Delfosse v Schmits*.

- that there must be limits to the development of case law, otherwise it might take into consideration minor inconveniences and purely subjective discomfort;
- that therefore, only defects which are objectively sufficiently serious to compromise the intended use of the building should be considered as defects;
- that in the case of condensation on double-glazed units, however annoying it might be, it does not constitute a defect which endangers the stability or strength of the building, or even an inconvenience serious enough to compromise its use or intended purpose;
- that therefore, in this case, there are no grounds for the application of ten-year liability.

DEFINITION OF 'PRODUCERS' (CONSTRUCTEURS)

On which professionals and trades does 'ten-year post-handover' liability rest?

They are:

(a) Most obviously, contractors,[7] architects, consulting engineers, technical supervisors.
(b) Technical inspectors (*contrôleurs techniques*).[8]

Insurance companies usually require the latter to act as 'consultants' to the client as an essential prerequisite to agreeing to cover the ten-year liability risk. These 'inspectors' merely 'supervise' and do not assist in preparing drawings or overseeing the works. The architect and contractors bear complete responsibility for their own professional work.

It should be noted that in Belgium no state or municipal public control office exists.

(c) Property developers – after a Law passed on 8 July 1971 (see Appendix II to this chapter) regulating the construction and sale of dwellings 'to be built or in the process of being built' extended ten-year liability (already applicable to developer–contractors) to 'developer–vendors' of houses or apartments to be built or in the process of being built, if the house is intended to be lived in at least partially and the contract of sale requires the buyer to make one or more payments before completion of the building.
(d) 'Concessionaires of works', i.e. contractors who are granted the right by public authorities to operate public works (for example, a refuse incineration factory), which they have undertaken to build (article 5 of the Law of 14 July 1976 and Royal Decree of 14 November 1979).

---

[7] Unless they construct a building for themselves (i.e. outside the framework of a contract) and then sell it in a completed state.
[8] See also Appendix I to this chapter.

(e) Holders of a 'public development contract', which arises where a public law body which does not have the necessary budget resources uses a private law natural or legal person for the finance, execution and possible design of a building which meets its particular requirements (or the requirements of a third party such as potential purchasers or tenants of subsidised housing or manufacturers at whose disposal it is advisable to put certain construction works in the framework of an economic development policy). The building is usually let on a lease with a purchase option or even immediate or future transfer of ownership (article 1–3 of the Royal Decree of 18 May 1981, *Le Moniteur belge* (*journal officiel*), 16 June 1981, p. 7664).

(f) Sub-contractors, liable for ten years with regard to the main contractor, who is considered to be the client insofar as they are concerned.[9]

But do sub-contractors have a possible direct liability with regard to the actual client? In the absence of a contract between them, the client could not sue the sub-contractor direct for contractual liability.

Could he then – if it is in his interest[10] – consider the sub-contractor responsible for defects as a third party and seek damages on the basis of quasi-delictual liability (liability 'in tort')?

Case law had always replied in the affirmative in the past, but since a judgement by the Cour de Cassation (Supreme Court of Appeal) of 7 December 1973, this is no longer the case.[11]

The Supreme Court held:

- that the executing agent (such as the sub-contractor) whom the main contractor substitutes for himself in order to carry out all or part of the contract, is not a third party with regard to performance of the contract and to the client;
- that therefore the client cannot invoke the liability of the sub-contractor in tort 'unless the defect attributed to him consists in the violation, not of a contractual obligation, but of a universal duty of care and if this defect has caused damage other than that resulting merely from poor execution of the contract'.[12]

An excellent example of this quasi-delictual or tortious liability of the sub-contractor to the client is provided by a judgement of the Cour de Cassation (Supreme Court of Appeal) of 20 October 1983,[13] given in a case where a company called Sterilco had requested the contractor Mallet to carry out some extension works of an autoclave cell, intended for use in the

---

[9] Flamme and Lepaffe *Le contrat d'entreprise* (1966), nos. 635–636.
[10] For example, because the main contractor has been declared bankrupt.
[11] See M. A. Flamme and Philippe Flamme *Le droit des constructeurs* (1984), no. 201.
[12] It seems 'other damage' should be construed as the loss of an 'advantage' or 'benefit' which the victim was entitled to expect from the contract, in brief, an 'extra-contractual damage'.
[13] *Pasicrisie* (Belgian Law Reports) 1984-I-182.

sterilization of medical and pharmaceutical products and where Mallet (the main contractor) had sub-contracted this work to Gallier, sheet metal workers.

Gallier's welder was busy welding when a fire broke out in the room he was working in and completely destroyed Sterilco's workshop.

Since it was established that the cause of the fire was carelessness on the part of Gallier's welder, who was working without having removed the particularly inflammable wooden partitions which enclosed the cistern and provided thermal insulation, the court allowed the claim by the client against the sub-contractor Gallier, who had to answer (Civil Code, article 1384, para. 3) for its agent's wrongful act, which constituted a violation of the universal duty of care and which had caused damage other than that resulting from poor execution of the contract.

With regard to suppliers who have contracts of sale with producers, the client cannot seek to make them liable in contract.

The client could only sue them on a quasi-delictual basis or in tort.

TENANTS

If a tenant suffers damage which is the fault of one of the 'producers' or one of their agents, he can seek reparation on a quasi-delictual basis (Civil Code, article 1382 ff.), i.e. in tort.

1.2   During the Guarantee (Maintenance, Defects Liability) Period

PRELIMINARY REMARKS

In Belgium, it is normal practice (and not the law) which is the origin of the handover operation being split into two phases ('provisional handover' and 'final handover') separated by a 'guarantee' period (see 1.5, however).

This means that if the contract only provides for one handover, in Belgian law – contrary to the French system – there is no 'guarantee period' (called 'guarantee of formal completion' in France).

1.2.1   DETERMINATION OF THE COMMENCEMENT OF THE PERIOD

This commencement is fixed by the date of provisional handover, when the client, usually assisted by advisers (architects, engineers, technical offices), declares his acceptance of the works, thereby acknowledging that they conform to the state of the art and the requirements of the contract.

It is up to the producer (usually the contractor) to claim this provisional handover when he considers the works to be complete.

It cannot be refused if the defects are of a minor nature. Any unjustified refusal gives the producer grounds to appeal to the courts, who may declare a handover date which is retrospective to the date the works were actually completed.

Provisional handover may be with or without qualifications.

It may even be tacit, i.e. inferred from circumstances (taking possession, payment of price) indicating an undoubted intention on the client's part to accept the works.

1.2.2 SIGNATORIES TO AND NAME OF THE DOCUMENT STATING THE COMMENCEMENT OF THE PERIOD

This document usually takes the form of a report which is signed by the producers whose works are accepted, as well as by the client, assisted by his advisers – usually his architect.

Obviously, handover granted to the contractor does not release the architect from his own contractual obligations.

If he has neglected to point out in the report certain defects which should be apparent or detectable by a professional man, he is liable to the client.

1.2.3 THE PERIOD'S DURATION AND TERMINATION

Belgian law remaining silent on this subject, it is the contract which stipulates the duration of the period, usually one year.

The end of the period is then marked by 'final handover', which usually takes the form of a second report, but which can sometimes be inferred from the actions of the client (for example, the repayment of a bond or bank guarantee).

This period may obviously be extended, but only by agreement between the parties.

1.2.4 TIME ALLOWED FOR DIFFERENT CATEGORIES OF PRODUCERS FOR MAKING GOOD, REPAIRING OR PAYING DAMAGES

As there are no statutory provisions on this point, these periods are decided upon by the client or, if the producer disagrees, by the court to which the client appeals.

1.2.5 KINDS OF DEFECTS AND DAMAGE COVERED BY THE GUARANTEE

During the guarantee period, the duty of the producer is essentially to 'maintain', which means carrying out all necessary works in order to put or maintain the building in proper condition and in good working order.

In addition to this obligation of maintenance, it is his duty to repair damage pointed out on provisional handover and mentioned in the report, as well as damage which was not apparent on handover but which was revealed during the period. However, he cannot be held responsible for repairs necessary due to a case of force majeure, either by the act or through the fault of a third party or by the client himself.[14]

It is by no means certain that a defect which has not developed into damage cannot give rise to the payment of damages. In fact, a defect which does not take the form of actual damage can certainly reduce the value of a building.[15]

There are no rules for determining the type and extent of damage which gives rise to a right of indemnification.

1.2.6 EXCULPATION OF PRODUCER

VISIBLE DEFECTS

Whether provisional handover exempts the producer from all liability for visible defects is the subject of lively controversy in Belgium.

Wrongly basing itself on what it believes to be usual practice, prevailing case law states that 'provisional handover aims only to record completion of the works and, being of a provisional nature, it does not in itself imply acceptance'.

However,

- on the one hand, a recent judgement by the Cour de Cassation[16] (Supreme Court of Appeal) finds that the parties are free to recognize a power of acceptance at the time of provisional handover with regard to defects which the client has been able to discover by that time;
- on the other hand, academic lawyers[17] have shown, at the conclusion of an in-depth investigation into the construction field, that the usual practice is the exact opposite of that imagined by the courts, namely, that

---

[14] Errors and omissions in the design, drawings, ground, materials specified; abnormal use of the works.

[15] Possibly insufficient proportions of cement, or the substitution of cheaper materials than those specified etc.

[16] 24 February 1983.

[17] See *Le droit des constructeurs*, no. 122 ff.

of the two handovers, it is provisional handover which assumes the greater importance and has serious consequences, particularly:

(a) the transfer of the 'custody' of the building and the risks of loss through force majeure;
(b) penalties for delays cease to run;
(c) the right to payment of the balance of the price;
(d) the right of at least partial reimbursement of the retention guarantee on the instalment payments and the bond;
(e) the covering of visible defects.

See 1.2.2 regarding the possible liability of the architect for negligence at the time of handover.

DAMAGE CAUSED BY AN UNKNOWN THIRD PARTY OR BY FORCE MAJEURE

Irrespective of whether acts of vandalism or events of force majeure (hurricane, fire, earthquake, flood etc.) are involved, the question is settled by article 1788 of the Civil Code, which relates to the situation where the producer supplies the material. It provides that 'if the goods perish in any way whatever before delivery, the loss shall be borne by the worker [i.e. the producer], unless the client delays in accepting the goods'.

In this way, therefore, the producer supplying the materials bears the risks of loss through force majeure up to the time of delivery, but on delivery (handover) – or from the moment the client fails to take over the works in spite of formal notice – the risks are transferred to the client.

USE OF EXPERIMENTAL TECHNIQUES OR TECHNIQUES INVOLVING INCREASED RISKS

The use of materials and innovative or high-risk procedures does not exempt the producer from liability.

This is not the case when there is interference by a manifestly competent client with whose orders a producer could legitimately comply, expressing strong reservations if necessary.[18]

It is important to add that even without interference by the client, the occurrence of damage arising from the use of new materials or procedures will raise the question of the apportionment of liability between producers.

Since the choice of materials and procedures is usually made at the 'design stage' of the works, it is the architect alone who will usually assume liability, thereby excluding the contractor, who is a mere executant.[19]

---

[18] See *Le droit des constructeurs*, no. 314 ff.
[19] Except, obviously, if the design defect is so obvious that the contractor should have pointed it out, expressed reservations, or even refused his co-operation.

LACK OF MAINTENANCE

A client, proprietor or tenant who fails in his duty of maintenance[20] or who uses the building or its equipment without complying with the appropriate instructions supplied by the producer, obviously commits a fault which will have the effect of obviating or at least reducing the producer's liability.

The recent publication – by the Belgian building research centre CSTC (Centre scientifique et technique de la construction) – of a *Maintenance Manual for Buildings (Carnet d'c..retien des bâtiments)* sums up the duties of proprietors and tenants and seems likely to assist judicial experts and the courts in establishing each party's liability.

'DUTY OF RESULT' OR 'DUTY OF CARE'?

This is an important factor with regard to proof of fault.

If the producer's obligation is to achieve a specific result and the quality of the works is inferior to that promised or there is bad workmanship or defects, since the producer has not achieved the promised result, he has obviously not fulfilled his obligation. Far from having to prove fault, the client can simply declare the existence of the shortcoming or defect, in which case the producer can only avoid liability by proving an outside cause (an act by the client or a third party; force majeure).

If, on the other hand, the producer only has an obligation to act with diligence, i.e. to take due care in achieving a given result – and this is often because the result desired is too uncertain – it is up to the client to prove fault by the producer.

Belgian case law is hesitant but appears to lean – much more so than in France – towards the obligation to act with diligence.

A judgement by the Liège Court of Appeal of 23 October 1974 (appeal dismissed by the Supreme Court on 26 February 1976) seems to be particularly clear:

'On being ordered to indemnify the victim of a traffic accident caused by an abnormally slippery road surface, the Belgian State was wrong to seek indemnification from the contractor who successfully tendered for the reprofiling works and road surface repairs, on the ground that it was the contractor who had an obligation to achieve a specific result, to ensure execution of the works without defects, which he failed to do. Secondly, being responsible for site traffic safety, he failed to give warning of the abnormality, rectify it and notify the public authorities.

The prevailing view of academic lawyers and case law is that, unless

---

[20] A judgement by the Tribunal de Bruxelles (Brussels Court) (Chamber 11 bis, 9 March 1983, Société Carrier Bénélux) refers to an obvious lack of maintenance and service of a heating installation.

there is an express stipulation to the contrary, the contractor generally has only an obligation to act with diligence. This applies particularly to contracts where the task of the contractor is limited to the execution of works designed by the client and in accordance with precise instructions imposed by him. This is a case in point, since there are instructions in the technical specifications giving precise details as to the nature and quality of the materials to be utilised and the way in which they are to be applied. It follows that although the contractual liability of the contractor is invoked under the guarantee, the State must furnish proof – which it has failed to do – of fault by the contractor in the execution of the works.'

In the same way, one could not reproach even a specialist contractor for not having found a process for the solution of a problem – that of condensation in double glazing which has simply been set with putty, a problem which the legal expert admitted was insoluble.[1]

However, case law finds that there is an obligation to achieve a specific result[2] with regard to:

- property developers who provide a complete construction service, material as well as administrative, thereby contracting to provide purchasers with an apartment which conforms to the type agreed;
- waterproofing works for cellars by a specialist contractor with specific knowledge of the works to be carried out and who had, in addition, made a site inspection;
- Thermopane glazing (insulating panes), the glazier contending in vain that he was only to carry out the installation 'to the best of his ability';
- achieving the heating levels stipulated in the contract;
- poor aesthetic result, and in any case not conforming to the sketches accompanying the drawings, achieved by the contractor in charge of the construction of a fountain.

Finally, it is essential to determine the parties' intentions and to analyse 'the content of the obligations undertaken', because one and the same contract often comprises a variety of obligations, some being for diligence, others being specific obligations or obligations to achieve a specific result.

With regard to the main criteria used by the courts to distinguish the two types of obligation, these are:

(a) The uncertainty of achievement of the desired result, i.e. whether this depends exclusively or not on the person who undertakes that the desired result will be achieved.
(b) State of technology.

---

[1] *Laboureur v Chambel-Sadacem*, 23 November 1976, Civil Brussels, Chamber 3.
[2] *Le droit des constructeurs*, no. 60.

For instance, the current state of science makes it necessary to assume that obligations for the supply of radar to the Régie des voies aériennes (Airways Authority) are relative.[3]

As regards the influence of 'standards' and technical 'regulations' published by public bodies (national or international) and even scientific publications and recommendations from research centres (such as CSTC) it is obvious that these standards and recommendations rapidly acquire the status of 'rules of the art', or accepted practice, so that failure to comply with them puts the producer at fault.[4] Compliance with such standards etc. does not necessarily exclude any fault whatever, but it gives rise to a very useful presumption of no fault by the producer.

(c) Degree of specialisation of the producer.

In a case where a timber building was constructed by the assembly of prefabricated parts by a specialist firm, virtually without the services of an architect, the specialist contractor taking on the drawing up of a description of the works, preparation of drawings and role of architect, it was stated that:[5]

'It is common sense that the contractor or producer who specialises in timber buildings is solely responsible for the quality and treatment of the materials he uses and for the proper execution of their installation; . . . in this type of construction, the contractor has an obligation to erect a building which has no defects relating to his speciality, especially the treatment of the timber to be used because this work requires specialist skills for which reason the client has entrusted him with the design as well as the execution of the project; . . . it is because of these special conditions that it seems natural not to employ an architect.'

(d) The active role played by the producer in the design of the project itself, for example in the supply of a turnkey project or a refuse incineration plant.[6]

(e) Examination of the contract clauses, any specific obligations,[7] the intention of the parties.

---

[3] Airways Control Department, Brussels, 27 February 1967, JT 418, notes by Flamme.
[4] A specialist plumbing contractor who, contrary to the rules of the art explained in the recommendations of the CSTC, failed to insulate the pipes bedded in a concrete screed, was liable for their corrosion and damage affecting a central heating installation. He could not invoke in his defence either a desire to effect savings by the architect, or errors in design by the latter or in the supervision of the execution of the works (Brussels, 8 May 1980, ED 1981, p. 141).
[5] An unpublished judgement by the Cour d'Appel de Bruxelles (Brussels Appeal Court) of 21 December 1970 in *SPRL Firme Vandeleene v Hemeleers*.
[6] Cour de Cassation, 19 November 1970, *L'entreprise et le droit*, no. 171, p. 142, notes by Flamme.
[7] Regarding, for example, the watertightness of a swimming pool or foundations, the waterproofing of a roof or the maintenance of water supply pipes.

For instance, in a case where the client complained of condensation in double glazing and the technical specification stated that this glazing should be protected against condensation by external evacuation or that condensation should be rendered impossible, the Brussels Appeal Court[8] stated:

> 'that the client's interpretation – that the contract allowed no manifestation of condensation – was not in accordance with the general character of the provision, which is designed to allow adaption to different methods of production; that its object is protection against condensation but it leaves to the contractor the choice between two methods in order to achieve this; that the existence of this alternative shows that the desired result was to ensure efficient protection against condensation, but not to render condensation impossible; that as the respondent correctly observed, in imposing evacuation, the clause acknowledged that condensation could occur; . . . that on the other hand, the tenderer should manage to eliminate any damp, by whatever means at his disposal.'

1.2.7 CONSEQUENCES OF PRODUCER'S NON-COMPLIANCE WITH UNEQUIVOCAL DUTY TO MAKE GOOD, REPAIR OR PAY DAMAGES

(a) If the producer can effect repairs, but refuses to do so, the client should usually apply to the court (Civil Code, article 1184) to order the recalcitrant producer to carry out the repairs or alternatively to rescind the contract with damages.

Since judicial procedures are lengthy and repairs often cannot wait, case law allows the client, subject to certain precautions,[9] to proceed with the necessary repairs himself, in anticipation of the court's decision and therefore at his own risk if the court eventually concludes that the producer is not at fault.

(b) If the producer is declared bankrupt, and on the assumption that he had liability insurance, the victim of the damage – i.e. the client or successive owners of the building – has a preferential claim on the indemnity the insurer owes to the insured.

The insurer can only obtain a valid discharge by paying the insurance indemnity to the victim.[10]

(c) If the building company is wound up, it is still considered to be in existence for purposes of liquidation, and the liquidators may

---

[8] 2 March 1979, *Laboureur v Chamebel-Sadacem*.
[9] Formal notice to effect repairs within a reasonable period, addressed to the recalcitrant producer, drawing up of a statement of the defective works.
[10] Cf. *Le droit des constructeurs*, nos. 254–255.

complete any contract, the non-completion of which might prejudice the company's interests.[11]

Otherwise the situation remains the same as in (a) and (b), above.

With regard to 'other producers', i.e. other professionals participating in the construction of the same building, the question is not clear. No doubt it amounts to asking what is the potential liability of other producers to the client if one producer has refused to carry out the requested repairs.

If the damage for which repairs are demanded is attributable to one producer, it is hard to see how the failure of one party could involve any liability on the part of other producers.

If, on the other hand, the damage is caused through the combined fault of several producers,[12] who contributed indivisibly to the whole of the damage, the client may obtain judgement against all of them jointly and severally, which means that each of them may be made liable for the whole damage, although he can later seek indemnification from the others for their respective share of the liability.

[1.2.7: *Not applicable*]

1.5   From expiry of the Guarantee (Maintenance, Defects Liability) Period

Under Belgian law – as has been seen (in 1.2) – no legal provision[13] exists which institutes a 'guarantee period' separating two handovers, the first 'provisional' and the second 'final'.

If the contract remains silent on this issue and there is therefore only one handover – which is more usual than is generally belived – there is no question of a 'post-guarantee' period and the 'ten-year guarantee period' provided for by articles 1792 and 2270 of the Civil Code (see 1.1.0 and 1.1.1) arises.

On the other hand, if two handovers are provided for,[14] the first constituting the beginning and the second the end of the 'guarantee' period, the question in Belgium is then whether the 'post-guarantee' period is exactly the same as the 'ten-year guarantee period' (Civil Code, articles 1792 and 2270).

---

[11] Articles 178 and 182 of the Commercial Companies Laws.

[12] Often a fault in execution by the contractor and, in the case of the architect or engineer, a fault in the design or supervision of the works.

[13] Except articles 5 and 9 of the Law of 9 July 1971 on the construction of housing and the sale of housing to be built or in the process of being built, when the purchaser or client is required to make one or more payments before completion of the works, cf. Appendix II to this chapter.

[14] Either by law (see preceding note) or more often by the contract (especially technical specifications covering public works contracts).

In fact, according to prevailing case law, provisional handover, due to the fact that it is only provisional, in no way constitutes acceptance and could not therefore start the ten-year statutory guarantee period running.

Under this system, therefore, the ten-year guarantee would begin to run at the date of formal handover and would only end 11 years after the completion of the works recorded by the provisional handover document.

In practice, however, the contracting parties usually stipulate in the agreement[15] a clause expressly providing that the ten-year period runs from the date of provisional handover and the Supreme Court has held (see 1.2.6) that such a provision[16] is valid.

In this situation – the most frequent today – producers' ten-year liability ends ten years after the date of provisional handover but during the first year of these ten years, i.e. during the year separating the two handovers, the client may:

- not only claim against producers under their ten-year liability for major defects endangering the stability of the building or its main elements;
- but also claim against them under their contractual liability in general laws where articles 1792 and 2270 of the Civil Code do not apply, i.e. even for minor defects or damage.

1.5.1 DETERMINATION OF THE COMMENCEMENT OF THE (POST-GUARANTEE) PERIOD DURING WHICH A PRODUCER MAY BE HELD LIABLE

The idea of a 'post-guarantee' period is not very helpful in Belgium, where it is often confused with the 'post-handover' period or the 'ten-year guarantee' period provided for by law (Civil Code, articles 1792–2270).

In fact, as we have seen above (see 1.5), either:

(a) where no guarantee period is provided for – the period called 'post-guarantee' is exactly the same as the ten-year guarantee period beginning on the date of (sole) handover of the works; or
(b) where the contract provides for a guarantee period (one year) separating the two handover operations – the 'post-guarantee' period coincides either, again,
  - exactly with the ten-year guarantee period commencing on the date of final handover, or
  - with the last nine years of this period of ten years if the latter has already commenced on the date of provisional handover, by virtue of a specific clause in the contract.

---

[15] Especially in the general technical specifications for public works contracts.
[16] In the absence of which there is therefore a great risk that ten-year liability (to be insured) terminates only 11 years after completion of the works.

1.5.2 NAME OF THE DOCUMENT STATING THE COMMENCEMENT OF THE POST-GUARANTEE PERIOD; ITS SIGNATORIES

As already seen in 1.2.2, this involves a document signed by the client, his advisers (architect, engineer) and the producers whose works are accepted or approved.

When the contract provides for two handovers, the final handover document is frequently 'forgotten', in which case final handover is sometimes inferred from the actions of the client and the expiry – without qualification on his part – of the one-year guarantee period.

1.5.3 PERIOD(S) OF LIMITATION

LIMITATION PERIOD FOR CONTRACTUAL LIABILITY

Articles 1792 and 2270 of the Civil Code lay down a period of ten years[17] for all categories of producers bound by contract to the client.

The parties are obviously free to extend the ten-year liability – for example, by stipulating a longer period in the contract. However, with regard to the question of whether it is possible for the parties validly to include a clause in the contract excluding or limiting ten-year liability, Belgian case law has replied in the negative (unlike the Dutch courts) on the ground that articles 1792 and 2270 of the Civil Code embody requirements of public policy.

However, two reasons for prolongation of the guarantee period cannot be ignored:

(a) Acknowledgement of liability: If, during the guarantee period (i.e. ten years) the producers 'acknowledge' their liability for the damage, e.g. by spontaneously carrying out repairs without requesting payment, a new period of ten years beings to run for the repair works.[18]
(b) Fraud: In the event of fraud (i.e. if the producers have intentionally carried out bad workmanship or fraudulently substituted one material for another) liability lasts for 30 years.[19]

LIMITATION PERIOD FOR CIVIL OR DELICTUAL (TORTIOUS) LIABILITY

Apart from their contractual liability to the client, contractors, architects, engineers, technical offices etc. have quasi-delictual liability (in tort) for 30 years from the date of the act generating the damage.

---

[17] This period concerns not only the duration of the guarantee, but also means that any action must be brought before the end of the ten years, otherwise the right of action will be lost.
[18] See *Le droit des constructeurs*, nos. 258–262.
[19] *Le droit des constructeurs*, no. 264.

As among themselves

The problem of a guarantee claim by the architect against the contractor or vice versa[20] arises not only when a joint and several judgement[1] is given against both of them, but also when one of them is ordered to make good the entire damage suffered by a client who, whether because of negligence, deliberate intention, prejudice or personal convenience, brings an action against only one producer.

This hypothesis is often confirmed in the situation where the client obtains judgement against one producer by reason of some fault originally on the part of another. The former then usually seeks indemnification from the latter, whose fault is the direct cause of the damage.

With regard to third parties

(a) Professional producers may incur their quasi-delictual liability for personal acts (Civil Code, article 1382) with regard to third parties,[2] sometimes in connection with failure to meet their contractual obligations to the client, sometimes independently of such failure;
(b) they must answer for damage caused to articles[3] of which they have 'custody', i.e. which are in their possession with powers of supervision, management and control (article 1384, para. 1);
(c) they are answerable for fault on the part of their employees (article 1384, para. 3 and article 1797). See example at 1.1.1 (f).

And also, although rarely, with regard to the client

Although contractually bound to the producer, the client may bring an action against him in quasi-delictual if, first, the fault consists in a breach of the universal duty of care and, second, if the damage caused is other than the loss of an advantage expected from the performance of the contract (see 1.1.1 (f)).

On the other hand, settled case law refuses to apply the ten-year limitation period to claims against architects, contractors and other advisers by a client who has to make good damage caused to third parties (tenants, passers-by, neighbours) through poor design or bad workmanship in the structure.[4]

For example, under article 1386 of the Civil Code, 'the proprietor of a

---

[20] Non-contractual claim, since the architect and the contractor are third parties in their personal relationships.
[1] When the fault of several producers has indivisibly produced the whole damage (see 1.2.7 *in fine*).
[2] For example, tenants, neigbours, passers-by etc.
[3] Site machinery etc.
[4] See *Le droit des constructeurs*, no. 385.

building is liable for the damage caused by its decay when it has occurred as a result of lack of maintenance or a defect in construction'. The proprietor obviously has a claim against the professional person (architect, contractor etc.) to whom the construction defect is attributable.

1.5.4 TIME ALLOWED (DIFFERENT CATEGORIES OF PRODUCERS) FOR MAKING GOOD, REPAIRING OR PAYING DAMAGES

See 1.2.4.

1.5.5 KINDS OF DEFECTS AND DAMAGE QUALIFYING FOR DAMAGES

Unlike defects and/or damage recorded at the time of provisional handover or manifesting themselves during the one-year guarantee period – which all, without exception, require repairs or compensation[5] – defects and damage covered by articles 1792 and 2270 of the Civil Code which are likely to give rise to the producers' ten-year liability must fulfil specific conditions (see 1.1.0 and 1.1.1), namely:

- affect a 'building' or 'major works';
- be sufficiently serious.

It follows that a defect which has not developed into damage will not generally be considered to endanger the stability of a building or its main elements and that its manifestation during the period of ten years following formal handover[6] should not usually give rise to the producer's ten-year liability.

We say 'should not usually . . .'. Why? Because a curious Belgian decision,[7] prompted by a concern for fairness and increased protection for consumers with regard to professionals, considers – by analogy with the law of sale of goods[8] – that as handover does not have the slightest effect of acceptance with regard to a hidden defect, after handover producers remain liable (even after final handover) for minor hidden defects, i.e. defects which do not meet the statutory conditions (especially that of seriousness) for the commencement of ten-year liability.

For these minor defects, the client may therefore seek to hold the

---

[5] Even if a defect has not manifested itself as damage (see 1.2.5).
[6] Or the last nine years of the ten-year period if this commenced on the date of provisional handover.
[7] See *Le droit ds constructeurs*, no. 126.
[8] Civil Code, article 1641 ff., concerning the warranty that the goods sold have no hidden defects.

contractor liable for the duration of 30 years (i.e. three times longer than for the most serious defects!) provided that he brings an action shortly after[9] discovery of the defect.

This case law is absurd but it prevails in Belgium, and has been approved by the Supreme Court in a judgement dated 25 October 1985.[10]

1.5.6 EXCULPATION OF PRODUCER

See 1.2.6.

As to whether the producer is under a duty of care or a duty of result (cf. 1.2.6), Belgian law – unlike French law which since 1978 has been based on a presumption of fault against producers – may be illustrated by a recent judgement given by the Antwerp Civil Court.[11]

After handover a storm damaged the asphalt roof and sliding doors of a recently constructed industrial workshop. In an action against the main contractor on the basis of ten-year liability, the client relied on a very brief expert report which merely described the damage and remained silent as to whether it was caused by a defect.

The main contractor replied that it was up to the client to prove that the damage was the result, not only of a defect, but of a defect giving rise to the producer's ten-year liability, i.e. endangering the stability of the building or one of its main elements.

The court dismissed the client's claim on the ground that he was manifestly confusing the dilapidation (or damage) to the roof and doors with the defect which was the cause of the damage and which he had to prove.

1.5.7 CONSEQUENCES OF PRODUCER'S NON-COMPLIANCE WITH UNEQUIVOCAL DUTY TO REPAIR

See 1.2.7.

1.6 Transfer of ownership and producers' liability

1.6.1 RULES

It is settled case law that the client's right of action for ten-year liability passes to the successive purchasers of a building, as articles 1792 and 2270 of the Civil Code attach statutory protection to ownership of the building.

---

[9] The length of the period is assessed case by case by the courts.
[10] No. 4758, *NV Metaalbouw Vandekerchove v NV Lannoy and NV Solvay.*
[11] 9th Chamber, 22 March 1985, *NV Herold v MBG etc.*

Where the ownership of a house or apartment to be built, or in the process of being built, is transferred in the framework of a contract of sale, the same Law (of 8 July 1971) which extended ten-year liability to the vendor of such a house or apartment provides that 'the ten-year guarantee by the vendor shall benefit successive proprietors of the house or apartment' but 'an action may only be brought against the original vendor'.

2 PROPERTY (OR MATERIAL DAMAGE) INSURANCE

2.1 General[12]

2.1.0 LAW OR CONTRACT OR BOTH

No legal obligation for this type of insurance exists in Belgium. It is relatively little used in the construction of housing, but is more often taken out for apartment buildings, offices, industrial buildings and civil engineering works.

We shall examine two types of policy here, i.e. 'All Risks' insurance and so-called 'Supervision' insurance.

The former may be obtained by annual agreement with an additional declaration at the opening of each new site. Therefore it usually covers contractors only and guarantees exclusively damage sustained by the works under construction. It is usually taken out for specific works and may therefore contain a section for 'civil liability'.

The latter is always obtained for a particular construction. Its main purpose is to cover the ten-year guarantee period. During this time, it also covers civil liability which the insured may incur as a result of loss or damage to the project.

2.1.1 TERMS AND CONCEPTS

'BUILDING' (BÂTIMENT)

All Risks policies usually define the item insured as being the works which are the subject of the contract, including materials, whether prefabricated or not, which are intended to be incorporated therein. In each case, specific conditions often describe in more detail the subject of the insurance cover.

In principle, fixed installations are included in the general concept.

For Supervision (*contrôle*) insurance, the works will be those inspected by the supervising office; parts not inspected will therefore in principle be excluded. The whole of the works carried out on site may therefore be

---

[12] This analysis was written in strict conformity with the questions asked. It is therefore not to be considered exhaustive insofar as certain questions are not raised in the Questionnaire.

covered inasmuch as they are necessary for the execution of the project insured. (See Appendix I to this chapter)

It will be noted, therefore, that the term 'building' (*bâtiment*) is not found in this type of policy. This word is generally used only in insurance for 'fire and various risks' (water damage, storm damage etc.) which is not the subject of this study.

POLICY HOLDER (BÉNÉFICAIRE)

The usual holder of a policy of this type is the person who actually suffers the loss or damage, i.e. in principle the client.

Certain All Risks ('TR': *tous risques*) policies expressly state that payment will be made to the policy holder.

Other, more sophisticated policies, state that all the parties involved on site shall be joint and several creditors and the insurer will choose, taking into account essentially the objective of repairs or rebuilding.

2.1.2 TECHNOLOGY AND INSURANCE TERMS AND CONDITIONS

The parties are free to contract according to their own wishes, merely respecting the relevant legal limits and regulations, which in practice are not very restrictive.

The reply to all the questions posed in (a) to (e) of the Questionnaire (see Appendix A) is therefore 'yes'.

Competition is free and therefore dependent on all the conditions of the market in question, though the Insurance Supervisory Office (Office de contrôle des assurances), a public body of the Ministry of Economic Affairs, supervises the general conditions of insurance issued by companies. A prudent insurer will therefore only contract in accordance with a preliminary study of the risk related to all the elements of which it consists.

2.1.3 INSURER'S TECHNICAL CONTROL

The distinction to be made here between All Risks policies and Supervision insurance is absolutely fundamental.

The former only rarely provide for the site to be inspected by a supervisory body, but this is compulsory in the latter. This difference is obviously justified by the fact that All Risks policies only rarely cover the post-handover period and this only for a very limited period, while Supervision policies cover precisely the ten-year period.

Moreover, insurers working in these two areas usually have available a technical department experienced in site matters or, failing this, will call upon specialist experts to inspect certain phases of execution of a specific construction, if necessary. Contracts expressly provide for the insurer's access to the site, as a safeguard.

Insured producers are required to enter into a contract with the technical office designated by the insurance company. This office's task consists, amongst others, in examining the drawings, supervising the execution of the works, participating in handover and giving technical assistance to the insurer in the event of a claim. The instructions of the supervisory body are mandatory.

2.1.4   INSURERS' POSSIBLE INFLUENCE UPON BUILDING TECHNOLOGY

No specific organization exists in Belgium for the collection and centralization of technological experience. The same applies to systematic information (data) from producers and clients.

Nevertheless, different bodies springing mainly from private enterprise undertake in practice to draw conclusions from actual experience. There is, among others, the National Committee for Action in Construction Safety (Comité national d'action pour la sécurité dans la construction), the Scientific and Technical Construction Centre (Centre scientifique et technique de la construction), the National Construction Federation (Confédération nationale de la construction) etc. All these institutions have acquired in the course of time a kind of official recognition de facto or de jure. They act in all areas of their sectors and in this way carry out various precautionary measures. Obviously, insurers specialising in this field give these bodies their active co-operation. To this is added the efforts of experts and supervisory bodies in whose interest it is to co-ordinate their initiatives in this direction.

It is to these different institutions that the authorities and bodies responsible for standardization will turn, if necessary.

2.1.5   NEW TECHNOLOGY

The insurability of a risk depends primarily on the wishes of the insurers. There is nothing to stop one or more insurer from covering an 'innovative' building, even in the absence of any statistics or technical data.

In practice, however, insurers who are approached with a view to covering a technologically new risk will, in most cases, instigate an in-depth appraisal of the potential risk by an expert. They will seek to spread the risk by turning to co-insurance and re-insurance of the cover. They will study the technical tests which will have been carried out prior to the adoption and implementation of the system or product.

All these preliminaries will be likely to influence the terms and conditions of the policy, although the insurer will allow himself every freedom in negotiations.

2.1.6 THE INSURANCE MARKET, GENERAL IMPRESSION

Generally, it may be said that cover for post-handover risks is not much sought after, at least for the whole of the period of contractual liability (ten years). Ten-year insurance is only usually taken out for major, high-risk projects, which makes it relatively expensive.

There are companies which specialise in the construction sector but generally it can be said that all the major companies, even foreign ones, currently operate in this sector.

2.2 During the Guarantee (Maintenance, Defects Liability) Period

See 2.1.1.

There is, in principle, in fact no 'damage' insurance in the French sense of the word, i.e. serving exclusively to cover the losses sustained by the client and/or proprietor of the building.

In Belgium the All Risks policy performs this purpose during the execution of the works, and possibly for a short period thereafter (one or two years). Insofar as the conditions of the policy are duly fulfilled, the insurer indemnifies the prejudiced party, usually the client, for any damage or losses affecting the insured property.

Therefore the insurer will possibly only exercise a right of recourse if the producer responsible does not have the capacity of an insured party (if, for example, the policy holder has not insured the sub-contractors), or if he is covered for civil liability by another policy having priority.

The system is the same for Supervision insurance, since this covers not only the client, but also the contractual and non-contractual liability of producers, during a maximum of ten years from the date of provisional handover.

2.5 After the Guarantee (Maintenance, Defects Liability) Period

2.5.1 NORMAL DURATION OF COVER; CANCELLATION

If the policy is for a specific project, which is always the case for Supervision and often for All Risks policies, the duration of cover will usually equal the period of site works, plus the agreed post-handover period.

In an ongoing All Risks policy, the duration will be freely determined by the contracting parties.

A provisional premium is usually fully or partly paid when the insurance contract is taken out, final settlement taking place on completion of the works, on the basis of their total cost.

Early cancellation may take place by agreement or automatically in the event of the insured party's bankruptcy or during legal settlement, delays in the payment of premiums, death or dissolution of the insured party and, finally, in the event of flagrant violation of his contractual obligations.

2.5.2 LEVELS OF PREMIUMS; EXCESS; FRANCHISE

Many criteria and classifications are involved in the working out of rates, which are at the same time extremely complex and very flexible in order that they may be adapted to the negotiation of each contract. These basic rates are, however, strictly confidential and it is hardly possible to give a detailed analysis here.

The type of project is obviously an important factor. The same is true of the specific risks which it entails by reason of its location, choice of materials, architectural design etc.

There are also special conditions for conversion works.

2.5.4 TIME ALLOWED FOR NOTICE OF CLAIM AND FOR PAYMENT OF DAMAGES

NOTICE OF CLAIM

Periods of time are always prescribed by policies so that a claim is properly made known to the insurance company. These periods of time are usually precise and expressed in terms of days or even hours.

Sometimes the text is rather more vague, such as 'The insured party must as a matter of urgency inform the company of any damage or loss likely to give rise to a claim and submit a detailed report in writing as soon as possible, including all relevant information, otherwise he may forfeit his claim.'

Other conditions are more specific and usually give a period for notice varying from three to eight days from the occurrence or knowledge of the loss.

If notice is not given within the prescribed period, the insured party must prove that he was unable to give notice, i.e. that he was prevented by an occurrence of force majeure. Therefore this involves good faith, i.e. independence in relation to the other party's wishes. If such proof is not duly provided, the insurer will be entitled to forfeit any right to indemnification. This does not necessarily mean that he will do so automatically. In fact, insurers are usually very flexible in this respect. In the majority of cases, it would require very serious negligence on the part of the insured party, which actually hampered the insurer's actions regarding determination of the causes and assessment of the loss, for the contractual condition to be relied upon.

Case law currently shows a clear tendency to require the insurer to prove the existence of loss in this respect or, at least, that the likelihood is sufficient. However, this does not affect the actual principle of forfeiture, only some of the details of applying it.

PAYMENT OF INDEMNITY

Some policies provide for periods of time for the payment of indemnity, but this is not a general rule. Where it does exist, this condition usually provides that payment will be made as the repairs or the rebuilding proceed.

Legally, any claim for payment of damages is barred three years after the occurrence of the event which gave rise to it. Most insurance policies shorten this period contractually and it usually varies from 3 to 12 months.

2.5.5 KINDS OF DEFECTS AND DAMAGE WHICH CONSTITUTE CLAIMS EVENTS

Defects are only covered for the damage they cause.

A mere defect as defined in the framework of the present study is not compensated for in principle. The general conditions of All Risks policies usually refer to the idea of damage or losses, which connects with the term '*désordre*' (damage) used here as a criterion. Some policies prefer to use the word '*dommages*' (losses). It all amounts to the same thing, in that the conditions usually exclude losses or damage resulting from failure to comply with accepted practice or legal, administrative or contractual provisions, as well as any loss consisting, inter alia, of lost profit, aesthetic damage, contractual penalties etc. The last-mentioned are, in other words, those not directly affecting either the works or the insured property.

If an incident is to be covered, it must be unusual or abnormal in some way. Therefore, ordinary shrinkage cracks, wear and tear, or slight deterioration due to lack of use or to wear and tear are not indemnified.

Certain Supervision policies, moreover, only insure the consequences of total or partial collapse or serious damage likely to endanger the stability of the works.

2.5.6 EXCLUSIONS

In addition to what has been stated in 2.5.5, regarding the limits of the concept of 'claims', it may also be useful to point out below the following exclusions usually found in the relevant policies, which are:

- damage to any valuables and documents;
- mechanical, electrical or hydraulic failures;
- damage arising from stoppages on site;
- damage arising from malicious intent, fraud, conscious and deliberate violation of laws or accepted practice;
- costs in connection specifically with the defective part of the works;
- damage affecting means of locomotion or transport;
- events of war, radioactivity, theft etc.

Consequently, damage from defects which might reasonably have been discovered earlier will not automatically be excluded unless they are, for example, the result of obvious bad faith. Losses caused by the actions of a third party will lead to a claim against the latter by the insurer.

Natural forces are not automatically excluded, nor in general are occurrences of force majeure.

Finally, it should be noted that any behaviour constituting serious fault, i.e. virtually intentional fault making the loss foreseeable and almost unavoidable, constitutes a reason for exclusion.

To complete the picture, it should be pointed out that some or all of the above-mentioned exclusions can be waived by special conditions.

2.6   Transfer of ownership and insurance cover

2.6.1 RULES

As a general rule, the transfer of ownership does not affect the cover, unless otherwise agreed.

3   LIABILITY AND PROFESSIONAL INDEMNITY INSURANCE

3.1   General

3.1.0   LAW OR CONTRACT OR BOTH

In addition to the two policies already mentioned, namely Supervision (*contrôle*) insurance and All Risks ('TR') we must consider here the most commonly used policy, usually taken out by all contractors, i.e. the 'Civil Liability (Operation)', also called 'Civil Liability (Construction)' or 'Civil Liability (Enterprises)' policy. This covers civil liability engendered by all the professional activities of the insured party, not only for all events occurring during execution of the works but, optionally, also for a certain period of time after handover which is fixed by agreement.

All Risks and Supervision policies will continue to receive attention in this chapter, since they usually include a 'civil liability' section:

- The first covers the financial compensation which insured parties would have to pay under articles 1382–1386 of the Civil Code (extra-contractual liability) for damage to third parties which is attributable to the execution (on site) of insured works. This is additional to the cover provided by 'civil liability' policies taken out by individual contractors. It has already been seen that these All Risks policies normally provide (see 2.2) for a post-handover, so-called 'maintenance' period.
- The second offers equivalent cover if the damage arises as a result of a loss which is insured under ten-year liability (contractual)[13] and occurs while it is in force.

Contractors, design offices and clients have no legal obligation to insure. However, this is not the case for architects. A Royal Decree dated 18 April 1985, which appeared in the *Moniteur belge* of 8 May 1985, approved and made compulsory the new Rules of Ethics established by the National Council of the Order of Architects (Conseil national de l'ordre des architectes). Article 15 of this legal text states as follows:

'Architects working alone or in partnership or with a company, shall insure their professional liability, including ten-year liability. This insurance may be included in the framework of overall insurance compulsory for any party involved in the act of building. The effect of this insurance will continue for a period of ten years from the date of handover, and this applies to works completed at the time of the insured party's death.'

This Royal Decree came into operation on 1 June 1985. Recommendations relating to the practical implementation of this obligation to insure were later approved by the Conseil national de l'ordre des architectes on 11 October 1985; and architects were given until 1 April 1986 to comply with its provisions.

The insurance market responded with very little enthusiasm to this unilateral initiative by architects and the companies agreeing to give cover are few. It may be that future European regulations will extend the compulsory principle to other professionals in this sector, but this is uncertain. The tendency seems to favour harmonization within a framework of controlled freedom of competition.

3.1.1   TERMS AND CONCEPTS

'BUILDING' (BÂTIMENT)

See 2.1.1.

Civil Liability (Operation) policies do not contain any additional specifications in this respect.

---

[13] See Appendix I to this chapter.

It should be noted here that with Supervision insurance it is now possible, on payment of an additional premium, to obtain ten-year cover for items which are not supervised, on condition that the claim affects only these items.

'INSURANCE POLICY' (POLICE D'ASSURANCE)

See 3.1.0.

As Civil Liability (Operation) insurance is taken out individually it goes with the producer and not the project. In practice, however, this distinction has no effect.

'BENEFICIARY – INSURED PARTY' (BÉNÉFICIARE – SOUSCRIPTEUR)

Where liability is concerned, the insurer obtains a valid discharge only by making payment to the victim. This follows from article 20, para. 9 of the Law of 16 December 1851 amending the mortgage system which provides as follows:

'Claims arising from an accident which are in favour of a third party injured by the accident or his successors shall be preferred in relation to the compensation which the insurer of civil liability must pay by reason of the insurance contract. No payment to the insured shall discharge the insurer until the preferred creditors have been satisfied.'

The beneficiary of the compensation can, by definition, never be the insured party, unless exceptionally the latter himself indemnifies the victim beforehand. Consequently the liability of the insured party to the third party beneficiary is covered.

3.1.2 TECHNOLOGY AND INSURANCE TERMS AND CONDITIONS

On the subject of All Risks and Supervision policies, we refer to what has already been stated in 2.1.2.

This is equally valid, *mutatis mutandis*, for the Civil Liability (Operation) policy. At most, it may be stated that usually rates are fixed with greater care.

It may be recalled that this type of policy is not limited to one site but covers all the policy holder's professional activities. Terms and conditions will therefore be negotiated according to rates based on statistics even if, finally, there is nothing to stop the contracting parties departing from these.

3.1.3   INSURER'S TECHNICAL CONTROL

See 2.1.3.

There is no technical supervision in Civil Liability (Operation) policies. The system is identical to the procedures for All Risks policies.

3.1.4   INSURERS' POSSIBLE INFLUENCE UPON BUILDING TECHNOLOGY

See 2.1.4.

Different companies' rates are in fact fixed according to classes and categories and these are not published. Rates and all information concerning them are strictly confidential.

3.1.5   NEW TECHNOLOGIES

See 2.1.5.

3.1.6   THE INSURANCE MARKET, GENERAL IMPRESSION

It is an indisputable fact that producers' cover after delivery is very inadequate, not with regard to the cover offered, but with regard to obtaining it. Producers are currently becoming more and more aware of the necessity to cover themselves against extra-contractual liability (see 1.5.3). On the other hand, insurance for ten-year liability has had very little success, mainly due to its high cost and the need for assistance from a supervisory organization (*bureau de contrôle*).

Consequently, in Civil Liability (Operation) policies, the 'post-handover' cover (usually called 'after-delivery'), amounts to a conventional Civil Liability (Products) cover, excluding anything which might fall under ten-year liability. This distinction will be clarified at greater length by analysing customary exclusions.

It is expected that the 'post-construction' insurance market will undergo major changes in the next few years, in legal as well as contractual respects. However, it is impossible to foresee in which direction this probable evolution will take place, but obviously France's example is attracting attention.

3.1.7 COLLECTIVE INSURANCE FOR PROFESSIONAL OR OCCUPATIONAL GROUPINGS. GRADATION OF INSURANCE CONDITIONS

We have already mentioned the case of architects. Even if one cannot actually speak of a standard form contract, obviously the directives of the National Council of the Order of Architects (Conseil national de l'ordre des architectes) are a strict framework into which the terms offered by the market will have to fit.

There are some private initiatives in the form of recommendations which may be mentioned in order to complete the picture, but they could never be called typical civil liability insurance contracts, properly speaking.

3.2 During the Guarantee (Maintenance, Defects Liability) Period

Insurance contracts do not usually take into consideration a 'guarantee' period, insofar as it can be considered to exist, since it is not legally necessary (see 1.2). Either the producer is only insured for civil liability 'during the works' and, in this case, the cover will end on completion of the works, or the producer has 'after-delivery' insurance, and he will then be covered for the period insured, which may be two to ten years or even more from the date of completion, see below.

There is still some uncertainty mainly due to the absence of legal regulation. Therefore, in the interests of clarity insurers usually fix the date of the end or commencement of the policy as being that on which any of the following events is the first to take place: completion or taking possession, or provisional handover. Some refer to the more vague idea of direct or exclusive control of the works.

3.5 After the Guarantee (Maintenance, Defects Liability) Period

3.5.1 NORMAL DURATION OF COVER; CANCELLATION

(a) The contracting parties agree on the duration of cover. It varies between one and ten years, or even more.
(b) Cancellation in case of no payment of premiums is normally a right which the insurer reserves.
(c) The insured party normally cannot cancel prematurely. The insurer's rights to cancel are analogous to those described under 2.5.1.

3.5.2 LEVEL OF PREMIUMS, EXCESS AND FRANCHISE

Complete freedom (some would say anarchy) is the rule concerning the contractual fixing of premium rates. It is impossible to give even an approximate idea of real levels, given also the extremely confidential nature of this type of information.

There are two methods of calculating the premium: it is linked either to the wages and salaries of the firm's personnel or to its total turnover. In the first instance, cover will be limited to damage attributable directly to the firm's personnel, while the second covers all the works carried out, including those entrusted to sub-contractors, with a potential right of recource by the insurer of the main contractor against these last. It will therefore immediately become clear that the premium for the post-handover period will almost always be calculated on the figure for turnover.

3.5.3 TIME LAPSES BETWEEN OBSERVATION OF DAMAGE AND INVESTIGATION

See 3.5.4.

3.5.4 TIME ALLOWED BETWEEN THE INSURER'S RECOGNITION OF A CLAIMS EVENT AND HIS PAYING OUT

Regarding the different periods of time referred to here in the Questionnaire, please refer to the reply given to 2.5.4. No other customary, legal or contractual rules exist apart from those mentioned there.

Obviously, it is always in the insurer's interest to proceed with an expert valuation as quickly as possible. This is all the more important since, when a new file is opened, the insurer must forecast and freeze a financial reserve which will be used to cover the claim if the insured party is ultimately found liable. It is natural, therefore, that the information gathered on the spot by the expert will serve as a basis for estimating this reserve.

3.5.5 KINDS OF DEFECTS AND DAMAGE WHICH CONSTITUTE CLAIMS EVENTS; ONUS OF PROOF

Apart from what has already been stated in 2.5.5, it may be added that most Civil Liability (Operation) policies try to define exactly what is meant by the word '*sinistre*' (claims event). There is no standard wording but, in construction, the most frequently used is often[14] something approaching the following:

---

[14] More and more frequently a claims event is defined as occurrence of damage (assimilation claims event – damage).

'A claims event is an event at the origin of the damage and which is likely to give rise to cover, the whole of the damage resulting from the same event or a series of identical events constituting one and the same claims event.'

Within these limits the insurer will guarantee reparation for personal injury and property damage caused by his insured party and attributable to him, contractually or extra-contractually. Intangible damage, otherwise known as economic or financial loss resulting from the interference with a right such as loss of profits, customers or production, will not always be indemnified in full.

The insurer's contribution in respect of intangible damage is usually limited to that arising as a result of some kind of tangible damage which is covered.

For the onus of proof, refer to 1.5.6 for the main points. In principle, the general law of Belgium follows the maxim *actori incumbit probatio*, namely that it is up to the person who claims to be the injured party to prove the validity of his claim. However, it must be said immediately that this principle has been considerably modified by the courts, some of which do not hesitate to refer to a defect merely because an item operates in an abnormal way.

This remains an area which is treated in a very flexible fashion. We should therefore approach it with caution.

3.5.6 EXCLUSIONS

As already mentioned, Civil Liability (Operation) insurance in its post-handover section should be interpreted as true Civil Liability (Products) cover. This is clearly seen when analysing the exclusions.

First, there are what one might call standard exclusions, in that they are found in all areas of insurance, such as gross negligence, foreseeability, drunkenness, radioactivity, wars and acts of terrorism, increased contractual liability etc.

More specifically, for the post-handover period, there are:

- damage suffered by products or works delivered or manufactured, as well as all related costs of removal, repair, replacement etc.;
- ten-year liability (which is usually covered, it will be recalled, only by Supervision insurance);
- damage caused by products known to be defective but which have not been withdrawn from the market;
- non-conformity (with specifications) or desired results not being obtained, it being understood that any other harmful effect of products or works remains covered.

APPENDIX I SUPERVISION INSURANCE

It will be noted that ten-year civil liability insurance has been mentioned in both sections. This is precisely because Supervision insurance can be interpreted both as property and liability insurance. Ten-year insurance is in fact a *ratione temporis* and *ratione materiae* limitation of contractors' and architects' contractual liability to the client. In this respect, it is, of course, linked to the item built.

In order that the guarantee may be invoked, the damage or event must occur within the ten years following handover. The claim must also be brought in the course of this period.

For insurance cover purposes, the insured works, i.e. the parts of constructions which will have been subjected to inspection by the agreed authority, will automatically be considered as 'fabric' or 'a building', within the meaning of articles 1792 and 2270 of the Civil Code. It is therefore possible to receive indemnification for damage which would not necessarily give rise to a producer's ten-year liability, if he has previously agreed to submit the damaged area to inspection.

Cover is valid up to the declared value which, at the time of handover, should normally correspond to the reconstruction cost of what is insured. After each claim, this amount will be reduced by the payments agreed by the insurer, but it may be reinstated, under certain conditions.

Indemnities are paid subject to the application of a depreciation factor and an agreed excess.

The ambit of this contract does not stop there. Different covers are provided for third party liability and for indemnification for damage to sections of the building not subject to inspection. Extra finishing items and equipment may also form the subject of cover by agreement.

APPENDIX II OUTLINE OF POSITIVE BELGIAN LAW CONCERNING PRODUCERS' 'POST-HANDOVER' LIABILITY FOR RESIDENTIAL BUILDINGS

This appendix will be limited to specific aspects of Belgian law regarding residential buildings.

First of all, it should be noted that Belgian law usually makes no distinction between residential and non-residential buildings.

However, a Law of 8 July 1971 (called the 'Breyne' Law, after the Minister who had it passed by Parliament) provides specific protection for clients and purchasers of houses or apartments to be built or in the process of being built, provided that two conditions are fulfilled:

- first, the house or apartment is intended to be used as housing (or for professional use and housing);

- second, the purchaser or client is required to make one or more payments before completion of the building.

This means that if the housing is sold after completion, the Breyne Law is not applicable and the vendor's liability is limited to the guarantee against hidden defects, which the buyer must claim on as soon as possible (assessed by the court) after discovery of the defect.

On the other hand, the agreements covered by the Breyne Law must ensure the client's or the purchaser's protection, by virtue of the procedures described below, i.e. mainly:

- extension of ten-year liability to the vendor;
- requirement of a completion guarantee (by a bank); and, above all,
- the technique of the 'regulated contract', i.e. the Act requires or prohibits certain clauses (otherwise the contract will be void), so as to restore the balance between the lay purchaser and the professional contractor-vendor who is presumed skilful.

1.1.0 LAW OR CONTRACT OR BOTH

'Property development', i.e. the fact of selling or promising to build, causing the construction or procuring a house or apartment 'to be built or in the process of being built' in consideration of payments in anticipation of the completion of the works, is therefore governed by a Law of 9 July 1971, which applies (article 1):

'to any agreement concerning the transfer of ownership of a house or an apartment to be built or in the process of being built, as well as to any agreement entailing an undertaking to build, to cause the construction of or to procure such a building, if the house or apartment is intended to be used as housing or is for professional and housing use and, by virtue of the agreement, the purchaser or client is required to make one or more payments before completion of construction.'

In order to prevent any attempt to evade this Law through subtle legal schemes, the Belgian legislature has not only decided (article 2) that all agreements covered by article 1 are (in whatever form) governed by the provisions of the Civil Code relating to the 'sale' or a contract for services – i.e. *locatio operis* – but also (article 6):

- that articles 1792 and 2770 (ten-year liability) are equally applicable to the vendor; and that
- 'the [ten-year] guarantee by the vendor benefits successive proprietors of the house or apartment', although an action can only be brought against the original vendor.

On pain of invalidity[1] (of the agreement or the clause which is against the law), agreements covered by the Law, as well as the promise of similar agreements must (article 7):

(a) State the name of the proprietor of the plot and existing structures.
(b) State the date of issue of the building permit and its conditions.
(c) Contain an exact description of the privately owned and common parts forming the subject of the agreement.
(d) Include an appendix with the drawings and precise technical specifications, signed by an architect authorised to practise in Belgium and, in the case of an apartment, also a copy of the basic deed in the form of a notarial act and the co-ownership regulations.

    The absence of these appendices in the notarial act may be covered by a declaration by the notary, which must be repeated in the act, to the effect that the documents are in the parties' possession.
(e) Specify the total cost of the house or apartment and the conditions of payment; state that the price may be revised.
(f) Fix the commencement date of the works, the period of execution or delivery and compensation for delay in execution or delivery; compensation must at least be equal to a normal rent.
(g) Indicate the method of handover.
(h) Contain confirmation by the parties that they have had knowledge of the information and documents mentioned in this section for 15 days.

Under the terms of article 10 of the Law:

• The vendor or contractor may not require or accept any payment in whatever form, before conclusion of the agreement referred to by article 1.

If, on conclusion of the contract, a prepayment or deposit is paid, the total must not exceed 5 per cent of the total cost.

The property developer or contractor may require, on signature of the notarial act, the payment of a sum which, taking into account the prepayment or deposit already paid, is equal to the cost of the land or the proportion of it which is sold, plus the cost of works carried out.

The balance of the cost of the works will not be due, in instalments, until the date when the notarial act is signed, and the instalments, in any case, may not be more than the cost of the works executed.

When a promise to contract is not followed by conclusion of the contract the contractual compensation owed by the purchaser or client may not exceed 5 per cent of the total cost: the contractual compensation, despite its

---

[1] The objection concerning invalidity may be raised either by the purchaser or the client before the deed of sale is signed before a notary or, in the case of a building contract, before provisional handover of the works.

lump sum nature, may be increased or reduced if it is adjudged that the amount is less or more than the loss actually sustained.

On the other hand, the Law (article 12) has considered it desirable to ensure a financial guarantee for buyers.

If a property developer or contractor is already recognized as a public works contractor and is therefore entitled to be considered reliable in principle, he may merely set up a bond equal to 5 per cent of the cost of the building.

If he is not recognized, he will be required to guarantee the completion of the house or apartment or to reimburse the sums paid to him in the event of cancellation of the contract due to non-completion.

This guarantee of completion must be the subject of an agreement by which a bank enters into a joint and several guarantee by the vendor or contractor to the purchaser or client to pay the necessary sums for completion of the house or the building of which the apartment forms part.

This guarantee expires on provisional handover of the works.

1.2.6 'DUTY OF RESULT' OR 'DUTY OF CARE'

Settled case law imposes on developers an obligation to achieve a so-called 'specific result' on the ground that, as they supply the entire construction service, physical as well as administrative, they have a contractual obligation to supply purchasers with an apartment of the agreed type (cf. *Le Droit des constructeurs*, no. 60).

1.5 'POST-GUARANTEE' PERIOD

The Breyne Law includes the only provision in Belgium (article 9) which makes a distinction between two handovers:

'Final handover (*réception définitive*) of the works may only take place one year after provisional handover (*réception provisoire*), provided that final handover of the common parts, including means of access, has already been effected, such that normal habitability is ensured.'

This legal provision was brought into effect by article 2 of the Royal Decree of 21 October 1971:

'2.§ 1. The handover of a structure built pursuant to an agreement referred to by section 1 of the same Act must fulfil the following minimum requirements.

Only a written document signed by both parties is proof of handover, provisional as well as final, of the works.

Refusal to accept handover is notified, with reasons, by registered letter, addressed to the vendor or contractor.

§ 2. However, subject to proof to the contrary, a purchaser or client who occupies (or uses) the property is presumed to have tacitly accepted provisional handover.

A buyer or client is presumed to have accepted the works provisionally or finally, as the case may be, if he does not respond to a written request from the vendor or contractor to effect handover on a given date and if, within fifteen days following notice served by the vendor or contractor through a bailiff he fails to appear on the date fixed in the notice, in order that handover may take place. This provision does not apply to handover of common parts of a building.

§ 3. A vendor or contractor who remains the owner of part of the building which he presents for handing over, shall not exercise any of the rights attached to co-ownership on handover of the common parts of the building.

If the validity of the provisional or final handover of common parts requires the presence of one of the co-owners and if he fails to appear within the reasonable period of time that the vendor or contractor has given him through a bailiff's notice, the court shall give a ruling on the matter of handover insofar as the defaulting co-owner is concerned.'

Denmark

Flemming Lethan

1 LIABILITY

1.1 General

1.1.0 LAW OR CONTRACT OR BOTH

The liability of consultants and contractors is not governed by statutory provisions. There is freedom in terms of contract, anybody can make a contract with any content (with some general exceptions, e.g. not contrary to public policy etc.). However, liability is mainly regulated by two documents: General Conditions for Works and Supplies for Building and Civil Engineering Works (AB 72) and General Conditions for Consulting Services (ABR 89). These are agreed documents in wide use in the building and consultancy sector. English translations of AB 72 and ABR 89 are available.

AB 72 is being revised, see later.

With regard to *the basis of liability and defect powers* (i.e. actions or omissions which may involve liability and the responses that may be necessary), AB 72 and ABR 89 refer to the general rules of Danish law. These are not set down in statutes but derive primarily from the practice established by courts of law and arbitration.

Individual contracting and consultancy can incorporate liability provisions which deviate from AB 72, ABR 89 or the general rules of Danish law.

Unless otherwise stated, the answers in 1.2.1–1.6.1 relate to the legal position of the contractor and not the consultant or supplier.

The consultant's liability is a professional liability for negligence. The supplier's liability is normally based on the Sale of Goods Act 1906, or on specifically worded terms of delivery drawn up by the individual firms or trades involved, and as far as building materials are concerned is normally a question of liability for negligence, approaching objective liability.

The liability of consultants and suppliers is of the same nature during the period of construction, the repair period and after expiry of the latter. With regard to the lapsing of liability, see 1.5.3.

As will be noted, the 'producer' concept is not particularly well suited to Danish conditions. A distinction is required between 'contractor', 'consultant' and 'supplier'.

Liability is mainly based upon the contracts and thus is a matter of liability in contract. In some cases – but this is not normal – claims can be based upon liability in tort.

The Danish rules are identical in residential and non-residential building.

1.1.1   TERMS AND CONCEPTS

'BUILDING'

A legal definition is required of what is meant by 'a building' because the Danish Tenders Act and, for example, AB 72 apply only to building (and construction) *activities*. There is no specific definition but this must be assumed from natural usage and a series of individual decisions.

Installations and other equipment will normally be considered part of the building. The Danish Land Registration Act determines that mortgage security in real property extends to piping, wiring, heating systems, household articles etc. installed in the building at the owner's expense for the use of the building. Similarly, mortgage security in a commercial property normally also includes the operating fixtures and equipment, including all types of plant and machinery, used by the commercial concern occupying the property.

The increasing sophistication of building equipment means that an ever-growing range of equipment is included under 'a building', e.g. sprinklers.

On the other hand, plant and installations outside the building are normally not included in the latter.

'PRODUCER'

In the context of this chapter this must be taken to mean:

- architects, engineers and other consultants, whether involving programming, designing, technical supervision, site management or special consultancy;
- contractors (design-and-build, main, individual trades etc.);
- suppliers of goods and services ordered by the building owner;
- the owner himself to the extent that he does work organized by him;

It does not include:

- suppliers (except of goods and services ordered by the building owner);
- sub-contractors (normally);
- sub-consultants (normally);
- building authorities and other public authorities;
- approval and monitoring bodies;
- lawyers;
- building administrators.

See also 1.1.0.

'TENANT'

There is no relationship in law between producers and tenants.

1.2 During the Guarantee (Maintenance, Defects Liability) Period

1.2.1 DETERMINATION OF THE COMMENCEMENT OF THE PERIOD

The repair period runs from the date of handing over of the building work as specified in AB 72.

AB 72 stipulates with regard to handing over of work:

'Upon completion of the work, the contractor shall give notice thereof in writing to the employer [i.e. building owner], and the employer shall within two weeks of such notice convene a meeting for handing over of the work, at which meeting the contractor shall attend personally or by deputy, otherwise he shall be liable to accept measurements by the employer and his opinion on the work.

The handing-over meeting shall constitute handing over of the works to the employer, unless he shall have shown the existence of defects and shortcomings to an essential extent during the meeting. If so, a new handing-over meeting shall be convened, when the contractor has given notice to the employer of the making good or completion.'

The handing over of building work is frequently marked by the drafting and signing of a special completion record. This is not, however, a stipulation of AB 72.

1.2.2 SIGNATORIES TO AND NAME OF THE DOCUMENT STATING THE COMMENCEMENT OF THE PERIOD

See 1.2.1.

1.2.3 THE PERIOD'S DURATION AND TERMINATION

AB 72 stipulates this rule:

'The making-good period is one year from a time fixed in special conditions. Where no such time is fixed, the period shall be taken to commence at the time of handing over.

If the defects and shortcomings require replacement work to a substantial extent, the guarantee period for the parts of the work concerned shall commence at the time when such work has been completed and written notice thereof has been given to the employer.'

The repair (or making-good) period expires without any special formalities under AB 72. In actual fact – and this will depend on the conditions of the contract involved – a special inspection is often conducted on expiry of the period. It is frequently attended by those consultants who have assisted the building owner during the construction phase. Participation in this inspection is considered part of the consultant's 'normal' service, i.e. payment is included in the consultant's standard fee.

1.2.4 TIME ALLOWED FOR DIFFERENT CATEGORIES OF PRODUCERS FOR MAKING GOOD, REPAIRING OR PAYING DAMAGES

The AB 72 system is as follows:

'For the making good of the defects and shortcomings ascertained during the handing-over meeting, the employer shall at the handing-over meeting fix a reasonable period of time. When the making good has been completed, the contractor shall give written notice thereof to the employer. If the making good shall not be completed before the expiration of the time fixed, the employer shall be entitled to have the defects and shortcomings made good at the expense of the contractor. If the employer, after having received the contractor's written notice on completion of the making good, shall not consider that the defects and shortcomings have been duly made good, he shall within two weeks notify the contractor in writing of the defects and shortcomings he claims to exist. If the contractor shall not hereafter promptly make good such remaining defects and shortcomings, the employer shall be entitled to have them made good at the contractor's expense. During the making-good period the contractor shall be liable to make good without payment any defects and shortcomings caused by errors ascribable to him.

Where the employer intends to refer to the contractor's liability under clause 1 hereof (i.e. the previous passage), he shall in a written notice to the contractor concerned fix a reasonable time for the making good. After the expiry of the time thus allowed, the employer may have the making good done at the contractor's expense. Where the defects and shortcomings require immediate making good, the employer, in the absence of the contractor, is entitled to carry out the making good at the contractor's expense. The employer shall not refer to defects of which he has notified the contractor in writing within reasonable time after he has or ought to have discovered the defects.

After the expiry of the making-good period, the employer shall be entitled to refer only to such defects and shortcomings as he could not have discovered through a customary and reasonable inspection of the work.'

1.2.5 KINDS OF DEFECTS AND DAMAGE COVERED BY THE GUARANTEE

During the repair (i.e. making-good) period the contractor is required under AB 72 to make good any defects which are the result of an error or omission on his part. He is liable according to a rule of negligence. A defect is deemed to exist when the contract work is not executed as agreed or as the building owner should be entitled to expect. No specific definition of a defect is given in AB 72 or anywhere else. The question must be settled in each specific instance. A lack of compliance with public building regulations will normally imply the existence of a 'defect'. A defect can mean 'damage' but defects are also often possible without damage, e.g. the incorrect execution of a plan, building too close to a neighbouring boundary line, incorrect choice of colours or installations with inadequate capacity.

The concept of damage is not limited by any special rules. It is determined by practice.

1.2.6 EXCULPATION OF PRODUCER

By way of introduction, a summary of the rules in AB 72 as they relate to contractors' liability would be useful.

The contractor bears objective liability for defects ascertained at the time the work is handed over. He has an obligation and right to make good the defect. If he fails to do so, the building owner can be awarded compensation for the cost of repairing the defect. In respect of defects discovered during the repair period (normally one year from handing over of the work), the contractor is liable under a rule of negligence. The contractor has a duty and a right to make good the defect. Apart from the right to compensation for the cost of making good the defect if the contractor fails to do so, the building owner has the general right to point out any defect.

In the case of defects ascertained after the repair period, the contractor is liable under a rule of strict negligence but only for latent (hidden) defects. He has neither a duty nor a right to make good the defect. The building owner has the general right to point out any defect.

In the case of defects in connection with the *materials* used in the contract, liability is probably rather stricter than for the performance of work.

In the event of a failure occurring in the building work which is ascertained during the making-good period, it is very likely that the contractor will be relieved of liability in the following instances:

(a) If the failure is not a consequence of errors or omissions attributable to the contractor. In practice, liability is to be strictest when failure is ascertained during the period immediately following handing over of the work, whereas liability tends to be alleviated the longer the period that passes after handing over of the work. The contractor will always avoid liability if failure is the result of an error in the owner's tender documents or in the work of another contractor. The contractor will also be relieved of liability if failure is the result of miscalculated building experiments (unless the contractor has experimented without the knowledge of the client or has made the experiment in a wrong way). The development risk normally rests with the building owner, which means that the contractor is not liable for defects in his work as long as he has acted professionally and carefully based on the information at the disposal of technical experts at the time he carried out his work. Unavoidable accidents are also the owner's risk.

(b) If a defect is obvious, the employer must complain quickly to the contractor. Otherwise the owner forfeits his right to point out defects.

(c) If a defect is due to the owner's lack of or incorrect maintenance of the building or his incorrect use of the latter.

(d) If the deadline for registering complaints is exceeded, see below.

As will be noted, the Danish rules are primarily a 'duty of care'. The 'duty of result' system is relevant only if the contractor has assumed a *guarantee* for specific qualities in particular parts of the building work (e.g. guaranteeing that the roof will be watertight for ten years). In such a case, the contractor would be liable, whether or not there was any question of errors or omissions.

Unless otherwise agreed, design-and-build contractors have the same liability as other contractors. As opposed to other contractors, the design-and-build contractor is also liable for design and co-ordination of design and implementation of the works. Design work performed by the design-and-build contractor must be evaluated in accordance with contractors' rules.

Inspection consultants are liable for failures in their inspection work according to a rule of negligence. If ABR 89 is applied, special rules regulate and in some cases reduce the liability. A limitation of the damages is accepted.

Normally liability cannot be put upon approval and monitoring bodies.

1.2.7 CONSEQUENCES OF PRODUCER'S NON-COMPLIANCE WITH UNEQUIVOCAL DUTY TO MAKE GOOD, REPAIR OR PAY DAMAGES

If the contractor can repair but refuses or omits to do so, the employer can proceed as described in 1.2.4.

The employer will usually try to meet the cost of making good any defect by setting off the amount against any amount he owes to the contractor or by claiming part of the guarantee sum deposited by the contractor. (According to AB 72 the contractor normally deposits a guarantee sum of 10 per cent of the contracting sum. This guarantee ceases when the making-good period has expired, except as regards the employer's claims of making good.)

If the contractor disagrees with the action taken by the building owner, the dispute can be settled by resort to law.

If the producer cannot repair, the employer will try to obtain cover by set-off or through the guarantee amount, and can also seek to recover payment by means of debt collection or by bringing a claim against the contractor's estate.

There is no point in discussing other producers' situations, as producers bear no reciprocal legal relationship, but are bound and entitled to bring their claims to the employer.

1.5 From expiry of the Guarantee (Maintenance, Defects Liability) Period

[1.5.1: *Not applicable*]

1.5.2 NAME OF THE DOCUMENT STATING THE COMMENCEMENT OF THE POST-GUARANTEE PERIOD; ITS SIGNATORIES

See 1.5.3.

Commencement is not postponed, even if damage occurs in the period. There are no special documents.

1.5.3   PERIOD(S) OF LIMITATION

TRADITIONAL SYSTEM

Theoretically the legal liability period for contractors is five years according to an Act of 1908, dealing not only with building contracts. But the period of limitation is suspended as long as the entitled party has not discovered he has a claim, and therefore the Act of 1908 has only little importance in respect to latent (hidden) defects in building.

CONTRACTORS' LIABILITY

It has been court practice since the 1930s that the statute of general limitation of 20 years stipulated in legislation from 1683 also applies to contract work. It is supposed that the period runs from the handing over of the work.

The 20-year rule can be waived by agreement.

CONSULTANTS' LIABILITY

In principle, this is subject to the rule just mentioned, but ABR 89 and its predecessor have been in effect since 1979, and under their provisions the period of liability is five years from the date of handing over of the building work (or, in the case of work executed later, completion of the work).

ABR 89 is an agreed document. It depends on the contract between the parties whether ABR 89 should be included as part of the agreement or not, but it is normal for it to apply. ABR 89 is a mandatory condition in all building work ordered or subsidized by the state.

Suppliers' liability is governed by the Danish Sale of Goods Act from 1906, under which liability expires one year after delivery. The one-year rule can be waived by agreement, and this is in fact often done in the form of shorter statutory or complaint deadlines in the delivery conditions observed by individual companies or trades. There are also many examples of the supplier offering liability for periods of five or ten years. With the exception of goods and services ordered by the building owner the liability of a supplier is normally no concern of the building owner, as it is the contractor who is liable to the building owner under contractors' rules.

NEW CONTRACTUAL SYSTEM FOR CONTRACTORS AND SUPPLIERS

Since 1986 the general tendering reservations, which contractors normally include in their tenders, have contained the following rule:

> 'The tender's/contractor's liability in respect of defects shall cease five years after the handing over of the work.'

This five-year liability system is only in force if it is agreed in the individual contract. The employer must decide whether he will recognize the reservation or not. If he will not, he runs the risk that he cannot get the tender.

The National Building Agency has decided in a Circular on Liability that the five-year liability reservation normally has to be recognized in state and state subsidized building, and in this way the five-year liability period is the normal rule here. The municipalities and the investing companies use the same system with some exceptions. Among other employers the reservation is normally recognized.

The five-year rule applying to suppliers manifests itself in a 'building supply clause' which provides that the supplier's liability for defective supplies shall cease five years after the handing over of the building work for the completion of which the supplies were effected. Furthermore, the clause stipulates that in the case of supplies to persons other than owners/employers and contractors, liability shall cease not later than six years after supplies being effected to the purchaser. This double period of five and six years is described below.

The building supply clause has been introduced in the following way: in the Circular on Liability, the National Building Agency commits all owners/employers of government or subsidized building projects to ensure the inclusion of the clause in agreements between contractors and their subcontractors concerning supplies for the execution of the contract.

This system ensures the elimination of the 'contractor's scrape', i.e. the situation in which the contractor is liable to the owner/employer but has no right of recourse against his own supplier because of the short supplier liability period.

As mentioned above there is a double period for supplier liability, i.e. five and six years.

The five-year period applies to the 'ultimate' or 'end' supplier, i.e. the supplier who enters into a supply agreement with the contractor or an owner/employer, and he will then be the person liable for a period of five years after the handing over of the project. However, since a supply to a building project often passes through several suppliers, e.g. importers, wholesalers and middlemen, it may be desirable to be able to apply the five-year rule further back in the chain of suppliers.

If no possibility existed of applying the five-year rule to these previous links in the chain, the above-mentioned 'contractor's scrape' would have been replaced by a 'supplier's scrape', since it would mean that whereas the supplier would be liable for a period of five years to the contractor or the owner/employer, he himself would normally only have a right of recourse against his own suppliers for a period of one year, as provided by the Sale of Goods Act.

In order to prevent the occurrence of this situation, a provision was included in the building supply clause to the effect that in the case of supplies for stocks or resale, liability shall cease not later than six years after delivery to the purchaser. This addition means that the previous suppliers in the chain live with a double limit: five years from the handing over of the building work, but never more than six years from the time of their own delivery. The extra year is meant to provide for periods of storage with the intent to maintain, to a considerable extent, the end supplier's right of recourse against previous links in the chain.

This rule presupposes that the supply clause is adopted in the contracts between suppliers.

DETAILS CONCERNING THE FIVE-YEAR RULE

The liability rule applies to hidden defects only, i.e. defects which the owner/employer could not have discovered by an ordinary and reasonably thorough inspection of the work.

For apparent defects the general rules for claims and complaints apply and this means that unless a claim is raised immediately upon the discovery of a defect, or immediately after the defect ought to have been discovered, the employer loses his right to raise such a claim.

The five-year limitation rule will not apply to defects caused by fraud or gross negligence. A probable consequence of this rule is that the right of an owner/employer to raise claims in respect of defects will be extended to run beyond the period of five years in cases such as some of those which have appeared over the last years and which involve examples of serious building damage. A definition of the difference between gross and ordinary negligence will have to be established through court practice.

Whenever a claim in respect of a defect is raised within the five-year period, the employer's right to raise claims in respect of defects will be preserved even though the actual clarification and allocation of liability is not made until after the expiry of the five-year period. This is the background for the rule which stipulates that five-year inspections of building constructions are to be made a few months prior to the expiry of the five-year period so as to allow a period of some months in which to raise claims without exceding the five-year period.

The rule limiting the duration of liability does not affect the basis of liability and the course open to the employer (the building owner) to raise claims of defects.

The above will not apply to experiments and pilot projects in the building sector. Such projects run a development risk which in principle rests with the employer.

The development risk may be described as follows: where a consultant or a contractor has acted with the necessary professional care and in accordance with the knowledge available to all members of the profession at the time of the performance, which is based on experience, he has fulfilled his commitments. Should defects nevertheless occur in the project because of the solutions chosen, such cases will fall under the development risk of the owner.

AMENDMENT OF AB 72

In the information given above reference was made to the contractual document AB 72. This document has been revised in 1988–92 and in October 1992. A new edition, called 'AB 92', has been issued and will be in general use from 1994.

The most important amendments concerning items which are discussed in this chapter are as mentioned below.

The following rules are transferred from the contractual (applying to building, not to construction) system described above:

- the building supply clause (meaning five or six years liability for suppliers of defective supplies);
- the five-year liability for contractors in respect of defects. There are three exceptions to the rule: (a) agreed guarantees, (b) fraud and gross negligence, and (c) essential failure to comply with an agreed quality assurance undertaking. In the last case the 20-year rule of liability is applied;
- inspection of the work one and five years after handing over;
- a special completion record is introduced;
- the making-good period of one year is abolished. The contractor has the right and the duty to repair defects during the whole five-year period of liability.

Other new rules:

- the handing-over system is rationalized;
- besides the right of making good and damages the employer gets a right of 'proportional reduction' in the contracting sum in case of defects;
- a rule of 'limitation of sacrifice' is introduced to protect the contractor against unreasonably large expenses concerning making-good efforts;
- a special set of rules for design-and-build contracts will be issued. Most of the AB 92 rules can be used also for these contracts, but some special rules are needed.

1.5.4 TIME ALLOWED (DIFFERENT CATEGORIES OF PRODUCERS) FOR MAKING GOOD, REPAIRING OR PAYING DAMAGES

See 1.2.4.

1.5.5 KINDS OF DEFECTS AND DAMAGE QUALIFYING FOR DAMAGES

See 1.2.5.

1.5.6 EXCULPATION OF PRODUCER

See 1.2.6.

1.5.7 CONSEQUENCES OF PRODUCER'S NON-COMPLIANCE WITH UNEQUIVOCAL DUTY TO REPAIR

See 1.2.7.

1.6 TRANSFER OF OWNERSHIP AND PRODUCERS' LIABILITY

1.6.1 RULES

Basically the system is as follows: if a successive owner has a right to bring complaints of defects against the vendor, he too has the right to bring the claims against an earlier vendor or against contractors, consultants or suppliers involved in the building work.

This right is subjugated to the limitations following from the contract between the earlier party and his contracting party.

2 PROPERTY (OR MATERIAL DAMAGE) INSURANCE

2.1 General

[2.1.0–2.1.1: *Not applicable*]

2.1.3 INSURER'S TECHNICAL CONTROL

DANISH BUILDING DEFECT FUND

The Danish Building Defect Fund is an independent institution set up by the state under legislation passed in 1985.

The Fund has two aims: directly, to ensure the repair of defects in buildings covered by the Fund, thereby avoiding deterioration or massive rent increases; indirectly, to prevent the occurrence of building faults by stipulating that buildings covered by the Fund must be quality assured, maintained in accordance with detailed plans, and subject to inspection five years after handing over to the owner.

All buildings for which a state subsidy is granted after 30 June 1986 are required to comply with Fund regulations. In return, such buildings have a legal right to cover under the Fund, provided the relevant conditions are satisfied. The Fund primarily covers the non-profit-making housing sector.

The Fund can be regarded as a type of property insurance contracted by building owners. Producers (architects, engineers, contractors and suppliers) do not contribute to premiums but become involved partly if they become liable to recourse by the Fund and partly through their own insurances, if applicable, as any cover provided by insurance is deducted from the amount covered by the Fund.

For every building a non-recurring premium is payable to the Fund equivalent to 1 per cent of the cost of the building. Half of the premium income is earmarked for the five-year inspection, the remainder is set aside to cover any building faults which may occur in buildings covered by the Fund.

In the event of defects occurring in insured buildings, there is an uninsured risk of 5 per cent of the cost of repair.

The Fund only covers damage which stems from the building's construction period. In the context the terms 'defect' and 'damage' mean rupture, leak, deformation, weakening or destruction of the building or part of the building, or other physical conditions which substantially impair the utility of the building or the purpose for which it was intended.

If a defect is attributable to the use of new or untried materials, structures, building components or methods, cover will be provided, if the work or the part of the work concerned has been properly planned by the building owner as an experimental or developmental project.

The Fund does not cover failures in the building which are not in the nature of defects. For example, materials and qualities which are not of the contracted standard, incorrect positioning of buildings on the site, inadequate capacity in installations or excessive energy consumption.

It is up to the building owner to prove that a defect has been ascertained which is coverable under the Fund.

There is no need to transfer the insurance, as the buildings covered by the Fund are not normally sold but permanently in the posession of public or semi-public owners.

The Fund provides cover for a period of 20 years from the completion (handing over) of the individual building. The Fund has recourse against any liable parties under current liability rules. In the case of the work of consultants, contractors and suppliers this means that the Fund normally has recourse for a period of five years after completion, provided that grounds for liability can be established.

The Fund need not wait for clarification of liability before covering repair.

It is a condition for buildings covered by the Fund that the building is quality assured during the period of construction and is subject to a detailed maintenance plan.

The Fund is not required, therefore, to approve building projects or research and development projects, and it is not intended that the Fund should exercise any independent intermediate control during the building process or once the building is in use.

Mention should be made of the fact that local authorities perform a supervisory function in connection with subsidized building and that this takes into account such things as maintenance.

FIVE-YEAR INSPECTION

There is one area in which the Fund does exercise direct supervision: five years after a building is completed and handed over to the owner (or rather, some months before expiry of the five-year period) the Fund arranges for an inspection of the building with a view to assessing its condition, including recording of defects and signs of any defects and, if possible, the reason for the defects. The idea behind the inspection is that most defects will have had the opportunity to crop up – at least in the form of an indication that something is wrong – when the building has been in use for five years. The five-year inspection must also be seen in conjunction with the fact that the liability of consultants, contractors and suppliers normally lapses five years after the handing over of a building under the Fund.

The Fund till now (January 1993) has only paid small amounts as cover for defects, but the impression of the Fund is that there will be increasing requests for cover.

The Fund has established an inspection system, consisting of an A-inspection (mainly visual and by means of usual basic technical tools), a B-inspection (a sort of indirect inspection, used if the A-inspection leaves uncertainty about the condition of the house, and based upon examination of drawings, control documents and maintenance plans) and a C-inspection (thorough and eventually destructive investigations by specialists and laboratories, used to clear up complicated cases).

The A- and B-inspections are carried out by private firms (architects and engineers), engaged by the Fund.

The experience of the Fund is that the system is effective.

The inspections have revealed certain defects, which are probably connected with the fact that the inspected buildings were built in the first years after the quality assurance reform.

A special building damage fund has recently been established for state subsidized building renovation projects.

The Fund has begun to publish details of the experience it has gained from the five-year inspections and from its defect claims, with a view to the future prevention of defects.

[2.1.4–3.5.6: *Not applicable*]

England, Wales and Northern Ireland

Liability

Jenny Baster

Property insurance

Non-residential buildings: *Peter Madge*

Residential buildings: *Padraic Doyle*

Liability and professional indemnity insurance

Peter Madge

Appendix I 'Normative bureaux'

Appendix II Building control authorities

Appendix III Testing laboratories

Appendix IV British Board of Agrément

Co-ordinated by Donald Bishop, Stephen Dunmore

and Andrew Fryer

The arrangements for post-construction liabilities and insurance in the UK are set out according to the standard format of the country overviews. There are in addition four appendices (I to IV) describing the roles and liabilities of 'normative bureaux', building control authorities, testing laboratories and the British Board of Agrément.

1 LIABILITY

1.1 General[1]

1.1.0 LAW OR CONTRACT OR BOTH

Producers can have post-construction liability under contract or in tort or both. As a general comment, there is no significant distinction between

[1] Three general references to current practice in England, Wales and Northern Ireland are given on p. 163.

liability arising after completion of construction and liability which may arise during the course of construction, except in the case of a contractor, which is dealt with in 1.2.1. There are also many statutory provisions and regulations which may impose obligations on different categories of producers, including a local authority, e.g. the Defective Premises Act, the Health and Safety at Work Act etc. Liabilities under a contract are governed by its terms.

It should also be noted that, leaving aside express terms of any particular contract entered into by a producer, there are terms which are implied by law into contracts. Implied terms affect different categories of producers in different ways. For example, there is generally an implied term in a building contract (i.e. a contract with a contractor for work and materials) that the materials will be reasonably fit for the purpose required and of merchantable quality and that the contractor will do the work in a good and workmanlike manner. In a design-and-build contract, there is also generally an implied term that the building will be fit for the purpose for which it is required. In any contract for the sale of goods (i.e. including contracts with suppliers), there is generally an implied term that the goods are of merchantable quality and that they are fit for the purpose required. In each of these examples, fitness for purpose involves an absolute obligation, i.e. strict liability.

Fitness for purpose is not a term implied in contracts which architects, engineers or other professional categories of producers enter into and they have a duty simply to exercise the reasonable care and skill of the ordinarily competent member of their profession, unless they agree to assume a different contractual duty.

As far as liability in tort is concerned, this is concerned principally with negligence or breach of a duty to take care. In order for a client or third party to succeed in an action against a producer for negligence, it must be shown that a duty of care is owed to the client or third party, the producer was in breach of that duty and the client or third party suffered damage as a result of that breach.

After a period in the 1970s and early 1980s in which the scope of the law of tort was greatly extended, there have been very significant restrictions since the mid 1980s. As a result of these recent developments, the circumstances in which a duty of care in tort arises, the scope of that duty and the damage recoverable for breach of the duty are now very much more circumscribed than before. Although there are still many uncertainties, the duty of care is unlikely to extend beyond a duty not to cause personal injury or damage to other property, but damage to the defective element itself or economic, i.e. purely financial, loss will not be recoverable in the absence of a special relationship of reliance. A recent decision by the House of Lords[2] has further restricted the grounds for actions in tort, e.g. once a defect is discovered before it causes personal injury or damage to other property,

[2] *Murphy v Brentwood District Council* [1990] 3 WLR 414.

there is normally no ground for an action in tort because the property can be safeguarded and the damage avoided. Where there is already a contractual duty the existence of a concurrent tortious duty is in question, but if one exists it is likely that the contractual duty will go some way to defining the limits of the duty in tort.

Examples of liability incurred by different categories of producers:

- Any producer who has a contract directly with the client (e.g. architect, engineer, quantity surveyor, or other consultant or contractor) may be liable for breach of that particular contract. The contract may be in a standard form or not. Any concurrent liability to the client in tort, if it exists, is likely to be circumscribed by the contractual duty, as explained above.
- A supplier may have liability in tort to the client in the very limited circumstances outlined above. In certain limited circumstances, although a supplier may contract with the contractor and not the client, it may be said that there is a collateral contract between the supplier and client, such that a supplier may be liable to the client for breach of that contract.
- A sub-contractor may have liability in tort to the client in the very limited circumstances outlined above and, in addition, it is not uncommon for a sub-contractor to enter into a direct contractual agreement with the client, so that he may also have liability in contract. Any concurrent liability in tort, if it exists, is likely to be circumscribed by the duty in contract.
- Local authorities have duties imposed on them by certain statutory provisions as mentioned above. During the period of the expansion of the law of tort, there were many decisions when a local authority was held liable in tort for breach of statutory duty or negligence or both, e.g. where its supervising officer approved foundations, drains etc. Again, this liability is now significantly circumscribed as a result of the developments explained above (see also Appendix II).

1.1.1 TERMS AND CONCEPTS

There is no universal definition of the term 'building'. It may have a specific definition given to it in specific contracts or in specific statutes. There is no distinction between new building and rebuilding in the general law of contract or tort. It would depend on the context whether fixed installations are included in the term 'building', but in the context of liability generally in respect of a building, one would understand such a term to include any part of the works carried out.

'PRODUCERS'

In general, the categories of producers may include:

- architect, engineer, quantity surveyor, construction manager or any other consultant, whether engaged directly by the client or engaged as a sub-consultant to another consultant, or appointed by the contractor;
- contractor, including management contractor;
- sub-contractor;
- supplier;
- local authority: as explained above, a local authority's liability in tort for breach of its statutory obligations is now significantly circumscribed.

'TENANT'

Different categories of producers may possibly have liability in tort to a tenant, although as described elsewhere whether there is a duty of care in tort and the scope of any duty is likely to be viewed very restrictively in the current state of the law. Any special reliance by the tenant on the relevant producers, however, may be of significance in establishing a duty of tort and this is most likely to arise in the case of an architect, engineer or other consultant.

COLLATERAL WARRANTIES: A UK PHENOMENON

For reasons that are not relevant to this chapter, the great majority of commercial leases in the UK make tenants liable for repairs and for the cost of reinstating the building at the end of the lease to the condition when it was taken over by the tenant; such leases are full repairing and insurance leases. Clearly, this can be an onerous liability, especially when tenants have to meet the cost of rectifying latent defects. The considerable extension of the law of tort from the mid-1970s enabled tenants to bring claims in tort directly against producers with whom they had no contract.

Now that decisions by the House of Lords have severely limited this route to restitution, owners and tenants without privity of contract with producers have sought to forge contractual links where none formerly existed by 'duty of care' letters or collateral warranties, which effectively give the owner or tenant the same contractual rights against producers as the clients have and which enable those rights to be assigned to successive owners or tenants.

Almost inevitably, collateral warranties increase the exposure to risk by all who give them. This is especially the case when only some of the producers can be coerced to give warranties, normally the principal consultants and the main contractor. They are then automatically 'in the firing line' for any defect in the building and are unable to seek contributions from sub-contractors or other producers who have not given such warranties. Moreover, some warranties attempt to impose onerous conditions which are not acceptable to the majority of professional indemnity insurers. Collateral warranties are one of the major and unresolved current problems.

1.2   During the Guarantee (Maintenance, Defects Liability) Period

1.2.1   DETERMINATION OF THE COMMENCEMENT OF THE PERIOD

As a general comment, the Defects Liability Period has specific relevance only for the contractor and not for the other categories of producers.

There is normally a defects liability provision in building contracts which provides that the contractor shall make good defects or repair and maintain works for a certain period after completion. To secure the contractor's obligations, part of the contract sum, called retention money, is retained by the client until the end of the period and is not released until the architect gives a certificate evidencing his satisfaction that the works accord with the contract. Such a defects liability provision forms part of the contract administration, obliging the contractor to put defects right within the Defects Liability Period. It does not affect any other rights there might be against the contractor. If he is in breach of contract, other losses may be recoverable, even though the defect may have been repaired.

Under the standard form of building contract, the period commences on issue of the Certificate of Practical Completion.

1.2.2   SIGNATORIES TO AND NAME OF THE DOCUMENT STATING THE COMMENCEMENT OF THE PERIOD

The Certificate of Practical Completion is normally signed by the architect.

1.2.3   THE PERIOD'S DURATION AND TERMINATION

The duration of the Defects Liability Period is fixed by the contract. The period expires without the necessity for a further certificate, although a certificate of making good defects would normally be issued by the architect under the standard form of contract when the defects had been made good.

1.2.4   TIME ALLOWED FOR DIFFERENT CATEGORIES OF PRODUCERS FOR MAKING GOOD, REPAIRING OR PAYING DAMAGES

Depends on contract.

1.2.5   KINDS OF DEFECTS AND DAMAGE COVERED BY THE GUARANTEE

The nature and extent of the obligations of the contractor vary according to the terms of the particular contract. Under the standard form of building contract, the contractor must make good 'defects, shrinkages, or other faults, due to materials or workmanship not in accordance with the contract or to frost occurring before practical completion of the works'.

The contractor's liability for damage generally is not removed by the existence of a defects clause and such a clause does not normally prevent a contractor also being liable for breach of contract.

The kind and extent of damages for breach of contract recoverable are subject to the general rules of law relating to damages, which provide that, in contract, damage is recoverable if it results either from the natural consequences of the breach or from special circumstances of which the party has had actual knowledge and which caused the breach to result in exceptional loss. The measure of damage is usually the actual monetary loss and the plaintiff, of course, has a duty to mitigate his loss.

1.2.6   EXCULPATION OF PRODUCER

In considering exculpation, as has been noted, the liability of producers, excluding the contractor, is not related specifically to the Defects Liability Period. The basis of liability of professional producers is negligence, in the sense of breach of contract or breach of duty of care in tort. The standard of care required is to exercise the reasonable skill, care and diligence of the ordinarily competent member of their respective profession. A professional producer will not be liable if he can show he exercised reasonable skill, care and diligence or, indeed, if the damage was caused by a third party, vandalism etc. and not by the producer's act. The standard is judged by the prevailing practice at the time of the design and construction of the building. If it was not then generally known that a particular practice was bad, the producer has not failed in his duty to exercise reasonable care. To the extent that damage is caused or contributed to by lack of maintenance or defective operation by the client, successive owner or tenant, such person will bear some liability.

A contractor may also be liable in negligence in the sense of breach of contract or breach of duty of care in tort. In some respects he is under a stricter duty inasmuch as certain implied terms referred to above will be implied into his contract. A design-and-build contractor is also under a stricter duty because there will be an implied term that the completed work is fit for the purpose for which it is required. In this respect, a design-and-build contractor has a higher duty than has a professional designer, unless the contract expressly provides for the contractor to exercise the normal professional duty of reasonable skill and care.

As has already been mentioned, the contractor is not generally relieved from liability at the expiry of the Defects Liability Period. It should be said that under earlier forms of the standard building contract a contractor would be relieved from liability once a Final Certificate had been issued because it was conclusive evidence that the works had been properly carried out and completed in accordance with the terms of the contract (subject to fraud etc.). Under the current form of standard building contract, the Final Certificate is evidence of compliance with the terms of the contract only in very limited areas where the quality is expressed to be to the architect's satisfaction. Under the standard form of building contract, a Final Certificate is issued as soon as is practicable but before expiration of three months from the latest of the end of the Defects Liability Period or completion of making good defects or receipt of necessary documents from the contractor.

If a producer uses experimental techniques, in order to discharge his duty to act with reasonable skill and care the producer should advise the client of the possible risk. If the client agreed to the risk, that might discharge the producer from liability.

The rules of limitation in respect of time for bringing claims (see 1.5.3) also operate to exculpate producers.

1.2.7 CONSEQUENCES OF PRODUCER'S NON-COMPLIANCE WITH UNEQUIVOCAL DUTY TO MAKE GOOD, REPAIR OR PAY DAMAGES

In general, if a producer fails to remedy defects, the client's recourse is to sue for damages. If the producer is in liquidation, unless that producer is insured, the client has no recourse against that particular producer. If the producer is in liquidation and carried insurance, the client has certain rights against the insurers.

1.5 From expiry of the Guarantee (Maintenance, Defects Liability) Period

1.5.1 DETERMINATION OF THE COMMENCEMENT OF THE POST-GUARANTEE PERIOD DURING WHICH A PRODUCER MAY BE HELD LIABLE

Liability commences when the cause of action accrues and commencement is not related to the guarantee period. The cause of action accrues in contract on the date of breach. In practice, where producers are under a continuing duty to supervise, review design or correct faults, liability in contract runs from the date of the Practical Completion Certificate. The cause of action accrues in tort when damage (not defect) occurs. The House of Lords decision in *Pirelli v Oscar Faber*[3] to this effect overruled previous cases where it was held that cause of action accrues when damage was discovered or reasonably discoverable. The Latent Damage Act 1986 confirms the principle that the cause of action in tort accrues when damage occurs but introduces an additional period where the damage was not discovered or reasonably discoverable and subject, finally, to a long stop (1.5.3 refers).

1.5.2 NAME OF THE DOCUMENT STATING THE COMMENCEMENT OF THE POST-GUARANTEE PERIOD; ITS SIGNATORIES

There is no formal certificate stating the commencement of the post-guarantee period.

1.5.3 PERIOD(S) OF LIMITATION

The relevant features of the law of limitation of actions in English law may be summarised as follows:

- A claim in contract must be brought within six years (12 if the contract is executed as a deed) of the breach.
- A claim in tort for negligence must be brought within six years of the cause of action arising, this being the occurrence of relevant damage, whether or not this is discoverable.
- A claim may also be brought in tort for negligence within three years of the damage first becoming reasonably discoverable.
- No claim for tortious negligence may be brought after the expiry of 15 years from the negligent act.
- No limitation applies where the relevant act has been deliberately concealed.
- Limitation periods do not protect a defendant from liability in contribution proceedings brought by other defendants.

1.5.4 TIME ALLOWED (DIFFERENT CATEGORIES OF PRODUCERS) FOR MAKING GOOD, REPAIRING OR PAYING DAMAGES

No specific time is laid down, either in contract or in tort, for any category of producer to make good, repair damage etc.

---

[3] *Pirelli General Cableworks Ltd v Oscar Faber & Partners* [1983] 1 All ER 65.

1.5.5 KINDS OF DEFECTS AND DAMAGE QUALIFYING FOR DAMAGES

See 1.2.5 – same principles involved.

1.5.6 EXCULPATION OF PRODUCER

See 1.2.6 – same principles involved.

1.5.7 CONSEQUENCES OF PRODUCER'S NON-COMPLIANCE WITH UNEQUIVOCAL DUTY TO REPAIR

See 1.2.7.

1.6 Transfer of ownership and producers' liability

1.6.1 RULES

THE LIABILITIES OF PRODUCERS IN RELATION TO SUCCESSIVE OWNERS

A successive owner may have separate rights in tort although these rights are now relatively limited as explained in 1.1.0. The successive owner, however, will not be deprived of any rights in tort because he was not owner at the time damage occurred to the building. Apart from rights in tort, it is becoming increasingly common for a successive owner, such as a first purchaser from the developer, to insist on acquiring contractual rights against the producers equivalent to the rights the client would have. These rights are normally conferred by collateral warranties given in practice by only some of the producers, in particular by the design professionals.

2 PROPERTY (OR MATERIAL DAMAGE) INSURANCE

NON-RESIDENTIAL BUILDINGS

Clients will insure their buildings for specific perils, e.g. fire, explosion, flood etc. Defects, in the sense of this enquiry, are not a peril normally covered by such insurance. Insurance for defects (in this sense) was not generally available in the UK, but since 1987 three insurers have been issuing latent defects insurance policies, i.e. ten-year non-cancellable policies covering latent defects. During the 1990s, other insurers have entered the field. There are now at least 12 underwriters offering material damage insurance against latent defects, including three composite insurers, Lloyds syndicates and mutual insurers. Some offer wider cover, e.g. for engineering services and for periods longer than ten years after Practical Completion. This is a rapidly evolving type of insurance, in part stimulated by the end of the sellers' market for commercial buildings and in part by the uncertain efficacy of collateral warranties.

RESIDENTIAL BUILDINGS

Details of insurance and guarantee schemes available from the Building Employers Confederation, the Federation of Master Builders and the National House Building Council for the construction or conversion of residential buildings are set out in the table opposite.

Mortgages for new houses and flats are very difficult to obtain unless the client or purchaser is safeguarded by an approved guarantee and insurance scheme. The National House Building Council is responsible for over 95 per cent of the guarantees within the private house building sector. It also offers warranties in the housing association sector in competition with the Housing Association Property Mutual, HAPM. HAPM began offering 35-year warranties to housing association members in 1991. Whilst these schemes can be used to safeguard consumers who employ builders to carry out conversions and other smallish works two other schemes are available for these purposes. The Building Employers Confederation was first in the field. As from September 1991 their scheme was acquired by Building Guarantee Scheme (UK) Belfast. The second is managed by the Federation of Master Builders. Both schemes apply to projects costing between £1 000 000 and £500 and thus cover the whole range of work likely to be ordered by consumers.

In addition, the majority of UK insurance companies provide cover against damage caused by 'subsidences, settlement or heave'.

Substantial claims are paid each year in respect of this cover for residential buildings of all ages. All policies have exclusion clauses; the most common relate to the original producer's negligence, the condition of the site on which the building stands and whether other insurance is in force covering the same risk.

Building societies and banks normally require the cover to be obtained by mortgagors. The normal franchise is £500 and the cost is 0.025 per cent of the building value (but the policy also covers other perils, e.g. fire).

3 LIABILITY AND PROFESSIONAL INDEMNITY INSURANCE

3.1 General

3.1.0 LAW OR CONTRACT OR BOTH

Liability is generally not imposed by statute law, i.e. by Acts of Parliament (except in the limited circumstances covered by the Defective Premises

Details of protection schemes operated by BEC, FMB, NHBC new homes, NHBC conversions and PRC homes. This table was compiled in 1991.

		BEC	FMB	NHBC: New homes	NHBC: Conversions	NHBC: PRC houses
1.	Scheme applies	All BEC members	FMB members on Warranty Register	NHBC members on Register	NHBC members or those on Conversions Register	NHBC members or those on PRC Register
2.	Are builders vetted?	No	Yes – 3 yrs as FMB member or as local authority tenderer	Yes	Yes	Yes
3.	Scheme funded: premium	1% of contract price (up to limit) (minimum £20 / maximum £250)	1% of annual turnover of members and annual registration fee	Average of 0.33% of selling price (up to limit of scheme) per dwelling plus annual registration fee	Dwellings offered by members	Approx 5% of cost of repair
4.	Work undertaken under scheme: Financial limit	Work in excess of £500 and less than £25 000 (plus small variations providing there is no architect)	Work up to £35 000 (separate insurance for larger jobs)	All homes		PRC dwellings in private ownership designated under the Housing Act 1985
5.	Warranty	'Reasonable care and skill and in accordance with specification' *plus design* if undertaken by builder	Faulty workmanship and materials	Compliance with NHBC's Requirements	Common law warranty of proper materials, in a workmanlike manner, and fit for habitation	
6.	Time limit starts	Practical Completion determined by builder	Render of Final Account by member paid within 21 days	Certificate of Completion by NHBC	Certificate of Completion by NHBC	Certificate of Completion by PRC inspector

Continued

	BEC	FMB	NHBC: New homes	NHBC: Conversions	NHBC: PRC houses
7. Time limit of scheme	(a) pre-completion (b) 6 months for all defects plus (c) 2 years for structural defects	(a) pre-completion (b) 2 year bankruptcy etc. cover	(a) pre-completion (b) 2 years for all defects (c) 8 years for major damage caused by structural defects and defects in drainage system	(a) pre-completion (b) 1 year for all defects (c) 5 years for structural defects	(a) pre-completion (b) 2 years for all defects in PRC repairs (c) 8 years for major damage caused by structural defect in PRC repairs
8. Financial limit on insurance	£5000 (may be increased for inflation)	(a) £7500 (b) None	(a) 10% of purchase price or £10 000 whichever is greater (b) Purchase price up to £500 000	(a) £5000 (b) £50 000 subject to scheme limit (c) per development	(a) £5000 (b) & £24 000 (c)
9. Who insures?	Norwich Union	Provincial Insurance plc	NHBC and Reinsurers	NHBC and Reinsurers	NHBC backed by Reinsurers
10. Guarantee or insurance	Guarantee: builder pays unless insolvent	Guarantee: builder pays unless insolvent or refuses to honour arbitration award; Board pays plus 2 year certificate for remedial work	Guarantee: pre-completion – insurance for 8 years against structural defects and defects in the drainage system, commencing 2 years after Practical Completion, with the developer guaranteeing this period	Guarantee for 1 year then insurance for 5 years	Guarantee for 2 years then insurance for 8 years against structural defect in PRC repairs

Continued

	BEC	FMB	NHBC: *New homes*	NHBC: *Conversions*	NHBC: *PRC houses*
11. Conciliation	Yes: obligatory unless both parties agree to arbitration	Yes	Yes: by agreement of parties	At request of purchaser	No
12. Arbitration	(a) Yes: but with builder only after conciliation but if client starts action in the courts insurance no longer applicable but builder cannot object	(a) Yes – but not conciliation	(a) Yes	(a) Yes	(a) Yes
	(b) Documents only arbitration but site inspection at arbitrator's discretion	(b) Documents only – hearing not on site	(b) Arbitrator determines procedure	(b) Arbitrator determines procedure	(b) Arbitrator determines procedure
	(c) Arbitration Rules	(c) Arbitration Rules	(c) No arbitration	(c) No Arbitration Rule	(c) No Arbitration Rule
13. Consequential loss excluded	Yes	Yes: 1 year for central heating etc.	From April 1986 alternative accommodation, removal expenses and storage insured		Yes
14. Exclusions on scheme applicable to builder and insurance	(a) Roof work unless complete replacement	(a) Roof work unless complete replacement	(a) Anything outside curtilage of home		Same exclusions as in NHBC Scheme for new homes
	(b) Landscape gardening	(b) Matters covered by Household Insurer's Policy	(b) Lifts, swimming pools		

Continued

	BEC	FMB	NHBC: New homes	NHBC: Conversions	NHBC: PRC houses
	(c) Swimming pools	(c) Wear and tear (d) Missing damp proof course	(c) Normal shrinkage, wear and tear (d) Matters covered by Household Insurance Policy (e) Work not carried out by builder (f) Normal shrinkage etc. i.e. covers only work done by builder		
15. Who pays fee?	Builder or client	Builder	Vendor	Vendor	Builder or client
16. Is there a standard form agreement?	Yes	No	Yes	Yes	Yes
17. Other matters covered by agreement	(a) Payment (b) Insurance of materials (c) Variations – minor building works agreement	None: but builder must have proper insurance for employer's public and product liability	Agreement covers only builder's performance in relation to workmanship, materials and design – all other matters covered by sale or building contract		
18. Approval from 3rd party needed for changes to scheme	Office of Fair Trading must approve	No	No – NHBC no longer applies to Secretary of State for approval in order to achieve exemption under Defective Premises Act	No	Yes: from insurers
19. Scheme introduced	October 1984	June 1981	1965	1982/83	1985

20.	Common law rights preserved	Yes	Yes	Yes	Yes	Yes
21.	Is scheme voluntary?	No	No, if builder is on warranty register	Yes in theory but mortgage finance difficult to obtain without NHBC cover	Yes	Yes

Note 1
This table refers only to residential buildings, i.e. to houses and to individual dwellings in multi-apartment blocks.

Note 2
The Defective Premises Act 1972 imposes a statutory duty in respect of design, workmanship, materials and fitness for habitation on any person connected with the construction of a dwelling or with its enlargement or conversion. This duty cannot be excluded by contract, but does not apply where there is an 'approved scheme' in force. The NHBC Insurance Schemes of 1974, 1975 and 1977 were approved under the Defective Premises Act. The 1979 Scheme and subsequent NHBC Schemes (1985, 1986, 1988 and 1989) have not been approved and residential buildings built under these Schemes are not therefore exempted from the provisions of the Defective Premises Act. The limitation period under this Act is six years from completion of construction, or a further six years from the date of any remedial work carried out by the producer.

Note 3
BEC : Building Employers' Confederation (the BEC scheme was purchased by Building Guarantee Scheme (UK) in September 1991; these details may not remain valid from this date).
FMB : Federation of Master Builders
NHBC: National House Building Council
OFT : Office of Fair Trading
PRC : Pre-stressed Reinforced Concrete

Note 4
No reference has been made in this table to the Housing Association Property Mutual.

Act); it arises from common law, i.e. by the general and unwritten law of the community. As far as insurance is concerned:

(a) Contractors take out and maintain liability insurance because they are required to do so by the standard forms of building contract and as a matter of commercial prudence and common sense. Such policies do not normally cover defects in the building and usually exclude the cost of making good defects, materials, workmanship and design. They cover injury and damage to third parties *arising out* of the construction. Such insurance is not imposed by law. Clients will often insist it is taken out, as will financiers of large projects.

(b) Similar considerations apply to professional indemnity insurance which may be a condition of the appointment by a client, or may be a general requirement of the professional institution relevant to the consultant, or may be mere commercial prudence, or any combination of these.

3.1.1 TERMS AND CONCEPTS

Liability insurance follows the producers, e.g. contractors' All Risks policies or architects' or engineers' professional indemnity policies.

3.1.2 TECHNOLOGY AND INSURANCE TERMS AND CONDITIONS

Do terms and conditions depend:

(a) Upon the latitude given to the insurer to collect technical information about the project in hand?
    No (except for latent defects insurance).
(b) On the geographical location of the building, its height, free spans, useful loads or other data (normally) determined in the brief?
    No (except for latent defects insurance).
(c) On the actual technology to be used (geotechnique, spans, wind bracing, materials etc.)?
    No (except for latent defects insurance).
(d) On the organization of the designer's or control bureau's office, or of the works, e.g. the levels of competence and the skill of different participants, including different suppliers (for consultants' professional indemnity policies)?
    Yes.
(e) On whether the insurer uses his own control bureau (see also 3.1.3 on the matter of technical control)?
    No (except for latent defects insurance).

(f) On the claims record of the insured party (see also 3.1.4)?
Yes.

3.1.3 INSURER'S TECHNICAL CONTROL

These questions are not relevant to the UK because each producer will
normally hold an annual insurance policy covering the relevant risks for all
his activities. Insurers seldom become involved in individual projects unless
they are unusually large or unusually risky or both.

3.1.4 INSURERS' POSSIBLE INFLUENCE UPON BUILDING TECHNOLOGY

From the standpoint of producers, liability insurers have no direct influence
on building technology, because such insurances normally provide annual
cover for the general categories of work of each producer. In the long run,
premiums for any producer will reflect claims made and, hence, the risks
entailed – but the link with technology is very weak.

3.1.5 NEW TECHNOLOGIES

This section is not relevant to the UK because producers normally hold
annual policies and little technical information is either sought by or disclosed to insurers. That is, insurers react to their experience in the relevant
market and to the claims record of individual firms.

3.1.6 THE INSURANCE MARKET, GENERAL IMPRESSION

In the UK, the market

(a) is national, rather than international;
(b) is competitive;
(c) is specialised as far as professional indemnity policies are concerned, in
    part because of the risks involved and, in part, because of the claims
    experience of insurers.

In 1986 a number of underwriters offering professional indemnity
insurance (to architects especially) withdrew from the market. Therefore,
temporarily at least, this kind of insurance was less – much less – competitive than in the past. Since 1988 the supply of this insurance has
increased and premiums related to turnover have fallen. For a variety of

reasons, it is unlikely that the 1990/91 rates will be held at their current relatively low levels.

3.1.7 COLLECTIVE INSURANCE FOR PROFESSIONAL OR OCCUPATIONAL GROUPINGS. GRADATION OF INSURANCE CONDITIONS

(a) So far in the UK, solicitors have made more progress than others in establishing models for professional indemnity insurance.
(b) As a consequence of the high (and then rapidly increasing) cost of professional indemnity premiums, the building professions actively investigated group insurance systems, approved policies etc. Architects in England and Wales (in the guise of RIBA[4]) have arranged a scheme, whilst those in Scotland achieved one many years ago. The Association of Consulting Engineers has a scheme which most members use. The RICS[5] is considering a compulsory scheme for surveyors.
(c) So far in the UK, continuing professional development is not yet mandatory on fully qualified professionals. The majority of the institutions require younger members to undertake CPD and most encourage all their members actively to participate in CPD. It is likely that the majority of the institutions will soon impose mandatory requirements on the majority, if not all, of their active members.

3.2 During the Guarantee (Maintenance, Defects Liability) Period

(See 1.2 if in doubt about the meaning of the words in brackets.)
Contractors will be covered against their liabilities during guarantee periods as an extension to contract works policies (policies covering damage to the works – e.g. by fire – during construction and which are a condition of contracts for building works).

3.2.0 EXISTENCE OF LIABILITY AND PROFESSIONAL INDEMNITY INSURANCE

This exists:

(a) for contractors, who are liable under the specific building contract;
(b) for consultants; their professional indemnity insurance will cover them to the extent liability attaches to them.

3.2.7 RIGHT OF RECOURSE AGAINST A PRODUCER

This remedy is not normal in the UK.

---

[4] Royal Institute of British Architects.
[5] Royal Institution of Chartered Surveyors.

3.5 After the Guarantee (Maintenance, Defects Liability) Period

3.5.1 NORMAL DURATION OF COVER; CANCELLATION

(a) Is the cover always/normally annual or does it run over more than one year?
Annual.
(b) Does non-payment of the premium necessarily entail cancellation of the insurance contract?
Yes, if delay is substantial.
(c) Is it possible for each of the parties to cancel the insurance contract prior to its normal expiry?
No, normally only the insurers have this as of right.

Contractors' public liability policies cover *damage occurring* during any policy.
Professional indemnity policies cover *claims made during* any policy.

3.5.2 LEVEL OF PREMIUMS, EXCESS AND FRANCHISE

Premiums are reviewed when any annual policy is sought or a current annual policy is extended. It is not possible to give examples because the range is wide and depends on, e.g., the experience of an insurance company in the market, the size of the premium and the claims record of the producer concerned.

It is not possible to obtain policies with zero excess (called, in the UK, the deductible) because no insurer would offer such a policy. Excesses are seen as essential to protect the interests of the insurer and avoid the costs of handling small claims.

Professional indemnity insurance policy premiums increased, on average, up to early 1988. Since then, they have fallen as the supply of PI insurance has increased; some practices have faced material increases; a few are finding it difficult to insure. Architectural and structural engineering practices have, in general, been affected by high premiums, although the rates, related to turnover, are very variable.

3.5.3 TIME LAPSES BETWEEN OBSERVATION OF DAMAGE AND INVESTIGATION

These questions are too complex for specific answers. The circumstances vary greatly: many proceedings are settled by negotiation, often because a consultant wishes to maintain good client relationships or because clients prefer to accept an offer rather than risk the uncertainties, delay and high costs of action in court. The details of these (many) cases are not made known.

In general, clients have six years from the date of *damage* to sue, whether or not the damage was *discoverable*. By virtue of the Latent Damage Act 1986, the period is three years from the date of the *discoverability* of damage, subject to an overall limit of 15 years from the *breach of duty*, which is itself an uncertain event in many instances (1.5.3 refers).

3.5.4   TIME ALLOWED BETWEEN THE INSURER'S RECOGNITION OF A CLAIMS EVENT AND HIS PAYING OUT

Determined by law (on insurance or other law)?

Insurers pay when claimants have proved the liability of the insured party. This often involves complex and highly debatable issues of fact and law. The evidence of experts often plays a material part in the proceedings, whether by negotiation or before the courts. In 1987, the average elapsed time between the registration of a writ and an award by an Official Referee was 31 months: actual elapsed times were very variable, as were the intervals between a claims notification and the registration of a writ.

3.5.5 KINDS OF DEFECTS AND DAMAGE WHICH CONSTITUTE CLAIMS EVENTS; ONUS OF PROOF

(a) Defects: It is probably rare that a defect not having led to damage will constitute a claims event. Have you any instance of this in your system? Not so: poor design, use of wrong materials and inadequate health and safety measures are all defects which would lead to valid claims.

(b) Damage: Are there, in your system, any fixed rules for determining the kind and the extent of damage which will constitute a claims event? If so, any reference would be useful.
No.

(c) Does it rest with the plaintiff (the client, successive owner etc.) to identify the individual producer who has committed the breach of duty which has led to observed damage (possibly an observed defect) in order to file a claim for damages against him? Or will the 'thing speak for itself' (is the *res ipsa loquitur* principle applied)?
A plaintiff must prove who was liable and the extent of their liabilities. In practice, plaintiffs may sue all producers 'in sight' leaving them to apportion the liability amongst themselves. In multi-party actions, the plaintiff can recover the full amount of the award and taxed costs from any defendant ('he who can, pays').

3.5.6 EXCLUSIONS

Professional indemnity policies cover liabilities for *negligence*, error or omission. These do not include contractual warranties or indemnities given under contract where liability may arise regardless of negligence, error or omission, although such warranties are becoming more frequent. Generally, insurers will not insure 'fitness for purpose' design responsibilities except, perhaps, work within the scope of the Defective Premises Act 1972.

Such PI policies may:

(a) exclude certain categories of work if the insurer concludes that the insured has not adequate experience in that area;
(b) permit a service only by individuals with the appropriate qualifications, e.g. building surveys can be done only by chartered surveyors, some structural surveys only by chartered structural engineers.

References (sections 1–3)

Professional Liability: Report of the Study Teams (HMSO, 1989) pp. 63–128 inclusive.
Capper, Phillip *General Statement: Tort and Statutory Duty in English Law* (Third Annual Conference, Kings College Centre for Construction Law and Management, London, 1990).
Winward, Friarton *Collateral Warranties* (Blackwood Scientific Publications, London, 1990).

APPENDIX I 'NORMATIVE BUREAUX' (UK)

BSI, the British Standards Institution, is the UK body charged with the production of national standards. It has this responsibility under Royal Charter. It is responsible for the production of British Standard specifications, test methods, codes of practice and other forms of standard for all industries including the building industry. British Standards are for voluntary implementation; many of them support the Building Regulations through reference in Approval Documents. BSI is the UK member body on CEN/CENELEC and 150/IEC.

No other UK body produces national standards, but there are other specialist standards produced at trade association or company level.

In the construction industry, the British Board of Agrément has a recognized specialist status.

Other bodies as well as BSI provide third party quality assurance services to the construction industry in response to demand. BSI's tasks are defined in its Royal Charter. BSI is an independent body.

1.1 LIABILITY

BSI's liability vis-à-vis the client is based in common law.

3 PROFESSIONAL INDEMNITY INSURANCE

BSI carries professional indemnity insurance for simple commercial reasons. The Institution needs to be able to meet claims against it.

3.1.6 THE PROFESSIONAL INDEMNITY INSURANCE MARKET

Cover is not now difficult to obtain and premiums are currently more stable.

3.5.1 INSURANCE COVER

BSI's cover is renewable annually subject to agreement between BSI and its insurers.

APPENDIX II BUILDING CONTROL AUTHORITIES (ENGLAND AND WALES)

The legislative framework of building control in England and Wales is contained in the Building Act 1984 and the Building Regulations 1985 made under it. The technical requirements in these regulations are expressed in a short functional form. The requirements are supported by Approved Documents prepared by the Department of the Environment which give practical guidance on how to comply with the functional requirements.

The function and duties of local authorities under the Act are carried out by their building control staff (who usually have some professional qualifications, though this is not a legal requirement). The Act also provides for the supervision of building work by private approved inspectors as an alternative to the local authority. This choice for developers is currently only available from the National House Building Council who now has 40 per cent of the residential building control market, so far for dwellings of up to eight storeys.

The procedures under the new system and the role of the local authorities are set out in a circular issued by the Department of the Environment.

A builder choosing to use local authority supervision is required before starting building work either to deposit full plans with the local authority or give them a simpler 'building notice' (this latter option not being available in certain cases). Where plans are deposited the local authority must check them for compliance with the building regulations and must either pass them or reject them within five weeks (or two months by agreement). The local authority may relax or dispense with a requirement in the regulations where appropriate. The builder must notify the local authority at certain stages of construction, and it is up to them to decide whether to inspect the work at those stages or any other.

The Act gives local authorities power to require work contravening the regulations to be altered or removed (but not if the work was executed in accordance with plans passed by authority).

The location of building control departments in local authority organizations varies. Some are located in planning departments, some report to the engineer and a few are part of the public health officer's department.

1.1 There is no document setting out a client's post-construction rights in relation to a local authority in the event of an omission or error. A person suffering damage as a result of negligence may have recourse to law, but his right of action against a local authority (or anyone else) is subject to certain limitations (see 1.5.3 for further details). In the majority of cases, a local authority is sued together with either the designer or builder or both but, as has been made plain, the grounds for actions in tort are now severely restricted.

1.2 In English law, an employer is held liable for the torts of his employees so long as the tort is committed when the employee is acting in the course of his employment. This is known as the principle of vicarious liability. There is an explicit statement of this principle in relation to the liability of local authorities for the tortious acts of building control officers in section 115 of the Building Act 1984.

1.5.3 The limitation period during which a person can bring an action for negligence against a local authority is the same as for other parties to the construction process. (The limitation periods in question are laid down in the Limitation Act 1980 and the Latent Damage Act 1986.)

1.5.6 Generally speaking, an action in the tort of negligence enables a person who has suffered injury to recover financial damages. But, first, the person must be able to identify a *duty of care* owed to the person who suffers the damage, a *breach of that duty* and the *damage* arising from that breach. The view taken by English courts of the duty owed by local authorities in latent defects cases has changed in recent years. In the mid-1980s, the courts began to adopt a new policy of caution, tending to restrict local authorities' liability to imminent threats to the health and safety of the occupier. Even this was rejected by the House of Lords in 1990.[1] Phillip Capper has commented on this case: 'Perhaps the greatest significance of that decision is the doubt expressed by their Lordships as to whether the building control functions of local authorities even attract the normal tortious liability in regard to actual physical injury and actual damage to other property.'

3　Individual building control officers do not need to take out professional indemnity insurance because their employers are liable in law for their professional actions.

APPENDIX III　TESTING LABORATORIES (UK)

1　THE TESTING LABORATORY'S LIABILITY

1.1.0　There is nothing defined in law and no model contract is known.
　Errors or omissions would (or should) be covered by the terms of individual contracts drawn up between the client and the testing laboratory.

1.2　The assumption is wrong – that is, a laboratory's liability could be called upon even during the guarantee period.

1.5　The assumption stated is presumed: there is no first hand experience.

2　THE IMPACT OF TECHNICAL TESTING UPON THE TWO KINDS OF 'CONSTRUCTION INSURANCE'

Insurers are becoming increasingly aware of the benefits of having a product or system testing. It is difficult for a testing laboratory to know whether insurance would be more expensive, or even refused, if testing was not carried out, although this is believed to be the case.

3　PROFESSIONAL INDEMNITY INSURANCE FOR THE TESTING LABORATORY

As a matter of commercial prudence, professional indemnity insurance is arranged. Occasionally, clients wish to confirm that such insurance is held, in which cases insurance may be a condition of obtaining a contract.

---

[1] *Murphy v Brentwood District Council* [1990] 3 WLR 414.

3.1.6 THE PROFESSIONAL INDEMNITY INSURANCE MARKET FOR TESTING LABORATORIES

Insurance companies are becoming increasingly reluctant to offer professional indemnity and (product liability) insurance. Therefore, the risk is spread around a number of insurance companies and such insurance is becoming substantially more expensive.

Details of an actual policy are:

Duration: 12 months
Cancellation: Insurers may cancel (giving seven days notice) in the event of a major change· of circumstances. In this eventuality, there will be a refund for the unexpired period of the policy.
               The insured cannot cancel during the term.

WORK VOLUME OF TESTING LABORATORIES

Technical testing is expanding in the UK. Except for fire testing, there is little evidence that 'construction insurers' are interested in developing technical testing.

One other important aspect is the growing interest of the construction industry in quality assurance. It has been publicly stated by at least two insurance companies that they will view more favourably manufacturing companies who have a quality management system that is registered with an independent certification body as conforming to the national Standard, BS 5750.

APPENDIX IV   BRITISH BOARD OF AGRÉMENT (UK)

(1)   The British Board of Agrément (BBA) issues agrément certificates for new and innovatory products and systems for use in the construction industry. Certificates can be used to demonstrate compliance with building regulations.

(2)   The Board is a 'company limited by guarantee' and the objects for which it is established are set out in its Memorandum of Association.

(3)   The Board was set up by the Government in 1966 as an independent body, to encourage innovation in the construction industry by providing reliable and independent assessments of new or innovatory products and systems. The Government pays grant towards the Board's runing costs, but these are largely covered by fees from clients.

(4)   The BBA's governing body is a Council whose members are appointed (normally for a three-year term) by the Secretary of State for the Environment. The members serve on a part-time basis. The Board

has a full time staff of about 90, headed by a Director who is also a member of the Council. An Assessor from the Department of the Environment sits on the Council and also on the BBA Finance Committee.

(5) Under the Memorandum of Association, the BBA has to comply with 'any direction of a general character as to the exercise of its functions' which may from time to time be given by the Secretary of State. A 'Memorandum of Arrangements', which was revised and updated in 1984, set out details of the BBA's financial and other relationships with the Department.

(6) The BBA has a contractual relationship with its clients. The responsibilities of the parties are set out in the contract. The BBA accepts responsibility for its assessments, but the question of liability has not been tested in the courts.

(7) The BBA does not take out professional indemnity insurance. That would be contrary to current Government rules for grant-aided bodies. For insurance purposes, similar rules apply to the BBA as apply to Government Departments.

France[*]

Jean Desmadryl

1 LIABILITY

1.1 General

1.1.0 LAW OR CONTRACT OR BOTH

Producers' post-construction liability has been regulated in France since 1804 by the Civil Code. It was reorganized by Law 78.12 of 4 January 1978.

The law defines producers (*constructeurs*) in the following terms (Civil Code, article 1792–1):

'The following are deemed to be producers of the works:
(1) Any architect, contractor, technician or other person bound to the client by a contract for works or services;
(2) Any person who sells, after its completion, works he has built or has had built for him;
(3) Any person who, although acting as a mandatory of the client [*le maître d'ouvrage*], carries out a task similar to that of an independent producer.' (Unauthorized translation.)

The ambit of this liability is therefore very extensive. All those involved are liable with the exception of sub-contractors and manufacturers (unless the latter are manufacturers of components).

Moreover, producers are automatically liable and they can obtain exemption only by proving that the damage (*désordre*) arose from a cause beyond their control or unrelated to their intervention (*cause étrangère*). Therefore a producer may be held liable without fault or negligence on his part.

So far as terminology is concerned, it should be noted that there can be a defect (*défaut*) without fault or negligence.

However, this liability does not cover any type of defect.

Under the terms of the law, it covers damage:

[*] Original text in French.

- which compromises the solidity of access roads, main conduits, foundation, load-bearing, enclosing or covering works or units of equipment which are inseparably connected with these works;
- or 'which renders the works unfit for their intended use' (unauthorized translation).

Other units of equipment are required to have a contractual warranty of not less than two years.

Moreover, a manufacturer of components is jointly and severally liable with the contractor who installed the component for the above damage.

Liability is a requirement of public policy, i.e. any contractual clause seeking to exclude or limit it is deemed by law to be void.

1.1.1 TERMS AND CONCEPTS

'BUILDING' (BÂTIMENT)

The word 'building' is defined by insurance contracts as being:

'A construction erected on the ground inside which humans are intended to move and which offers at least partial protection against the assault of natural external elements' (unauthorized translation).

It should be noted that the liability defined in 1.1.0 concerns all construction works, whilst the idea of a building is more restrictive; the definition of the word 'building' is necessary in French law since the statutory obligation of insurance only concerns buildings, other construction works being subject to the same liability but not to the obligation of insurance.

For building works, the area of liability is the same in the case of new construction and rebuilding works.

'PRODUCERS' (CONSTRUCTEURS)

The definition is given by law, see 1.1.0.

The term refers only to parties bound to the client by contract; under French law sub-contractors are therefore not producers. Nevertheless they are often found liable. The same applies to manufacturers of materials and equipment.

However, this liability is not within the ambit of the Law of 4 January 1978.

A tenant has no rights vis-à-vis producers.

1.2 During the Guarantee (Maintenance, Defects Liability) Period
(*Période de parfait achèvement* – formal completion)

1.2.1 DETERMINATION OF THE COMMENCEMENT OF THE PERIOD

The commencement of all liability periods is handover of the works.

Handover (*réception*) is the act by which the client declares his acceptance of the works with or without reservation. It is therefore a written document, drawn up by the client.

The parties contractually bound to the client must obligatorily be called to be present at handover.

In the event of a dispute, handover is decided in court.

A change in case law after the publication of the Law of 4 January 1978 resulted in the recognition of tacit handover. However, in most cases handover still gives rise to a written document.

1.2.2 SIGNATORIES TO AND NAME OF THE DOCUMENT STATING THE COMMENCEMENT OF THE PERIOD

(See also 1.2.1).

Producers are requested to sign a list of reservations attached to the Certificate of Practical Completion (*procès verbal de réception des travaux*), primarily when the client has expressed reservations (see 1.2.5) and when time limits must be set in this Certificate (see 1.2.4).

1.2.3 THE PERIOD'S DURATION AND TERMINATION

The duration of the formal completion period (*période de parfait achèvement*) is one year. It is fixed by law and cannot be extended. No formal procedures are required at the end of this period. However, once the works necessary in respect of the reservations are effected, a statement of removal of reservations must be drawn up by both parties.

1.2.4 TIME ALLOWED FOR DIFFERENT CATEGORIES OF PRODUCERS FOR MAKING GOOD, REPAIRING OR PAYING DAMAGES

Periods of time necessary for the execution of the remedial works related to the qualifications stated are fixed by agreement between the client and the contractor concerned.

1.2.5 KINDS OF DEFECTS AND DAMAGE COVERED BY THE GUARANTEE

Any defect, even if it does not manifest itself strictly as damage, may fall under a guarantee if it is a failure to conform to construction codes or standards or the stipulations of the contract.

1.2.6 EXCULPATION OF PRODUCER

Any defect giving rise to a reservation or qualification must be made good.

However, the guarantee does not extend to works necessary to remedy the effects of normal wear or abnormal use.

Visible defects must be pointed out at the time of handover, but hidden defects (*défauts cachés*) may become manifest in the course of the one-year guarantee period and the contractor concerned will be liable to remedy them.

Producers therefore are under a duty of result within the meaning of the Questionnaire, Appendix A.

1.2.7 CONSEQUENCES OF PRODUCER'S NON-COMPLIANCE WITH UNEQUIVOCAL DUTY TO MAKE GOOD, REPAIR OR PAY DAMAGES

The law provides that if obligations are not fulfilled within the periods stipulated, after formal notice (*mise en demeure*) to execute the works is unsuccessful, they may be carried out at the cost and risk of the defaulting contractor.

If the contractor no longer exists (has become insolvent, for example), his insurer must meet the cost of remedial works because all contractors must now be insured and the premium paid at the time of construction covers damage occurring during the whole of the liability period.

1.5 From expiry of the Guarantee (Maintenance, Defects Liability) Period

1.5.1 DETERMINATION OF THE COMMENCEMENT OF THE (POST-GUARANTEE) PERIOD DURING WHICH A PRODUCER MAY BE HELD LIABLE

The date of commencement of guarantee periods (ten years or a minimum of two years) is now the same as that of the formal completion period or guarantee (*période de parfait achèvement*), i.e. handover. In addition, this date, once fixed by the Certificate of Practical Completion, cannot be deferred.

1.5.2 NAME OF THE DOCUMENT STATING THE COMMENCEMENT OF THE POST-GUARANTEE PERIOD; ITS SIGNATORIES

Since this is the same document as that of the commencement of the formal completion period (*période de parfait achèvement*), please refer to 1.2.2.

1.5.3 PERIOD(S) OF LIMITATION

The limitation period is ten years from the date of handover for damage affecting the works listed under 1.1.0, as well as units of equipment which are inseparably connected with these works. It is also ten years for damage which renders the building unfit for its intended use.

For units of equipment not mentioned in 1.1.0, the limitation period is the same as the warranty period stipulated in the contract, but not less than two years. Limitation periods are the same for all producers as defined in 1.1.0.

[1.5.4: *Not applicable*]

1.5.5 KINDS OF DEFECTS AND DAMAGE QUALIFYING FOR DAMAGES

The kind of damage likely to require reparation is defined by law. Please refer to the statutory definition given in 1.1.0.

There is no right to reparation for a defect which has not developed into damage. It is this very point which distinguishes the formal completion period (*période de parfait achèvement*) from the guarantee period.

In the first case, as indicated in 1.2.5, any defect, even if it does not manifest itself as damage, as well as a defect of any kind whatever, including those not covered by the ten-year liability period (solidity of certain works) gives right to reparation.

1.5.6 EXCULPATION OF PRODUCER

The rules in this area are different again from those concerning liability for formal completion and are more restrictive.

The law provides that the producer may exonerate himself if he proves that the damage arose from a cause beyond his control or unrelated to his intervention (*cause étrangère*).

1.5.7 CONSEQUENCES OF PRODUCER'S NON-COMPLIANCE WITH UNEQUIVOCAL DUTY TO REPAIR

As producers are presumed liable, any dispute can only concern the nature of the damage or the cause beyond their control which may exonerate them.

If a dispute arises it may be settled by technical expertise, which is usually arranged by the parties' insurers. In the event of litigation, the problem must be referred to the courts.

So far as the producers as defined in 1.1.0 are concerned, only the one who executed the works which gave rise to the damage is liable. However, often more than one producer is liable, for instance if the architect and one or more contractors have each carried out a particular job within the causal chain of the different works.

Since insurers are subrogated to the producers' rights and liabilities, they always have to cover the cost of repairs.

1.6 Transfer of ownership and producers' liability

1.6.1 RULES

Please refer to 1.1.0 and 1.1.1.

The producers' guarantees are transferred in full to any successive owner during the guarantee periods.

Tenants do not have the benefit of these guarantees.

2 PROPERTY (OR MATERIAL DAMAGE) INSURANCE

See 3.

3 LIABILITY AND PROFESSIONAL INDEMNITY INSURANCE

Law 78.12 of 4 January 1978 imposes a duty to insure on any person[1] who may incur liability as described in 1.

---

[1] A Law of 31 December 1989, amending the Law of 4 January 1978, removed the obligation of public agencies to take out damage insurance, whereas originally the 1978 Law had only made possible derogation case by case from this requirement.

In addition, the same Law requires a client to take out damage insurance (*assurance dommage-ouvrage*). This insurance is of a particular type and is not exactly property insurance, as it is merely a way to pre-finance the liability insurance.

This is how the name usually given to the system of construction insurance in France arises, i.e. two-step (*à double détente*) insurance.

In effect, under the Law, this damage insurance covers, 'without seeking to establish liability, the payment for remedial works . . . damage of the nature for which producers are liable on the basis of article 1792' (unauthorized translation).

This damage insurance takes effect after the end of the formal completion guarantee period, i.e. one year after the date of handover.

However, it covers the payment for necessary remedial works in the following cases:

- before handover, if a formal notice has had no effect and the contract concluded with the contractor is terminated by reason of his failure to fulfil his obligations;
- after handover, if a formal notice has had no effect, and the contractor has not fulfilled his obligations.

Because of this special arrangement we have brought together sections 2 and 3, which were intended to deal with property insurance and liability insurance respectively. We have headed it 3 since the French system of construction insurance is more in the nature of liability insurance than of property insurance.

3.1 General

3.1.0 LAW OR CONTRACT OR BOTH

This is indeed insurance regulated by legislation. Furthermore, for building works liability insurance is mandatory; for damage insurance it is compulsory only in what concerns the ten-year liability and fitness for the intended purpose.

3.1.1 TERMS AND CONCEPTS

The insured parties are those described in 1. The beneficiary for the producers' liability insurance is the client's damage (*dommage-ouvrage*) insurer and for the damage insurer it is the client and his successors.

3.1.2   TECHNOLOGY AND INSURANCE TERMS
AND CONDITIONS

Since 1983, insurers have been subject to a system of capitalization and they could therefore fix different premium rates according to the characteristics of each construction operation.

In fact, they only do this for damage insurance. As in the past, liability insurance contracts continue to be renewal contracts, i.e. they cover all the activities of the insured party (architect, technical office, building enterprise etc.). However, these contracts are negotiated according to the characteristics of the enterprise which, for contractors, are defined by an organization on which insurers rely, the OPQCB, which describes and classifies building contractors. This classification is mainly based on the size of the firm and the competence of the person in charge of it.

A firm's insurance contracts limit the total value of construction operations insured as well as the nature of the works which can be carried out in the framework of an insurance contract.

These insurance contracts include excesses and exclusion limits. However, the part of the cost of remedial works which, because of the excess, remains in charge of the insured party cannot be invoked against the beneficiary.

At each setting-up of a site, the insured party must declare to his insurers the technical characteristics of the construction to be carried out and, at the closure of the site, the total value of the works executed.

3.1.3   INSURER'S TECHNICAL CONTROL

Since the 1978 Law, 'technical inspectors' (*contrôleurs techniques*) now intervene at the client's request; their task is to prevent different technical hazards likely to be encountered in the execution of the works.

Technical inspectors are government recognized; no one may act as a technical inspector unless he is recognized as such. The activities of technical supervision are incompatible with certain other activities, particularly design, execution of the works and expert reports (for instance, for tribunals or arbitration).

Most insurers concede a reduced premium when there is intervention by a technical inspector.

The intervention of a technical inspector is obligatory for major works.

3.1.4   INSURERS' POSSIBLE INFLUENCE UPON
BUILDING TECHNOLOGY

Insurers are grouped under the auspices of an Association (Association française des assureurs construction) which possesses a small technical service. This service operates primarily in order to help insurers assess the extent of the risks they are asked to cover, especially those concerning new technology.

The aim of the Agency for the Prevention of Damage and Improvement of the Quality of Construction (L'Agence pour la prévention des désordres et l'amélioration de la qualité de la construction or Agence qualité construction) is in fact to collect knowledge of construction malfunction and to elicit from this recommendations and advice necessary in order to avoid construction damage. It plays a role in the dissemination of know-how and can warn organizations responsible for issuing regulations, standards and technical advice. The construction insurers are members of this agency.

3.1.5   NEW TECHNOLOGIES

All new technology is the subject of technical advice written by groups of experts appointed by a Joint Ministerial Committee (Commission interministrielle).

Insurers decide under what conditions they will insure these new technologies from the information contained in the technical advice (*avis techniques*).

However, we should not lose sight of the fact that the obligation of insurance borne by the producers has its counterpart in the obligation of insurers to insure all construction works; this last duty is also regulated by law.

Only the setting of insurance premiums is therefore left, and an Office for Valuation of Construction Insurance (Bureau central de tarification construction) can set premium rates.

3.1.6   THE INSURANCE MARKET, GENERAL IMPRESSION

Since 1 January 1983, insurance companies have effectively competed. This was a goal desired by the government in order to break the de facto monopoly which previously existed.

However, as turnover in the area of construction insurance is very small in relation to total insurance turnover (some 2 per cent), many insurers do not venture into this complex and difficult market.

A greater number of insurers handle damage (*dommage-ouvrage*) insurance in order to satisfy the requirements of their regular clients who might undertake construction on a one-off basis.

[3.1.7: *Not applicable*]

3.2   During the Guarantee (Maintenance, Defects Liability) Period

3.2.0   EXISTENCE OF LIABILITY AND PROFESSIONAL INDEMNITY INSURANCE

Liability insurance must cover the formal completion period (*période de parfait achèvement*). As mentioned, in principle damage insurance also takes effect at handover (*la réception*), but it will be called upon during the period of formal completion (*la période de parfait achèvement*) only if a formal notice to remedy has had no effect (see also p. 175).

3.2.7   RIGHT OF RECOURSE AGAINST A PRODUCER

Obviously, if the producer agrees to proceed with remedial works insurance plays no part. However, since he has taken out liability insurance, it is in his interest to call upon this insurance and receive the compensation for the repair works necessary insofar as the damage is covered by the insurance contract.

3.5   After the Guarantee (Maintenance, Defects Liability) Period

3.5.1   NORMAL DURATION OF COVER; CANCELLATION

As in the case of damage insurance, liability insurance is effective for the entire period of statutory liability. It cannot be cancelled and continues to run even if the insured party ceases to exist.

The premium is calculated once and for all, but payment may be in instalments.

3.5.2   LEVEL OF PREMIUMS, EXCESS AND FRANCHISE

It is difficult to cite premium rates, given the competitiveness of the extremely active French insurance system.

However, it is possible to say that liability insurance premiums correspond to some 2 per cent of the value of the works, and damage insurance is less than 1.5 per cent.

With reference to excesses, major firms have taken out policies with excesses of 350 000 to 500 000 francs.

[3.5.3: *Not applicable*]

3.5.4 TIME ALLOWED BETWEEN THE INSURER'S RECOGNITION OF A CLAIMS EVENT AND HIS PAYING OUT

In the case of damage insurance, the insured must give notice of events likely to be covered by the policy at the latest within five days following his awareness of the event.

The damage insurer must, within 90 days of receiving notification of the event, notify the beneficiary of the proposed amount of indemnification for remedial works. The insured then has 15 days to accept or refuse this proposal.

If accepted, the insurer must pay out the indemnification within 15 days.

If the amount is higher than a figure freely fixed in the contract, payment is made in instalments, the first being within 15 days and the remainder so that the insured is never in the position of having to make advance payments for the repair works.

For liability insurance, these periods are not imposed on the insurers; but policies usually stipulate a maximum period for declaration of a claim.

3.5.5 KINDS OF DEFECTS AND DAMAGE WHICH CONSTITUTE CLAIMS EVENTS; ONUS OF PROOF

The ambit of liability and damage insurance policies is exactly the same as the area of statutory liability defined in 1.1.0, not more, not less.

3.5.6 EXCLUSIONS

As we have seen in 1.1.0, the only way in which the producer may be released from his liability is by proving a cause beyond his control.

In addition to this, the insurance contract does not cover damage resulting:

- from an intentional or fraudulent act by the insured party or a policy holder;
- from the effects of normal wear, poor maintenance or abnormal use;
- from a cause beyond the producer's control, particularly directly or indirectly, fire or explosion, unless the fire or explosion is the consequence of claims events covered by the contract;
- from very high winds, cyclones, flooding, earthquakes and other natural phenomena of a disastrous character;
- from acts of foreign war;

- from acts of civil war, acts of terrorism or sabotage;
- from direct or indirect effects of explosion, release of heat or radiation from transmutation of atom nuclei or radioactivity (unauthorized translation).

The insured party (the producer) will not be covered in the case of deliberate or unreasonable failure to comply with accepted practice ('DTU', *documents techniques unifiés*, codes of practice). But this will not exclude the insurer's payment of indemnification to a beneficiary.

Transfer of ownership and insurance cover

Damage insurance is automatically transferred to successive owners of the building. No formal procedures are required in this connection, but lawyers take great care to ensure that mention is made of the existence of damage insurance when buildings are sold during the liability period.

Additional information

So as to give further effect to the provisions of the Law of 4 January 1978, two important procedures have been set up:

- sole policy for a single building site (*police unique de chantier* or PUC);
- settlement agreement.

SOLE POLICY FOR A SINGLE SITE (PUC)

All those involved in a construction operation can agree to take out a sole insurance policy with the same insurance company.

This policy, which is called a 'sole site policy' (*police unique de chantier*: PUC), takes the place of all the policies which must be taken out by the different parties, including the client. It relates to matters for which insurance is compulsory but may also contain non-compulsory coverage.

This procedure greatly facilitates the payment of any claims. It is quite frequently used for very large construction operations. The negotiation of such a policy is sometimes rather difficult.

SETTLEMENT AGREEMENT

To the same end, insurers have drawn up a settlement agreement which lays down a fixed scale of apportionment of liability of the different parties and for different hypotheses of damage. The scale is used to apportion the indemnity among the different liability insurers. In the event of a dispute, the insurers have undertaken to respect the decision of a conciliation panel.

Italy[*]

Nicola Sinopoli

Before broaching the subjects in this chapter, it should be noted that Italian legislation makes no distinction between residential and non-residential buildings. The points made here are therefore applicable both to residential and non-residential buildings.

We would also add that Italian legislation on relations between clients and contractors lags well behind the current changes which have taken place in the field of construction techniques and the organization of the construction process. This is a constant source of disputes regarding liability and guarantees and such actions are brought before the courts if the parties cannot reach agreement. The cases usually lead to appeals and even reach the Supreme Court of Appeal.

1 LIABILITY

1.1 General

1.1.0 LAW OR CONTRACT OR BOTH

The contractor's liability is prescribed by law. This liability cannot be altered by contract.

Italian law (the Civil Code) only lays down the principle that the contractor is liable.

In Italy technical experts (designer, engineer, supervisor of works etc.) are only criminally liable for death or personal injury, and they have no civil liability for defects. They are not liable for damage even if the project is manifestly wrong: any defects in the project have to be discovered by the contractor.

Only very recently have the first principles of professional liability been laid down by two Laws, but these principles do not modify the Civil Code.

---

[*] Original text in French.

The professional liability introduced by the new Laws is not a type of civil liability for construction defects, but liability for the quality of the project in relation to safety requirements and efficient operation. These new Laws are Law 13/1989 concerning 'architectural barriers' for disabled persons and Law 46/1990 concerning the safety of technical installations.

The principles of liability laid down by these Laws are very different and may be summarised as follows:

- the Law on 'architectural barriers' compels the professional persons involved to guarantee, when signing projects, that standards have been complied with (dimensions of staircases, lifts, bathrooms etc.). In this case liability is not civil, but criminal. If the standards are not complied with, the professional person is punished because he signed a false statement in an official document;
- the Law on the safety of technical installations imposes responsibility for complying with safety standards on the client and producers. Failure to do so entails an administrative sanction (fine) or, for professional persons, withdrawal of their licence after they commit three offences. The Law provides for an 'implementing regulation', which has not yet been adopted, to lay down details of sanctions.

A regional Law of Emilia-Romagna (Regional Law 33/1990) concerning building standards laid down by local authorities (municipal regulations) provides for a different principle of liability. By their signature, professional persons guarantee that standards have been complied with. If, on official inspection, it is found that any aspect of the building fails to meet the criteria specified by the building standards, there is once again a false statement in an official document and a criminal sanction. This Law affects the question of defects because municipal regulations cover, among other things, the quality of construction (efficiency of installations, insulation, imperviousness to damp etc.).

The Civil Code (article 1662) states that the contractor has an obligation to carry out a construction according to 'the state of the art' or accepted practice. This means that the contractor must follow the technical regulations relating to the features of the construction requested of him.

1.1.1 TERMS AND CONCEPTS

'BUILDING'

No universal definition of the word 'building' exists. Italian law refers both to construction and to building. Each contract may furnish its own particular definition. Given the general definition of the term 'building', no distinction is made by law between new construction and reconstruction, or between residential and non-residential construction.

In the final analysis, 'building' means anything referred to in the contract between the contractor and the client.

'CONTRACTOR'

No universal definition of the word 'contractor' exists. The law refers to *l'appaltatore* (the contractor). The contractor is the person who, by organizing the necessary resources and undertaking the management at his own risk, carries out a construction or provides a service for financial reward.

'PURCHASER'

Italian law attaches importance to this party. Article 1490 of the Civil Code gives the purchaser protection as against the vendor.

1.2   During the Guarantee (Maintenance, Defects Liability) Period

In Italy, there is no provision for a guarantee period proper for building works.

Generally, contracts which are tendered for stipulate that maintenance of the works is the responsibility of the contractor from the date of completion until acceptance without reservation by the client.

This period can therefore be considered the guarantee period.

1.2.1   DETERMINATION OF THE COMMENCEMENT OF THE PERIOD

The period commences on the date of completion of the works.

1.2.2   SIGNATORIES TO AND NAME OF THE DOCUMENT STATING THE COMMENCEMENT OF THE PERIOD

The document is called a Certificate of Practical Completion (*verbale di ultimazione*); it is signed by the contractor and the Director of Works (*direttore del lavori*).

1.2.3   THE PERIOD'S DURATION AND TERMINATION

The duration depends on the time when *collaudo*, acceptance or handover without reservation takes place.

Acceptance entails inspection of the works by the client from the viewpoint of quality and to ensure that they conform to the project and the contractual conditions. The formal document is called the Acceptance Certificate (*certificato di collaudo*) and must be signed by the contractor, the Directors of Works, the client and the inspector (*collaudatore*).

The client may waive the inspection only if he is a private individual.

Clients in the public sector are obliged to carry out inspections. For minor projects carried out in the public sector, the Certificate of Acceptance is replaced by a Completion Certificate (*certificato di ultimazione*), signed by the contractor and the Director of Works.

Taking into account the importance attributed to the purchaser by Italian law, there is another date to consider: the date when the contract of sale is signed by the vendor (contractor, developer or client) and the purchaser.

Certain periods are fixed by law with reference to handover:

(a) Generally, handover must take place within six months of the date of formal completion of the works. For certain types of work or special installations, the commencing date of the period may be deferred (for example, heating installations are usually handed over in winter).

(b) Once started, handover must be carried out within the stipulated period. It must commence during a period of two months from the time the inspector (or inspectors if more than one) receives the completion documents for the works; it must be terminated within the following 120 days. This period may be extended if, during the process of handover, defects are discovered which are to be made good by the contractor at his own cost and during the prescribed periods. These periods stop the running of the handover period. During the handover process, the inspector or inspectors state their opinion on reservations which the Director of Works or the contractor may have made, during or at the completion of the works. Such reservations may be of an economic or technical order (in other words, they may concern the way the works were executed and aspects related to their quality).

1.2.4 TIME ALLOWED FOR DIFFERENT CATEGORIES OF PRODUCERS FOR MAKING GOOD, REPAIRING OR PAYING DAMAGES

These periods are fixed by the inspector (or inspectors) according to the nature of the defects found.

The guarantee covers all defects which show, in the opinion of the inspector, that the work has not been executed according to 'the state of the art' or accepted practice. See also 1.5.5 on this point, where different types of defects are illustrated and commented upon.

[1.2.5: *Not applicable*]

1.2.6 EXCULPATION OF PRODUCER

During handover, the contractor is held responsible (see also 1.5.6) for the rectification of all defects recorded by the inspector (or inspectors). We are therefore faced with a sort of obligation to achieve a specific result on the part of the contractor and a wide-ranging authority accorded to the inspector by law.

1.2.7 CONSEQUENCES OF PRODUCER'S NON-COMPLIANCE WITH UNEQUIVOCAL DUTY TO MAKE GOOD, REPAIR OR PAY DAMAGES

If the contractor refuses to carry out or delays the remedial works, by law they may be carried out by another contractor at the cost and risk of the defaulting contractor.

If the contractor is insolvent or no longer exists, it is the client's responsibility to carry out the works since insurance is not obligatory in Italy: see 2 and 3.

1.5 From expiry of the Guarantee (Maintenance, Defects Liability) Period

1.5.1 DETERMINATION OF THE COMMENCEMENT OF THE (POST-GUARANTEE) PERIOD DURING WHICH A PRODUCER MAY BE HELD LIABLE

The commencing date of the periods during which a contractor is liable for construction defects is the same as that of the guarantee period, i.e. formal completion of the works.

1.5.2 NAME OF THE DOCUMENT STATING THE COMMENCEMENT OF THE POST-GUARANTEE PERIOD; ITS SIGNATORIES

This is the document mentioned in 1.2.1 and 1.2.2.

1.5.3 PERIOD(S) OF LIMITATION

Italian law stipulates two different limitation periods:

(a) two years from the date of completion of the works for defects or differences from the project (*vizi*) (see 1.5.5): Civil Code, article 1667;

(b) ten years from the date of completion of the works for major defects (see 1.5.5): Civil Code, article 1669.

Italian law provides for two other periods which are relevant in the reporting of construction defects:

- a period of 60 days from the date of discovery for defects mentioned in point (a);
- a period of one year from the date of discovery for major defects mentioned in point (b).

These periods are applicable in client/contractor relations.

See 1.6 concerning relations with the purchaser of a building.

1.5.4 TIME ALLOWED (DIFFERENT CATEGORIES OF PRODUCERS) FOR MAKING GOOD, PEPAIRING OR PAYING DAMAGES

Italian law does not lay down precise periods. It merely states (Civil Code, article 1668) that the client may require differences and defects to be rectified at the contractor's cost or seek a pro rata reduction in price, while reserving the right to compensation in the event of fault on the contractor's part. The law adds that if the differences or defects are such as to render the building totally unsuitable for its intended use, the client may seek cancellation of the contract.

In practice, time limits and the type of compensation are decided:

- by agreement between the client and the contractor;
- by the ordinary courts; or
- by arbitration.

We should underline the fact that, in Italy, cases which go to court usually take a very long time (even over ten years). They involve very complex procedures which we cannot broach here.

On the other hand, it is quite usual to refer disputes to arbitration. The contractor and the client each select an arbitrator (technical expert or lawyer, according to the nature of the dispute). By agreement, the two arbitrators designate a third arbitrator, who becomes the chairman of the arbitral tribunal. The arbitrator's award is accepted by the parties since there is no possibility of appeal.

1.5.5 KINDS OF DEFECTS AND DAMAGE QUALIFYING FOR DAMAGES

Italian law identifies three types of differences which may give rise to defects or damage for which the contractor is liable (see also 1.5.3):

- differences in construction, i.e. the execution of a building which differs in part or totally from the project or contractual clauses. These differences need not necessarily be the cause of damage; the mere fact of their existence (and that they have not been authorised) may lead to an action by the client;
- building defects (*vizi*), i.e. work which is not in accordance with standard practice;
- 'major' defects (*vizi gravi*) which affect the very stability of the building or its intended use. Article 1669 of the Civil Code gives an implicit definition of 'major' defects as follows: 'with regard to buildings or other immovables which are by nature intended to last a long time, if, in the ten years following completion, the works suffer damage or are entirely lost as a result of a defect in the ground or in construction, or if there is a manifest risk of collapse or major defects, the contractor shall be liable to the client . . .' (unauthorized translation).

Judgements delivered by the courts over the years have not attempted to settle once and for all the meaning of the adjective 'major', which is a rather ambiguous term, nor have they clearly defined the kinds of defects for which the liability period is ten years. They merely specify a number of defects which are covered by article 1669.

In addition to statics errors which threaten the safety of a building (poor judgement of the load-bearing capacity of the foundation soil or the execution of the bearing structure in and above ground), the Supreme Court of Appeal has held that the following defects should be considered 'major':

(a) inadequate number of water pipes for the provision of drinking water for the tenants' requirements;[1]
(b) inadequate thickness of floors causing deflection and abnormal pressure on the walls, with no danger of collapse of the building, but which made support works necessary;[2]
(c) execution of the floors of a building with materials based on defective clay, causing rusting of the metal reinforcement and crumbling of the rendering and hollow blocks;[3]
(d) poor pointing of bricks on a third of the surface, causing the penetration of damp;[4]
(e) buckling and splitting of vertical sewage pipes made of plastic, allowing used water to seep into the building;[5]
(f) penetration of rainwater in the external walls of the building, rendering the apartment in question inhabitable or barely habitable;[6]

[1] Trib. Milan, 9 July 1970.
[2] Cass., 29 July 1976, n. 2928.
[3] Cass., 21 April 1976, n. 1426.
[4] Cass., June 1977, n. 2321.
[5] Cass., 4 May 1978, n. 2070.
[6] Cass., 11 January 1979, n. 206.

(g) omission of a floor beam, causing a static inadequacy, a loadbearing structure obviously influencing the future stability of the section of the building to which it corresponds;[7]

(h) damp patches in the apartments, caused by poor execution or mistakes in the sealing brickwork and the external cladding;[8]

(i) cracks in paving caused by lack of binding agent in the lagging;[9]

(j) construction defects in solar panels, so serious as to permit water to penetrate the building and especially the lower apartments;[10]

(k) poor damp-proofing, causing damp in the outer walls as well as the partition walls, due to inadequate insulating stonework;[11]

(l) construction defects in conduit for the removal of water, so that seepage from this conduit reached different floors and basements of the building, which in turn would have interfered with normal use of the parts for common use by the joint owners if they had not carried out the necessary replacements and repairs at their own cost;[12]

(m) damp caused by poor damp-proofing works;[13]

(n) major construction defects in chimney pipes for central heating installation;[14]

(o) construction defect in septic tanks serving a building and inadequate size of cesspools, by reference to municipal regulations. Such a defect causes a threat to the stability of the building by reason of leakage of waste water into the ground surrounding the foundation plinths of a nearby column, causing its lean concrete to become damp;[15]

(p) poor tar spraying of patios, shown by the presence of damp patches in the intrados;[16]

(q) poor insulation of north-facing rooms, carried out with inadequate rendering, causing condensation;[17]

(r) construction defects in garages and cellars of a building (which are by nature structures covered by the guarantee laid down by article 1669) shown by the presence of damp in the ceilings and floors in rainy conditions.[18]

To sum up, the Supreme Court of Appeal tends to define as 'major' defects which relate to water, damp or condensation in the interior of buildings. For example, no case concerning defects caused by lack of

---

[7] Cass., 5 February 1980, n. 839.
[8] Cass., 18 February 1980, n. 1178.
[9] Cass., 18 February 1980, n. 1178.
[10] Cass., 8 July 1980, n. 4356.
[11] Cass., 2 December 1980, no. 6298; Cass., 26 April 1983, n. 2858.
[12] Cass., 12 November 1983, no. 6741.
[13] Cass., 28 February 1984, n. 1427.
[14] Cass., 7 May 1984, no. 2763.
[15] Cass., 23 August 1985, n. 4507.
[16] Cass., 23 August 1985, n. 4507.
[17] Cass., 23 August 1985, n. 4507.
[18] Cass., 23 August 1985, n. 4507.

thermal or acoustic insulation has yet been brought before the Supreme Court of Appeal.

What is astonishing in Italian legislation is that these decisions are taken by judges who listen to the opinions of technical consultants, but who are not obliged by law to take them into consideration.

1.5.6 EXCULPATION OF PRODUCER

See also 1.2.6. In Italy, a contractor has an obligation to achieve a specific result. In other words, the law presumes that he knows his own trade and its standard practice. This means that he is always liable for defects, whether serious or minor, and any differences in the work when completed.

The law provides for only one case where the contractor is not responsible: this is where the contractor acts as a *nudus minister*, or mere servant, i.e. where the client has expressly instructed the execution of a particular job. However, the contractor must prove in such a case that he was opposed to the execution of the work and that he was obliged to carry it out at the client's risk.

Finally, the contractor is not liable for minor defects which are known to the client (or purchaser) and accepted by him.

The contractor is not liable for acts of vandalism and sabotage or occurrences of force majeure. The last-mentioned must be proved and their action upon the buildings must be exceptional.

1.5.7 CONSEQUENCES OF PRODUCER'S NON-COMPLIANCE WITH UNEQUIVOCAL DUTY TO REPAIR

If a contractor refuses to fulfil his guarantee obligations, the client may have remedial works carried out by a third party and seek damages from the defaulting contractor (see also 1.2.7).

If the contractor is insolvent or no longer exists, the client carries out the work at his own cost.

Where defects are attributable to several contractors, they are all jointly and severally liable. The client may also seek damages from each of them.

1.6 Transfer of ownership and producers' liability

1.6.1 RULES

Under Italian law (Civil Code, article 1669) the contractor's liability possesses characteristics of statutory liability as a principle of public policy, i.e. laid down in the public interest. Therefore it can be called upon by anyone affected by a construction defect. The periods of two years for minor defects and ten years for major defects (see 1.5.5) are also applicable for subsequent owners. Naturally, the commencement date of guarantees (with reference to contractor's liability) is always the completion of the works.

On the other hand, where subsequent owners are concerned, the periods within which defects must be notified change. These periods are as follows:

- eight days from the date of handover of the building for visible defects or from the date of discovery for hidden defects;
- one year from the date of discovery for 'major' defects mentioned in article 1669 of the Civil Code. The relations between purchaser and vendor, purchaser and contractor and contractor and vendor are not further elaborated on here; the potential combinations can be very complex and the situations covered by law extend beyond the framework of this book.

2 PROPERTY (OR MATERIAL DAMAGE) INSURANCE

Under Italian law this type of insurance is not compulsory and there is no specific provision for it.

Some contractors are insured but these are entirely private policies, often very different from each other, depending on the insurance company concerned.

3 LIABILITY AND PROFESSIONAL INDEMNITY INSURANCE

Here also, Italian law makes absolutely no provision. Where insurance exists, it is a private contract between a professional and an insurance company.

Japan

Kunio Kawagoe and Hiroyuki Nakamura

1 LIABILITY

1.1 General

The total investment in overall construction reached approximately 61.5 trillion yen (410 billion US dollars) for 1987, which was less than the EC + EFTA (not completed) level of 67 trillion yen, but more than the US level of 60 trillion yen as shown in the figure on p. 192. For 1989 the Japanese investment is expected to amount to 73 trillion yen and for 1990 to reach 77 trillion yen with less then 40 per cent for civil engineering projects and more than 60 per cent for the building and housing projects.

The private client usually directly contracts with a general contractor to build the building as a 'design-and-build' project, or orders a design office to design the building and contracts a general contractor to build. It is rare to order by tender.

The Government and other public bodies nominate contractors by tender. Negotiation between tenderers, although not legal, is commonly believed to be an effective means of eliminating unsuitable contractors thus assuring quality.

Many clients and contractors believe that it is not possible to make drawings and specifications in such a way that every contractor will produce the same quality of construction. The client wants to procure good quality buildings, and the contractor wants to achieve good building quality at a suitable price. But the legal tender procedure means that the lowest bid will get the contract.

In addition, tenders are obliged to negotiate for a suitable price and for the selection of a suitable company to get the contract (we call this negotiation DANGO). Many years ago some tenders used violence to get the contract by means of the Negotiation. People believed that the Negotiation was a bad custom, but now it has become non-violent. Some people still detest the the Negotiation but many people secretly allow it.

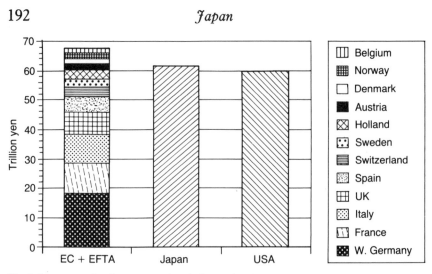

Total investment in the construction industry in 1987. (Exchange rate: $1 = 150 yen, £1 = 250 yen.)

According to the Public Accounts Act in Japan, only three types of bids are used for public works. They are:

(a) competitive bidding system;
(b) nominated bidding system; and
(c) direct selection and negotiation system.

In Japan, the direct selection and negotiation system is almost always used in bidding.

There are four reasons why public bodies apply the direct selection and negotiation system for their works. They are:

(a) the imbalance in the contract between the owner and the contractor;
(b) the licensed system in the construction industry;
(c) the target estimate; and
(d) the Japanese mentality.

First, the imbalance in the contract between the owner and contractor means that when a defect is found the contractor is always liable and the owner, in this case a public body and a supervisor, are not. In such circumstances, the contractor has risks which tend to create a big gain or loss and seldom a moderate profit. Consequently, contractors tend to share the risks by means of the Negotiation.

Second, there are too many contractors for the public body to evaluate whether bidding contractors have the technological ability to carry out the work to required quality levels. The public body, therefore, tends to nominate contractors who have a good record. In that sense, the Negotiation contributes to guaranteeing the quality of the work.

Third, in the Japanese bidding system, the target price is the upper limit of the bid.

Finally, the Japanese dislike making decisions by non-negotiated vote, so that it looks as if the decision has been made unanimously; before coming to a decision, they prefer to have negotiations between all the parties concerned.

About one hundred years ago the contract procedure between a client and a contractor based on right and duty similar to the western system was formally established by law, but in fact the traditional contract system based on personal reliance has remained in practice.

The Ministry of Construction and local government have the Committee for Adjustment of Construction Work Disputes to solve quickly building works disputes concerning the contract between the client and the contractor or between the contractor and the sub-contractor.

The number of arbitrations is few, about 250 disputes per year are solved by the Committee.

Most trouble occurring in post construction has been solved by negotiation between the client and the contractor and judgement by the court is not so common: the big contractors retain a sum of money for potential dispute negotiations as a means of self insurance.

1.1.0 LAW OR CONTRACT OR BOTH

The liability of the contractor is clearly defined both in Japanese civil law and the Standard Forms of Contract.

LAW: JAPANESE CIVIL CODE, ARTICLE 638

The contractor shall bear responsibility for constructions on the property for a five-year period after transfer (handing over) of the bond. However, this period of responsibility shall be extended to ten years if the property includes masonry, soil, brick or steel constructions.

STANDARD FORMS OF CONTRACT

(a) Shikai-Rengo Yakkan, article 23 (Standard Forms of Contract – generally adopted for civil works):

'(1) If any defect is found in the work, the Owner may demand repair thereof by the Contractor, specifying a reasonable period therefor, or claim compensation for damages against the Contractor either in lieu of such repair or in addition to such repair. However, if the defect is of minor importance and excessive expense is required for the repair, the Owner may not demand such repair by the Contractor.

(2) The guarantee period for defects provided for in the preceding paragraph (1) shall be one year in the case of wooden buildings, and two years in the case of stone, metal, concrete or similar buildings, or any other permanent structures or ground works from the date of delivery mentioned in two paragraphs (1) and (2) of Article 22, provided, however, that if such defects have been caused by intent or by gross negligence on the part of the Contractor, the one year period shall be five years, and the two years shall be ten years.

(3) The Contractor shall not be liable for defects in equipment and fixtures for utilities, interior decorations, furniture, or other similar items, unless the Supervisor demands the Contractor repair or replace such defects immediately upon inspection at the time of delivery, provided, however, that the Contractor shall guarantee against latent defects for one year from the date of delivery.

(4) Notwithstanding the provisions of paragraph (1), if and when the Owner finds any defect in the work at the time of delivery, the Owner may not demand repair or damages unless a written notice is given to the Contractor without delay; provided, however, that the same shall not apply, if the Contractor was aware of the defect.

(5) If the work has been destroyed or damaged by the defect mentioned in paragraph (1), the Owner may not exercise the rights mentioned in paragraph (1) unless it is within the periods prescribed in paragraph (2) and not later than six months from the date of destruction or damage.

(6) The provisions of the preceding five paragraphs (1) through (5) shall not apply to those defects, destruction, or damage of the work which have been caused under any of the items of paragraph (4) of Article 14; provided, however, that this shall not apply in case the matter falls under paragraph (5) of the same Article.'

(b) Public Works Standard Forms of Contract, article 37 (generally adopted for public works).

(c) Contract document as codified by FIDIC (Fédération Internationale des Ingénieurs-Conseils), the international federation of consulting engineers (generally adopted for works based upon foreign capital).

Note that the three contracts mentioned above are those prevailing in construction in Japan, however, other contracts may be used if the owner and the contractor so agree.

1.1.1 TERMS AND CONCEPTS

These are clearly defined in the Building Standard Law of Japan.

'BUILDINGS' (article 2 (1))

Structures fixed on the ground having roofs as well as columns or walls; gates or fences attached thereto; structures used as grandstands; or structures used as offices, stores, show houses, warehouses or other facilities similar thereto established in underground or elevated structures (excluding those facilities which are provided within the site of a railway or tramway for the operation and maintenance thereof, as well as overbridges, platform sheds, storage tanks and other facilities similar thereto). Building equipment shall be considered as part of a building.

'CONSTRUCTION' (article 2 (13))

To newly construct, add, rebuild, or relocate a building.

'MAJOR REPAIR' (article 2 (14))

Repair to one or more of the principal building parts of a building exceeding 50 per cent of all the parts of that building.

'MAJOR REMODELLING' (article 2 (15))

Remodelling of one or more of the principal building parts of a building exceeding 50 per cent of all the parts of that building.

'BUILDING EQUIPMENT' (article 2 (3))

Facilities provided in or on a building for the purpose of electricity supply, gas supply, water supply, drainage, ventilation, heating, cooling, fire extinguishing, smoke exhaust, or waste disposal; including chimneys, elevator equipment and lightning rods.

'PRODUCER'

Parties who shall be liable both under the law and under the contracts after the delivery of the building are as follows:

(a) contractor;
(b) clerk of works;
(c) designer (including structural, mechanical and electrical engineer).

A field representative, an engineer in charge (superintendent or chief engineer) is required to be stationed at the site during construction according to the Construction Contractors Law. Therefore, he is responsible for inspection of the materials employed and their use.

'TENANT'

The tenant may/can claim compensation or demand repairing of the defects during the Defects Liability Period: Civil Code, article 423.

1.2 During the Guarantee (Maintenance, Defects Liability) Period

1.2.1 DETERMINATION OF THE COMMENCEMENT OF THE PERIOD

COMPLETION CERTIFICATE, CERTIFICATE OF INSPECTION OF THE BUILDING

The Defects Liability Period generally commences on the date of the issuing of the Completion Certificate of the building just after the inspection made by the relevant authorities which receive the above-mentioned Completion Certificate. Although the Defects Liability Period often commences from the issue of the Completion Certificate after inspection by the relevant authority, it should be noted that the commencement date is dependent on the type of contract employed.

1.2.2 SIGNATORIES TO AND NAME OF THE DOCUMENT STATING THE COMMENCEMENT OF THE PERIOD

For example, the Certificate of Inspection, the signatories to which are a building official and an inspector who inspects the building.

1.2.3 THE PERIOD'S DURATION AND TERMINATION

(a) It may be defined both in law and in contract.
(b) Ten years (or longer for water-proof works) according to the Shikai Rengo Yakkan.
(c) None specifically (the relevant provision is usually included in the Contract Document).
(d) The parties are not required to observe the Civil Code (article 639) and it is not, therefore, compulsory. The parties may/can prolong the period by mutual understanding and agreement concluded in another specific

contract. For example, Shikai Rengo Yakkan provisions reduce the burden of the liability of the Contractor in comparison with civil law in Japan.

1.2.4 TIME ALLOWED FOR DIFFERENT CATEGORIES OF PRODUCERS FOR MAKING GOOD, REPAIRING OR PAYING DAMAGES

No time is specifically allowed.

If any defect is found in the work, the owner may demand repair thereof by the contractor, specifying a reasonable period therefor, or claim compensation for damages against the contractor, either in lieu of such repair or in addition to such repair.

1.2.5 KINDS OF DEFECTS AND DAMAGE COVERED BY THE GUARANTEE

DEFECT

(a) None.

   If the defect is of minor importance and excessive expenses are required for the repair, the owner may not demand such repair by the contractor (Shikai Rengo Yakkan, article 23; Civil Code, article 634).

(b) There is no rigid rule to define the kinds of defects and damages covered by the guarantee.

   It may be settled by mediation or conciliation under the Committee for Adjustment of Construction Work Disputes.

1.2.6 EXCULPATION OF PRODUCER

None.

According to the Shikai Rengo Yakkan, in the absence of clarifying instructions from the clerk of works, the contractor shall not be liable where:

(a) Any part of the drawings or specifications is unclear.
(b) Discrepancy exists between drawings and specifications.
(c) It is deemed inappropriate to perform the work in accordance with the drawings, specifications or instructions of the clerk of works.
(d) Errors or omissions exist to drawings or specifications.
(e) Any discrepancy exists between performance conditions in the design documents and the actual circumstances.

(f) Any unforeseeable condition is found at the jobsite, which constitutes an obstacle to the work.

To avoid liability, the contractor must notify these conditions to the clerk of works, unless the clerk of works is already aware of them.

In addition, when the contractor notifies the client of the above circumstances, the clerk of works is exempt from liability.

1.2.7 CONSEQUENCES OF PRODUCER'S NON-COMPLIANCE WITH UNEQUIVOCAL DUTY TO MAKE GOOD, REPAIR OR PAY DAMAGES

If the producer (the contractor) refuses to make good, repair or pay damages, although he could do so, the Maintenance Guarantee is paid as compensation to the client. The Maintenance Guarantee in this sense comprises the Performance Bond and the Retention Money (or the Bond in lieu of the Retention.) The former is the bank guarantee which may be paid to the client on breach of the contract. The latter is the amount of money which is reserved at each payment of the contract sum by the client to the producer. However, when the producer is not able to make payment, or the client cannot specify the debtor (the obliger), the client will have no way other than repairing or paying damages at his own expense.

1.5 From expiry of the Guarantee (Maintenance, Defects Liability) Period

It is the custom, after the expiration of the Defects Liability Period, for the Special Agreement between the producer and client to deal with the specific obligations/responsibilities of repairing or payment of the damages or relevant procedures.

The content of these may vary from one agreement to the other and it is, therefore, impossible to refer in detail to the circumstances listed in 1.5.1–1.5.7.

1.6 Transfer of ownership and producers' liability

1.6.1 RULES

The Civil Code, article 638 states that for major defects the Defects Liability Period shall be ten years, with all claims required to be notified to the contractor within one year of discovery. Under Shikai Rengo Yakkan form of contract, the liability also extends to defects, the period for this being one year for timber works and two years for concrete works. However, where these defects have been caused by intent or gross negligence by the contractor, the periods shall be extended to five years and ten years respectively.

2 PROPERTY (OR MATERIAL DAMAGE) INSURANCE

There is no property insurance at the moment except for fire insurance.

3 LIABILITY AND PROFESSIONAL INDEMNITY INSURANCE

There is only one kind of liability and professional indemnity insurance in Japan and the insurance was introduced from the United States in 1983. The insurance does not protect individual architects but any architect's office, which is a member of the Japan Federation of Architects Offices Association. The insurance covers defects resulting from unclear and imprecise drawings or specifications provided by the office and also covers defects resulting from erroneous instructions given during supervision by the office. The term of validity is the duration of the guarantee period which is specified in the contract mentioned in 1.2.3.

The insurance had protected 1038 architects' offices who were members of the Association by the end of 1990. The contract of insurance is voluntary for the architect's office.

There are about 117 500 architects' offices in Japan and 16 353 of them belong to the Association.

The Netherlands

D. E. van Werven

1 LIABILITY

1.1 General

1.1.0 LAW OR CONTRACT OR BOTH

INTRODUCTION

The Netherlands is a civil law country. Its Civil Code is under revision. Several parts of it (amongst others on property law and on general contract law: Books 3, 5 and 6) came into force on 1 January 1992. At this moment it is expected that it will take several more years before revised sections on specific contracts (such as construction contracts) will come into force.

The section on construction contracts that applies until that time is based on the French Civil Code and dates from 1838.

From article 306 of Book 3 of the new Civil Code it follows that every debtor is liable during a period of 20 years. This rule applies unless a different period is provided for by law or contract.

Dutch law is based upon the principle of freedom of contract.

Furthermore, the principle of good faith plays an important role in Dutch contract law. 'Good faith' is considered to be a source of law itself. Not only can it complement the contents of a contract, it can also have restrictive effects: e.g. as a result of the principle of good faith a party cannot always rely on an exemption clause (see 1.5.3).

I CONSTRUCTION CONTRACTS

A Civil code

The Dutch Civil Code contains a small section (articles 1640–1652) on contracting. The definition of contracting in this section not only covers the construction of buildings but also covers, e.g., the undertaking by a tailor to make a bespoke suit.

Within this section two articles apply to the construction of buildings only. One of these is article 1645 of the Civil Code (Burgerlijk Wetboek) which deals with the liability of the contractor after completion. Although its text resembles article 1792 of the French and Belgian Civil Codes, it is given a different interpretation in The Netherlands.

For more details on this article and its relation to the general 20-year liability period, see 1.5.3.

B   Standard conditions

Due to the fact that the rules in the section on the construction contract are minimal and unclear, standard conditions for these types of contracts were drawn up at an early stage.

Standard conditions for the construction contract were presented by the Ministry of Waterways as early as 1839.

These standard conditions have, of course, been amended regularly.

Of the various sets of standard conditions that are to be found in the construction industry today, the following are the most important:

1   Uniform Administrative Conditions for the Execution of Works 1989

The *Uniforme Administratieve Voorwaarden voor de uitvoering van werken 1989*, hereinafter referred to as 'UAV 1989', originally date from 1968 and, after lengthy consultation with the building industry, were adapted by a joint resolution of the Minister of Transport and Communications, the Minister of Defence and the Minister of Housing, Town and Country Planning. Originally intended to serve as standard conditions for public building contracts of the central and local government, the UAV are now widely used in the building industry.

The UAV were amended in 1989. A broad selection of representatives from the building industry have been consulted on the reform. In these consultations full agreement has been reached on the new UAV 1989.

The UAV are based on two principles. First, that the client (or a designer on his behalf) provides the contractor with the design or specifications. Second, that the client's agent supervises the execution of the works on behalf of the client.

2   Model Contract for Residential Building with Matching General Conditions 1991

The *Model koop-/aannemingsovereenkomst met bijbehorende algemene voorwaarden 1991*, hereinafter referred to as 'the model contract', was originally drawn up in 1974 and amended in 1976, 1989 and 1991. The background of the 1991 revision was the coming into force of the new

Dutch Civil Code. The central government, consumer organizations, organizations of building companies and developers and professional organizations of notaries and of estate agents all took part in the drafting of these conditions.

The model contract is exclusively written for the situation in which a client, a layman in building matters and not assisted by any professional, buys a piece of land from a developer and commissions the developer to build a house on it. Thus, the model contract combines a purchase contract (the building lot) and a construction contract (the construction of the house).

In this situation the developer takes the initiative towards building and takes care of both the design and the execution, whereas the client acts as a mere consumer.

The model contract includes a guarantee scheme. In a nutshell, this scheme provides the client with a claim against a guarantee fund if, as a result of insolvency of the developer:

- the building is under threat of not being completed;
- the developer is not able to repair certain defects that appear after completion.

Use of the model contract is compulsory if the dwelling is subsidized.

II COMMISSIONS TO ARCHITECTS AND ENGINEERS

A *Civil code*

The Dutch Civil Code does not contain a specific section on contracts between clients and architects or engineers. General contract law therefore applies to these contracts, which means that, unless otherwise agreed, an architect or engineer is in principle liable for breach of contract during a period of 20 years.

As there are no specific provisions for contracts between clients and architects in the Dutch Civil Code, standard conditions were drawn up for this type of contract as well. Professional organizations of architects presented their conditions for the first time around the beginning of this century.

At present the following standard conditions are used:

B *Standard conditions*

1 Standard Conditions 1988 for the Legal Relationship Client and Architect

The text of the *Standaardvoorwaarden 1988 rechtsverhouding opdrachtgever/ architect SR 1988*, hereinafter referred to as 'SR 1988', has been agreed upon by the Royal Society for the Advancement of Architecture, the

Institute of Dutch Architects (BNA) and representatives of contracting authorities and of public housing corporations.

2   Regulations Concerning the Relationship between Clients and Consulting Engineering Firms RVOI 1987

The *Regeling van de verhouding tussen opdrachtgever en adviserend ingenieursbureau, RVOI 1987*, hereinafter referred to as 'RVOI 1987', were agreed upon by the Royal Institution of Engineers (KIVI) and representatives of contracting authorities. The above-mentioned standard conditions of course only apply to a contract if the parties have agreed upon their applicability.

All standard conditions mentioned above contain provisions on post-completion liability of the contractor, developer, architect and engineer. These provisions often differ from general contract law.

Various other standard conditions exist in Dutch building practice, especially in the field of the execution of works, e.g. there are specific standard conditions for small building activities and for installation.

Arbitration

All the above-mentioned standard conditions (UAV 1989, Model contract, SR 1988 and RVOI 1987) have in common that they contain an arbitral clause. As a result of this, a substantial proportion of construction disputes in The Netherlands are settled by arbitration.

The three most important permanent arbitration courts are:

- the Court of Arbitration for the Building Industry (de Raad van Arbitrage voor de Bouwbedrijven, RvA) to which reference is made in the UAV 1989 and in the model contract;
- the Architectural Arbitration Institute (Arbitrage Instituut Bouwkunst, AIBk) to which the SR 1988 refers;
- the Disputes Committee of the Royal Institute of Engineers (Commissie van geschillen van het Koninklijk Instituut voor Ingenieurs), to which reference is made by the RVOI 1987.

1.1.1   TERMS AND CONCEPTS

'BUILDING'

Article 1 of the Housing Act (Woningwet) defines 'building' as the placing, the entire or part erection, the renovation or alteration and the extension of a structure. The word 'structure' in this definition has a wider meaning than the substantive 'building': every building is a structure but not every structure is a building, e.g. tunnels, bridges and dams are structures, not buildings. Furthermore, not every construction activity results in a structure. Roads and parking bays, for example, are not considered structures and therefore, in the sense of the Housing Act, their construction is not considered to be the result of 'building'.

However, it should be kept in mind that the Housing Act is written with the purpose of enabling communities to regulate building activities (amongst others) by issuing building permits. Thus the definition of 'building' in the Housing Act can probably not be applied generally.

Of more importance is that the relevant section on construction contracts in the Civil Code applies and the UAV 1989 can apply to every construction activity, thus not only to the construction of buildings, but also to the construction of civil engineering works such as tunnels, bridges and roads.

'PRODUCER'

According to the vocabulary used in this book, the term 'producer' is used to denote a party who may be liable towards the client after the taking over or completion of the building.

First and foremost, parties with whom the client has a contractual relationship are producers in this sense.

If the contractor is presented with the specification of the works, the client will have commissioned an architect and/or another consultant. In that case, *architect*, *consultant* and *contractor* are producers.

When the client is a layman in building matters and is not assisted by an architect, he will normally do business with one party, a developer. This developer is taking care of both the design and the execution of the works. Irrespective of whether the developer carries these tasks out himself or contracts them out, the *developer* is a producer as defined in the List of Terms, Appendix B. This is often the situation in the residential building sector.

A *supplier* will generally only be a producer if there is a contractual relationship with the client. If there is no contract between the client and the supplier, apart from liability on the basis of the implemented EC Directive on product liability, a liability of the supplier based on tort will only exist in very special circumstances. This also applies to the *sub-contractor*, with whom the client will usually not have concluded a contract.

To be mentioned separately is the situation in which the client has nominated a supplier or a sub-contractor, or has prescribed the use of a certain material.

Nominating a supplier or sub-contractor does not result in a contractual relationship between the client and the nominee. However, there are circumstances in which it is considered not to be reasonable to hold the main contractor liable for a fault by a nominated sub-contractor/supplier or for defective materials which have been prescribed by the client. In that case the client should bear the risk of a fault of the nominated sub-contractor/

supplier, and the client should be able to claim against the nominated sub-contractor/supplier.

The UAV 1989 give the solution to this problem with regard to nominated sub-contractors in clause 6, paras 26 and 27:

'**6.26** The contractor may sub-contract specific parts of the works, provided he has obtained the prior written approval of the employer's agent for the choice of such parts and the sub-contractors to be engaged for that purpose; nevertheless, the contractor shall remain fully liable to the employer in respect of such parts.

6.27 If the engagement of any particular sub-contractor is or has been prescribed by or on behalf of the employer, the contractor's obligations to the employer with respect to the work carried out by that sub-contractor shall be limited to the obligations for which the contractor can hold that sub-contractor liable under the terms of the sub-contract as accepted or approved by the employer.

If the nominated sub-contractor fails to perform or does not perform within the agreed time or performs improperly and the contractor has taken all reasonable steps in order to obtain performance and/or compensatory damages, the employer shall reimburse the contractor for the additional costs of execution incurred by him, insofar as the sub-contractor has not compensated the contractor for such costs. In consideration for such reimbursement the contractor shall, forthwith upon his request of the employer, assign to the employer his claim against the nominated sub-contractor up to the sum so reimbursed by the employer.'

The UAV also include provisions dealing with the case in which a supplier is nominated or a material is prescribed (clause 5, para. 4):

'**5.4** The employer shall be liable for functional unfitness of:
(a) any materials prescribed by him;
(b) any materials to be procured from a supplier nominated by him, unless the contractor had a choice in respect of such materials.
'Functional unfitness of materials' means inherent unfitness of the materials for the purpose for which they are intended according to the specification.'

This all means that, as a client may sometimes have a contractual claim against a *nominated sub-contractor* or against a *nominated supplier*, they can sometimes be considered 'producers' as the term is used in this book.

Article 100 of the Housing Act states that there is a *building and housing inspectorate* in every community. One of the tasks of this municipal service is to inspect whether the building is being carried out in accordance with the building permit and the applicable building regulations.

As there is no contract between the municipal building control authority and the client, liability of the inspectorate can only be based on tort. Such a tortious liability is generally only accepted if the inspectorate has been seriously negligent in carrying out the inspection.

Generally the *tenant* has no rights vis-à-vis a producer.

1.2 During the Guarantee (Maintenance, Defects Liability) Period

1.2.1 DETERMINATION OF THE COMMENCEMENT OF THE PERIOD

1.2.2 SIGNATORIES TO AND NAME OF THE DOCUMENT STATING THE COMMENCEMENT OF THE PERIOD

1.2.3 THE PERIOD'S DURATION AND TERMINATION

I CONSTRUCTION CONTRACTS

A Civil code

The section on the construction contract in the Dutch Civil Code uses the terms 'completion' and 'acceptance', but the way in which the acceptance can take place is not indicated.

The acceptance is therefore form-free and can take place tacitly.

A Certificate of Completion (or a similar document) is not mentioned in the section on the construction contract in the Civil Code, and neither is a Defects Liability Period.

B Standard conditions[1]

1 UAV 1989

(a) **Provisions on inspection, approval, completion and acceptance**

Clauses 9 and 10 of the UAV 1989 define in detail the procedures relating to the inspection, approval, completion and acceptance of the works.

Basically the system is as follows:

- upon written application by the contractor to the client's agent, the works will be inspected (clause 9, para. 1);
- after the inspection the contractor shall be given written notice, stating whether the works have or have not been approved. In the latter case such notice shall specify the defects by reason of which the approval has been withheld (clause 9, para. 2);

---

[1] As, in practice, Defects Liability Periods are exclusively agreed by clients and contractors/developers, only the UAV 1989 and the Model Contract for Residential Building are taken into consideration here (not the SR 1988 or the RVOI 1987).

- if the works have been approved, they shall be considered completed and accepted (clause 10, para. 1).

Three further remarks should be made:

- In accordance with para. 7 of clause 9, minor defects that can be repaired before the due date of any subsequent payment/instalment, shall not constitute a reason for withholding approval, provided that the works are in such condition that they are fit for use.
- Clause 9 of the UAV 1989 also deals with the situation that the client's agent fails to carry out the inspection or – when the inspection did take place – fails to dispatch the written notice to the contractor, stating whether the works have been or have not been approved.

   For both situations the UAV contain provisions to the effect that, after some formalities, the works are deemed to be approved. If the works are deemed to be approved they are considered completed and accepted as well. See clause 9, paras 5 and 6 and clause 9, para. 1).
- The written notice to be given to the contractor, stating whether the works have been approved or not, in practice often bears the title 'Report on delivery' (*Proces-verbaal van oplevering*) and is usually signed not only by the client's agent, but also by a representative of the contractor.

   In such a case, the *Proces-verbaal van oplevering* can be considered as a kind of Certificate of Completion: the works are accepted; the minor defects mentioned in the document are to be repaired.

   For the consequences of the completion and acceptance see 1.5.6.

(b) Provisions on a Defects Liability Period

If the construction contract refers to the UAV 1989, a Defects Liability Period (*onderhoudstermijn*) does not automatically apply. Clause 11, para. 1 only mentions the possibility of the specifications providing for such a period.

   The reason for this is that the specification and duration of a Defects Liability Period depends on the nature of the works. It is, for example, not usual to agree a Defects Liability Period when the work consists of dredging.

   If the specification provides for a Defects Liability Period, that period shall begin immediately after the date on which the works are considered completed and accepted in accordance with the provisions of paras 1 and 2 of clause 10.

   The duration of a Defects Liability Period normally varies between 3 and 12 months.

   The only event which the UAV 1989 expressly mention with regard to the expiry of the Defects Liability Period is that, upon the expiration date of the Defects Liability Period, the works shall be inspected again in order to ascertain whether the contractor has fulfilled his obligations. However, even if that inspection does not take place, the Defects Liability Period just expires by course of time.

2   Model Contract for Residential Building 1991
(a)   **Provisions on completion and acceptance**

The model contract states (in article 14, para. 4) that the day on which the keys of the house are handed over to the client will be considered to be the day of completion.

Before this handover a report of possible defects or shortcomings has to be drawn up and signed by or on behalf of both parties.

The developer should give written notice of the proposed completion date 14 days in advance.

In comparison to the UAV 1989, the model contract contains no explicit provisions on inspection and approval of the works.

(b)   **Provisions on a guarantee period and on a Defects Liability Period**

The model contract mentions two periods, both starting at the moment of completion. First, there is a guarantee period of six months. The developer is required to repair any defect or remove any shortcoming which comes to light during this period. The client is required to report in writing any defect or shortcoming before the expiry of the guarantee period.

Second, the model contract mentions a Defects Liability Period of three months during which the developer has to repair or make good defects and shortcomings specified in the report which was drawn up before handover of the dwelling.

In both cases the period expires by the course of time. Reference is made to para. 1 of articles 15 and 16 of the model contract.

1.2.4   TIME ALLOWED FOR DIFFERENT CATEGORIES OF PRODUCERS FOR MAKING GOOD, REPAIRING OR PAYING DAMAGES
I   CONSTRUCTION CONTRACTS

A   *Civil code*

In the section on the construction contract in the Dutch Civil Code no provisions are made for the time allowed for repairing defects during a Defects Liability Period. General contract law therefore applies.

Under general contract law defects should be repaired within a reasonable time.

B   Standard conditions[2]

1   UAV 1989

Paragraph 2 of clause 11 of the UAV 1989 refers to the general contract law rule by stating that repairs shall be carried out by the contractor within a period of time to be set reasonably by the client's agent.

2   The Model Contract for Residential Building 1991

According to the model contract, the defects and shortcomings mentioned in the report that is compiled on the occasion of the completion, should be repaired as soon as possible, but in any event within a period of three months after completion.

Defects and shortcomings that come to light during the six months Defects Liability Period should be repaired or made good immediately after they have been reported in writing by the client to the developer.

1.2.5   KINDS OF DEFECTS AND DAMAGE COVERED BY THE GUARANTEE

This will be discussed in 1.5.5–1.5.7. As will be seen, there is no difference between the situation during the Defects Liability Period and the situation after the expiry of this period.

1.2.6   EXCULPATION OF PRODUCER

I   CONSTRUCTION CONTRACTS

Standard conditions

Again, as Defects Liability Periods in practice are exclusively agreed between clients and contractors, this question primarily will be answered on the basis of the applicability of the UAV 1989 and of the Model Contract for Residential Building 1991.

1   UAV 1989

In clause 11 of the UAV 1989 a distinction is made between defects coming to light during the Defects Liability Period and damage occurring to the works during the same period.

As to *defects*, para. 2 of clause 11 states that the contractor shall be required to repair them, with the exception of defects for which the client is

---

[2] As, in practice, Defects Liability Periods are exclusively agreed by clients and contractors/ developers, only the UAV 1989 and the Model Contract for Residential Building are taken into consideration here (not the SR 1988 or the RVOI 1987).

responsible under para. 2 of clause 5 and for which he is liable under para. 3 of clause 5.

Those paragraphs amongst other things deal with the client's responsibility for the construction and construction methods, prescribed by him or on his behalf, and with the client's liability for damages cause by materials supplied by the client.

With regard to *damage* to the works occurring during the Defects Liability Period, para. 4 of clause 11 states that such damage shall be for the account of the client except for damage which is the result of unsatisfactory work done by the contractor.

2   Model Contract for Residential Building 1991
The situation to which the model contract applies differs dramatically from the one covered by the UAV 1989.

In the first situation, the client has no essential influence on the building process, whereas in the latter situation it is the client who takes the initiative towards building and provides the contractor with the specifications.

Therefore, in the model contract the provisions on liability during the guarantee period (and after that) are simpler than in the situation covered by the UAV 1989. As a result of this, the model contract offers the developer only limited possibilities to escape from liability for defects occurring during the Defects Liability Period.

Paragraph 2 of article 15 of the model contract states that with regard to shortcomings in appearance, resulting from the nature of the applied materials, the client has no claim, unless those shortcomings are the result of either the application of inferior quality materials, or incompetent application of materials by the developer.

1.2.7   CONSEQUENCES OF PRODUCER'S NON-COMPLIANCE WITH UNEQUIVOCAL DUTY TO MAKE GOOD, REPAIR OR PAY DAMAGES

See 1.2.5.

1.5 From expiry of the Guarantee (Maintenance, Defects Liability) Period

1.5.1 DETERMINATION OF THE COMMENCEMENT OF THE (POST-GUARANTEE) PERIOD DURING WHICH A PRODUCER MAY BE HELD LIABLE

1.5.2 NAME OF THE DOCUMENT STATING THE COMMENCEMENT OF THE POST-GUARANTEE PERIOD; ITS SIGNATORIES

I CONSTRUCTION CONTRACTS

A Civil code

Article 1645 of the section on construction in the Dutch Civil Code deals with the liability of the contractor after completion. The period of ten years mentioned in that article starts at the moment of completion. As mentioned previously, the Dutch Civil Code does not provide for a Defects Liability Period.

For the completion itself refer to 1.2.1–1.2.2.

B Standard conditions

1   UAV 1989
In clause 12 of these standard conditions some important rules are given on the liability of the contractor after completion.

The main rule is that after completion and acceptance of the works the contractor is no longer liable for defects.

Two important exceptions are given in para. 2 of clause 12: first, when the situation of article 1645 of the Civil Code applies, and second, when there are hidden minor defects, caused through the fault of the contractor, his suppliers, or his sub-contractors or his staff or labour.

If no Defects Liability Period is agreed, the periods related to the aforesaid liabilities start at the moment of completion and acceptance. If a Defects Liability Period is agreed, the liability periods referred to in clause 12 start on the day following the expiry of that Defects Liability Period.

As to the completion itself and possible documents to be drawn up, refer to 1.2.1–1.2.2.

2   Model Contract for Residential Building 1991
Pursuant to article 16 of the model contract a guarantee period of six months and a standard Defects Liability Period of three months applies. See 1.2.1–1.2.3. At the moment of expiry of the guarantee period – six months after completion – the contractual post-guarantee liability periods (for hidden defects and for serious defects) start to run. No further formalities, such as e.g. a repeated inspection, are required.

II COMMISSIONS TO ARCHITECTS AND CONSULTING ENGINEERS

A Civil code

If no contractual provisions are made, architects and consulting engineers will be liable for breach of contract during a period of 20 years.[3]

B Standard conditions

The standard conditions SR 1988, frequently used by architects, contain several articles (articles 55–63) relating to their liability. Normally the liability period starts the day on which the building project is completed and accepted. Reference is made to article 61 of SR 1988.

On this point para. 7 of article 16 of the RVOI 1987 stipulates more or less the same.

1.5.3 PERIOD(S) OF LIMITATION

I CONSTRUCTION CONTRACTS

A Civil code

The Civil Code contains a general 20-year liability period applicable to claims based on contract or tort. This general rules applies, except where other provisions are made in law or contract.

In the section on the construction contract, article 1645 of the Civil Code states that if a building collapses (or partly collapses) by a defect in the construction, the contractor is held responsible for a period of ten years after completion.

Prima facie this article implies that the contractor is not liable if the building collapses later than ten years after completion. In fact, this is only *one* of the interpretations of article 1645, but not the leading theory, which is based on a decision of the Dutch Supreme Court (Hoge Raad), given in the early 1930s.

The Supreme Court held that article 1645 does not affect the general (20-year) liability period, but only stipulates a reversal of the burden of proof to the disadvantage of the contractor during the first ten years.

This means that if the building collapses within a period of ten years after completion, the burden of proof rests on the contractor to prove that the collapse was not caused by any fault on his part.

The point of view that article 1645 does not have any influence on the general (20-year) liability period has as a consequence that:

(a) a contractor can be held liable for a collapse of the building up to 20 years after completion, provided that, from ten years after completion,

---

[3] The assumption is made that article 1645 of the Civil Code does not apply to architects.

the client proves that the collapse, whether total or partial, was caused by a fault committed by the contractor; and

(b) a contractor can be held liable for latent defects for 20 years after completion, provided that the client proves that they were caused by a fault on the part of the contractor.

Three additional remarks should be made on the controversial article 1645.

First, the Supreme Court has widened the article's field of application by ruling that if a building threatens to collapse, the article applies as well. Some learned authors support a broader interpretation of the term 'collapse'. Until now the Supreme Court has, however, not yet been given the opportunity to give a decision on this point.

Second, in The Netherlands it is generally accepted that the liability of the contractor under the Civil Code can be limited or even excluded by contract. Contrary to Belgium or France, article 1645 is not considered to be imperative.

This means that by way of contractual provision, not only the general liability of the contractor for latent (minor) defects can be limited (or even excluded), but also the liability for defects, which lead to the (partial) collapse of the building. Nevertheless, it should be kept in mind that the principle of good faith plays a very important role in Dutch general contract law and that one cannot go too far in limiting or excluding one's liability. The following quotation from an article by Professor Jan M. van Dunné[4] may illustrate this:

> 'In Dutch law the question whether a party can rely on an exemption clause is considered to be a matter of construction, in the light of the principle of good faith and the surrounding circumstances.[5] . . . So it all depends on the measure of negligence of the party seeking reliance on the clause; he will not succeed, according to Dutch law, in the case of gross negligence or wilful breach of contract. In a leading case it was held by the Hoge Raad that circumstances which should be taken into consideration in the case of an exclusion clause in a standard form contract, are: the measure of fault in relation to the interests of the parties, the type of contract and its terms, the position of the parties in society and their relation to each other, the way the clause had been negotiated, the measure in which the other party was conscious of the meaning of the clause.'[6]

This has resulted in an article of the new Civil Code (article 6:248) that explicitly states that a contractual clause does not apply as far as its

---

[4] J. M. van Dunné *Selected Problems of Construction Law: International Approach* (Fribourg, London, 1983) p. 140.

[5] HR 20 February 1976, NJ 486, *Pseudo-birdpest* case.

[6] HR 19 May 1967, NJ 261, *Saladin v HBU*.

applicability would be unacceptable with regard to the principle of good faith (justice and equity).

Third, the leading theory is that – apart from cases of fraud or gross negligence – the contractor can only be held liable on the basis of article 1645 of the Civil Code if the defect which caused the collapse was hidden to the client or his agent.

B Standard conditions

1 UAV 1989

As stated before, the main rule, according to para. 1 of clause 12 of the UAV 1989 is that, after completion and acceptance, the contractor is no longer liable for defects.

However, two important exceptions to this main rule are given in paras 2, 3 and 4 of clause 12, which read as follows:

'(2) An exception to the provision in para. 1 shall apply:
   (a) If the event referred to in section 1645 of the Civil Code occurs;
   (b) If the works or any part thereof contain any hidden defect caused through fault of the contractor or his suppliers or subcontractors or his staff or labour and the contractor is notified of such hidden defect within a reasonable period of time after it has been discovered.
(3) Any defect as referred to under (b) in para. 2 shall be regarded as a hidden defect only if despite close supervision during the execution of the work or at the inspection of the works as referred to in para. 2 of clause 9 such defect could not reasonably have been discerned by the employer's agent.
(4) Actions on account of hidden defects cannot be brought after five years have elapsed since the date referred to in para. 1.'

As in the UAV 1989 reference is made to 'the event referred to in section 1645 of the Civil Code', again the question arises as to which type of event article 1645 relates. There is a tendency in the decisions by the Court of Arbitration for the Building Industry (RvA) to apply the article not only to situations in which a building collapses or threatens to collapse but also if a latent defect affects the solidity of the building.

In two recent cases, the arbitrators even held that the article can be applied to situations in which, as a result of a defect that does not affect the solidity of the building and provided that high repair costs are involved, the building becomes unfit for its purpose.[7]

Furthermore, the question arises whether the reference to article 1645 in

---

[7] Court of Arbitration for the Building Industry (RvA), 21 January 1988, Bouwrecht 1990, p. 867 and RvA, 25 July 1990, Bouwrecht 1990, p. 867 (with a comment by H. O. Thunnissen).

the UAV 1989 has the effect that the interpretation by the Dutch Supreme Court of the nature of the contractor's liability also applies if the UAV 1989 govern the contract.

The answer is affirmative with regard to the reversal of the burden of proof during the first ten years after completion. Thus were decided several cases by the Court of Arbitration for the Building Industry.

The answer is probably negative with regard to the situation after the expiry of the first ten years after completion. Normally, when the UAV 1989 do not apply, for hidden defects, the contractor can be held liable, as any other debtor, for up to 20 years after completion. But when the UAV 1989 govern the contract, it is arguable that the main rule of clause 12 of the UAV 1989, stating that after completion the contractor is no longer liable for defects, entails that any exception to that main rule should be interpreted restrictively. This would have the effect that under the UAV only the ten-year period mentioned in article 1645 applies and not a general liability period of 20 years.

As to the liability for (minor) hidden defects the situation is quite clear with regard to the length of the liability period. The duration of this period is five years. Moreover, actions in relation to these defects cannot be brought after the elapse of the five-year period.

2   Model Contract for Residential Building 1991

As in the UAV 1989, the main rule in the model contract is that, after the expiry of the six-month guarantee period, the developer is no longer liable for defects.

As in the UAV, there are important exceptions to this main rule.

The first exception applies if the dwelling or any part thereof contains a serious defect. For such defects, the developer is liable for 20 years after the expiry of the above-mentioned guarantee period. Actions cannot be brought after this period.

Paragraph 3 of article 16 of the model contract defines a serious defect (to which the 20-year liability period applies) as a defect which either affects the solidity of the construction or an essential part thereof, or makes the building unfit for its purpose.

Please note that in this model contract no reference is made to the controversial article 1645 of the Civil Code, and that the term 'serious defect' is given a rather broad interpretaion.

The second exception to the main rule applies if the dwelling or any part thereof contains a hidden defect and the contractor is notified of such a hidden defect within a reasonable period of time after it has been discerned. The developer is liable for hidden defects for a period of five years after the elapse of the six-month guarantee period.

The definition of a hidden defect is to a large extent similar to the one found in clause 12, para. 3 of the UAV 1989. Again, the model contract states that claims cannot be brought after this five-year period.

The third exception to the main rule applies if a claim against a developer is based on the guarantee scheme which is included in the model contract. Generally speaking, according to this scheme, the developer guarantees that the dwelling complies with the technical requirements mentioned in the scheme itself and in the model building regulation prepared by the Association of Netherlands municipalities. Furthermore, the guarantee includes that the applied construction materials and parts are and will stay sound and fit for their purpose. The guarantee scheme mentions a general guarantee period of six years, which starts three months after completion. If a developer is unable or refuses to repair defects, to which the guarantee scheme relates, the client has a claim against a guarantee fund. This guarantee system, set up in 1976, was inspired by the British NHBC system.

II COMMISSIONS TO ARCHITECTS AND CONSULTING ENGINEERS

A Civil code

Unless special contractual provisions are made, architects and engineers are liable for breach of contract during 20 years. However, in most cases, standard conditions apply to these commissions.

B Standard conditions

The SR 1988, very frequently used by architects, limits the liability of the architect to a period of five years after completion of the works (SR 1988, article 61).

A similar provision is made by the RVOI 1987, which frequently applies to commissions to consulting engineers. Paragraph 7b of article 16 of the RVOI 1987 mentions a five-year liability period, starting at the moment of the completion of the project.

1.5.4 TIME ALLOWED (DIFFERENT CATEGORIES OF PRODUCERS) FOR MAKING GOOD, REPAIRING OR PAYING DAMAGES

With the exception of clause 46, para. 1 of the UAV 1989 (see 1.5.7), no specific provisions are to be found in standard conditions. The main rule of general contract law therefore applies, which states that repairs by a producer should be done within a reasonable time.

1.5.5   KINDS OF DEFECTS AND DAMAGE QUALIFYING FOR DAMAGES

First, refer to 1.5.6.

Second, it is assumed, at least by the Court of Arbitration for the Building Industry, that if a fault by the contractor leads to a material defect, the contractor can only be ordered to repair if damage results from that defect.[8]

1.5.6   EXCULPATION OF PRODUCER

I   CONSTRUCTION CONTRACTS

General

A party cannot always rely on a contractual exemption clause. This is a result of the principle of good faith which can have a restrictive effect on the contents of a contract. The new Civil Code explicitly states that a contractual clause does not apply as far as its applicability would be unacceptable with regard to the principle of good faith. In case these exemption clauses are part of general conditions that apply to a contract between producer and consumer, the Dutch Unfair Contract Conditions Act (part of Book 6 of the new Civil Code) can have as a result that some of these conditions can be considered void.

The nature of the construction contract

Before discussing the cases in which a contractor or developer is not liable for defects which come to light after completion (or after the expiry of a Defects Liability Period), it may be useful to say something about the nature of the construction contract. Often the question is put whether the contractor owes the client a duty of care or a duty of result. A duty of care implies that the contractor is not liable for defects as soon as it becomes clear that he has taken care to a reasonable extent. A duty of result implies that a contractor is liable for any defect, except if the defect is caused by force majeure.

The leading theory in The Netherlands is that the answer to the question as to what type of duty results from a construction contract depends on the degree of influence which the client exercises on the building process.

Therefore, if the building is done on the basis of a design and/or specifications provided by the contractor, and without any relevant influence or

---

[8] Court of Arbitration for the Building Industry, 19 June 1985, Bouwrecht 1985, p. 801. In a case in which it was not certain if a building defect would lead to damage in the future, the contractor was ordered to provide a client with a guarantee covered by insurance: Court of Arbitration for the Building Industry, 6 August 1986, Bouwrecht 1987, p. 68.

supervision by the client, the construction contract will lead to a duty of result.

On the other hand, if the work is carried out in accordance with a design and/or specifications provided by or on behalf of the client, or under supervision of the client or his agent, or if the client has any other relevant influence on the building process, the construction contract is of a mixed nature in this sense, that parts of the contract may impose a duty of care on the contractor and other parts of the contract may lead to a duty of result.

To put it in a different way: the more freedom the contractor has in determining his working methods, and in choosing materials, sub-contractors and suppliers, the more likely it is that the construction contract leads to a duty of result. On the other hand, the tighter the contractor is bound by the client regarding these matters, the more likely it is that the duty resulting from the construction contract should be considered as a duty of care.

The principle that the extent of liability varies with the extent of influence that a party exercises on the building process has been of great importance in the drawing up of standard conditions for construction contracts in The Netherlands.

But even if standard conditions do not apply and the contract is exclusively controlled by the provisions in the section relating to construction contracts in the Civil Code and by general contract law, this principle helps to determine whether – after completion – a contractor can be held liable for defects.

A Civil code

As mentioned before (see 1.5.3), unless agreed otherwise a contractor can be held liable for defects for 20 years after completion, provided that the client proves that the defect is caused by the fault of the contractor or is due to a circumstance for which the contractor bears the risk. A special case, however, exists if the defect leads to the collapse of the building within ten years after completion. In that case, the liability of the contractor is presumed but the contractor has the opportunity to prove that the collapse is not caused by his fault, or otherwise does not result from a circumstance for which he bears the risk.

The section in the Civil Code on construction contracts does not, however, clearly indicate for which circumstances each party bears the risk or, in other words, how the risks are allocated between the parties. However, the principle described above can help to determine this in a particular case.

B Standard conditions

1 UAV 1989
The principal effect of completion and acceptance is that, whereas in principle *before* completion the contractor bears the risk of defects and

damage occurring to the work, *after* completion this risk rests with the client. Thus, at the time of completion and acceptance, the risk for defects and damage to the work shifts from the contractor to the client. This can be considered to be a general rule of construction law.

This rule is implemented in clause 12, para. 1 of the UAV which states that after completion, with the exception of two cases, the contractor is no longer liable. Those two exceptions are:

(a) the case in which the event referred to in article 1645 of the Civil Code occurs; and
(b) the case in which the work contains a hidden defect, caused through the fault of the contractor, his suppliers, his sub-contractors or his staff or labour.

As to the burden of proof and the duration of the liability period, see 1.5.3.

A central element in both cases is that the contractor is only held liable if the defects are caused by a fault of the contractor or are due to a circumstance for which the contractor bears the risk.

The answer to the question as to which party (client or contractor) bears which risk is given in detail by the UAV.

It should be kept in mind that the UAV is written for the situation in which the client, or a designer on his behalf, provides the design or specifications, and the client's agent supervises the execution of the work on behalf of the client.

As a result:

(a) The client is responsible for the construction specified by him or on his behalf, including the effect that soil conditions may have on such constructions (clause 5, para. 2). Please note that if the described construction contains an obvious fault or defect, the contractor is required to issue a warning (clause 6, para. 14). If a contractor fails to give such a warning, the liability can be divided between the client and the contractor to the extent that each party has been negligent. Whether such a duty to warn exists depends on the specific circumstances of the case, e.g. the expert knowledge of the contractor.
(b) Apart from fraud or gross negligence on the part of the contractor, he is only held liable for defects that were hidden. Defects that could reasonably have been discerned by the client or his agent are considered to have been accepted if during the execution or at the inspection no comments regarding those defects were made by the client or his agent. In its decisions the Court of Arbitration has ruled that, amongst other things, the degree of supervision that took place, and the expertise of the client's agent, should be taken into consideration to determine if a defect in the execution could reasonably have been discerned by the client's agent or not.

If the client wants to exercise more influence on the building process than merely by specifying the construction and having the works superintended, he will soon be interfering in matters that relate to the actual execution of the work (which traditionally is the domain of the contractor).

If the client does so, the risk to be borne by him will increase.

The following provisions in the UAV may illustrate this:

- the client is liable for prescribed construction methods, including the effect that soil conditions may have on such construction methods (clause 5, para. 2);
- the client is liable for any damage caused by defects in the materials that he has supplied (clause 5, para. 3);
- the client is liable for 'functional unfitness' of:
  (a) any materials prescribed by him;
  (b) any materials to be procured from a supplier nominated by the client, unless the contractor has a choice in respect of such materials.
  'Functional unfitness of materials' means inherent unfitness of the materials for the purpose for which they are intended according to the design or specifications (clause 5, para. 4, quoted in 1.1.1).[9]
- to a large extent the client bears the risk if a nominated sub-contractor fails to perform properly. Reference is made to clause 6, paras 26 and 27 as quoted in 1.1.1).

It may be clear from the above that if the UAV apply and the client chooses to exercise considerable influence on the building process, the scope of the liability of the contractor narrows.

The duties of the contractor, resulting from such a construction contract, will be duties of care to a considerable extent. Although the client will not always appreciate that, it is the price he pays for interfering in matters concerning the execution of the work.

2 Model Contract for Residential Building 1991
The situation is entirely different if the model contract applies.

Here the initiative of the building is taken by the developer who also takes care of both the execution and the design. The client has no substantial influence on the building process; he acts as a consumer.

This construction contract therefore results in a duty of result of the contractor. As a consequence the provisions in the model contract concerning the liability of the contractor after the guarantee period are simple.

Provided that a defect is not caused by force majeure or by lack of maintenance on the part of the client, according to the wording of article 17, para. 2, the developer is liable for serious defects and for hidden defects as soon as they occur. It is not necessary that the defect is caused through the fault of the developer or due to a circumstance for which the developer bears the risk. Such a requirement would be useless as practically speaking

---

[9] In this clause the theory of the Court of Arbitration, developed in the so-called *Monoliet* case, was codified. For the *Monoliet* case see RvA, 10 November 1977, *Bouwrecht* 1978, p. 72 (with a comment by H. O. Thunnissen). On this theory see J. M. van Dunné *Selected Problems of Construction Law: International Approach* (Fribourg, London, 1983) p. 127.

a defect can only be caused by the contractor or by somebody for whom he is liable.

Again, the client has no substantial influence on the building process.

II COMMISSIONS TO ARCHITECTS AND ENGINEERS

If, in the relationship between client and contractor, the risk for a defect is to the account of the client, and damage occurs, the client will try to seek recourse against another party with whom he has a contract. This could, for example, be an architect, an engineer or a supplier. Their liability will briefly be dealt with below.

A Civil code

As mentioned several times before, there are no specific provisions in the Dutch Civil Code with regard to these commissions. Therefore, unless otherwise agreed, an architect or a consulting engineer is liable for breach of contract during a period of 20 years.

B Standard conditions

The standard conditions, used by architects (the SR 1988) and by consulting engineers (the RVOI 1987), limit the liability of these producers in practically the same manner. Therefore, only the SR 1988 are discussed.

Article 55 of the SR 1988 states that the architect is liable to the client for the loss suffered by the latter as a direct result of a serious error by the architect.

The use of the word 'direct' entails a certain restriction of the liability whereas normally a breach of contract will lead to liability for damage to the extent that it was foreseeable at the time of concluding the contract.

It may seem that the term 'serious error' implies a limitation as well, but this is not the case, as in para. 4 of article 55 a serious error is defined as follows:

'(4) In reading this article a culpable fault shall be understood to mean: a fault, which a good and careful architect, in the relevant circumstances and with regard to a normal attentiveness and a normal practice of the profession, should avoid.'

It is a principle of the SR 1988 that if on certain technical matters advice should be given by a consulting engineer, this will lead to two parallel contracts: a contract between client and architect and a contract between client and consulting engineer.[10] Therefore article 58, para. 1 of the SR 1988 states:

'(1) If certain parts of the building project are designed by third parties, or designed and executed by these, the architect is not

[10] To the latter the RVOI 1987 frequently will apply.

responsible for the design, or for the design and the execution thereof.'

Nevertheless, the architect can be held liable for incorporating the advice of other advisers and matching it with the architectural design. Article 23 of the SR 1988 reads as follows:

'If the client, in bringing about the design for a building project, has drawn in consultants or other advisory third parties, the standard services in the commission for a building project also include incorporating and fitting in at every stage of the architectural design the advice of these consultants and other third parties.'

Furthermore, the SR 1988 contain two other restrictions as to the liability of the architect.

First, there is a limitation in time, as article 61 states that each form of liability of the architect expires after five years from the day the building project is delivered (which means completed and accepted).

Second, there is a financial limitation, following from article 60, of which paras 1 and 2 read as follows:

'(1) With full commissions for a building project, the compensation to be refunded by the architect for each commission is always limited to a sum of one hundred and fifty thousand guilders, or, if the fees of the architect are higher than one hundred and fifty thousand guilders, the compensation equals the fees of the architect with a maximum of one million five hundred thousand guilders.

(2) With commissions other than a full commission for a building project, the compensation to be refunded by the architect for each commission is always limited to a sum equal to the fees of the architect with a maximum of one million five hundred thousand guilders.'

The liability of the consulting engineer is limited by the RVOI 1987 (article 16) in roughly the same manner. There is, however, the following difference. According to the SR 1988 the architect is obliged to conclude a professional third party insurance with respect to his liability for errors as meant in article 55. The RVOI 1987 do not contain such an obligation, but in practice an engineer will often insure his professional liability.

Finally, it should be added that the limitations of liability of architects and engineers, under the SR 1988 and the RVOI 1987, are still under criticism.

There is no jurisprudence yet on the SR 1988 or the RVOI 1987 regarding the limitation of liability, but it must be presumed that in specific circumstances the principle of good faith will prevent an architect or engineer relying upon these exemption clauses. Reference is made to para. 1.5.3.

In fact, in two cases where the old RVOI (the RVOI 1971) applied, the

Disputes Committee of the Royal Institute of Engineers held that, as a result of the extent of negligence by the engineer and as a result of the extent of the damage occurred, the engineer could not rely on the exemption clause in the RVOI 1971.[11] However, similar decisions have not yet been given by the Architectural Arbitration Institute to which the SR 1988 refers.

Supply contracts

If general conditions apply to supply contracts, they often contain limitations of the liability of the supplier. For example, liability is often limited to the value of the supply and, furthermore, there are frequently exclusions with regard to specific types of damage.

1.5.7 CONSEQUENCES OF PRODUCER'S NON-COMPLIANCE WITH UNEQUIVOCAL DUTY TO REPAIR

If a producer does not comply with the unequivocal duty to repair, the creditor may choose one of the remedies for non-performance, the most important of which are damages and dissolution of the contract.

These remedies are, of course, not of much use if a producer is bankrupt or if the claim is not otherwise recoverable and no guarantee is given by a guarantor (e.g. a bank).

Three remarks should be made on this:

(a) Only the UAV 1989 explicitly states in clause 46, para. 1:

> 'If the contractor fails to perform any of his obligations and the employer gives notice of default, such notice shall be given in writing and shall allow the contractor a reasonable amount of time within which his default may be remedied. In urgent cases the employer shall be entitled, even before the stated period of time has expired, to take such measures at the contractor's expense as the employer may deem apropriate for the benefit of the works.
>
> If the contractor continues to default in the performance of his obligations the employer shall be entitled to complete or have others complete the works at the contractor's expense, without prejudice to the employer's right to compensatory damages.'

(b) If the Model Contract for Residential Building 1991 applies, and a defect is covered by the guarantee scheme connected thereto, the client has a claim against the guarantee fund.

---

[11] Disputes Committee of the Royal Institute of Engineers, 8 December 1981, Bouwrecht 1986, p. 940 and 11 September 1987, Bouwrecht 1989, p. 788 (with a comment by H. O. Thunnissen).

(c) Unless otherwise agreed the Dutch system does not imply joint and several liability of producers, e.g. an architect will not be held liable towards a client for breach of contract by a contractor, and the contractor will not be held liable against the client for breach of contract by the architect.

1.6 Transfer of ownership and producers' liability

1.6.1 RULES

In The Netherlands learned authors discussed for some while whether, according to the preceding Civil Code, a successive owner might have a contractual claim against a contractor. Most authors were in favour of such a possibility. The main reason for accepting this is that the contractor should not be brought into a more advantageous position by the transfer of the ownership.

It was, however, not until 1986 that a civil judge – in this case in the Court of Appeal of The Hague[12] – expressly accepted the possibility of a contractual claim by the successive owner against the contractor.

Although the case before the Court of Appeal of The Hague concerned an action based on article 1645 of the Civil Code (see 1.5.3), the commentator noted that the decision certainly did not exclude the possibility of a claim, based on other grounds than article 1645, such as a claim based on minor hidden defects.

In fact, one year later, the Court of Arbitration for the Building Industry[13] also held that – in accordance with the aforesaid decision of the civil judge – a contractual action by a successive owner against a contractor is possible.

The legal theory, accepted in both decisions, is that with the transfer of ownership an eventual claim against a contractor is transferred as well. Such claims are connected with the capacity as owner. In the new Civil Code this theory is explicitly accepted (article 6:251).

2 PROPERTY (OR MATERIAL DAMAGE) INSURANCE

Property insurance as the term is used in this book is unknown in The Netherlands.

It is impossible for the owner of a building to insure his property against damages arising from construction defects.

---

[12] Court of Appeal The Hague, 5 September 1986, Bouwrecht 1987, p. 385 (see the comment by H. O. Thunnissen).
[13] Court of Arbitration for the Building Industry, 24 September 1987, Bouwrecht 1988, p. 142 (with comments by W. H. Heemskerk and H. O. Thunnissen).

A type of insurance that is often used in The Netherlands, however, is the Construction All Risks insurance (CAR) which is also used in many other countries, but which is not considered to be a property insurance.

Most of these CAR insurance policies also include damage occurring during the Defects Liability Period, provided that this damage originates from the actual construction. Moreover, the possibilities to extend this insurance coverage have increased significantly in recent years; in other words, extending the insurance coverage period up to three or five years is not unusual any longer.

In Holland there is, apart from the CAR insurance, a growing interest in the phenomena of insured guarantees. A well-known example is the guarantee scheme connected to the Model Contract for Residential Building with Matching General Conditions 1991 (see 1.5.3). If the contractor is unable (or unwilling) to comply with the agreed guarantee scheme, the owner of the dwelling has a claim against a guarantee fund. However, this scheme cannot be considered as a property scheme either.

3 LIABILITY AND PROFESSIONAL INDEMNITY INSURANCE

Professional indemnity insurance is not imposed by law upon producers of buildings.

This type of insurance is of interest only to parties such as architects, consulting engineers and contractors involved in the design of buildings.

As seen in 1.1.0 commissions to architects and engineers are usually governed by standard conditions (the SR 1988 and the RVOI 1987) that limit the liability of these producers.

According to the RVOI 1987, consulting engineers are not obliged to insure their professional liability following from these standard conditions. In practice, however, most architects do have such an insurance coverage.

According to article 6, para. 3 of the architects' standard conditions (the SR 1988) insurance of the architects' professional liability (as defined in the SR 1988) is compulsory.

Norway

Hans Jakob Urbye
Revised and updated by Jan Einar Barbo

1 LIABILITY

Non-residential buildings

1.1 General

1.1.0 LAW OR CONTRACT OR BOTH

In Norway, there is freedom in terms of contract, i.e. any person can make a contract with any party, and with any content. Contracts should be fulfilled according to their terms.

From this basic principle, there are several reservations:

(a) there are limitations regarding the *competence* of the parties involved, i.e. persons being under age, insane etc. cannot, as a main rule, enter into legally binding contracts;
(b) there are limitations regarding *how* contracts are made. Contracts made under compulsion or by any other unacceptable behaviour from one of the parties involved, are not valid. This is regulated in a special Act (1918) concerning the procedure for making contracts.
(c) there are limitations regarding the *contents* of contracts. The Act mentioned above stipulates that courts may partly or wholly declare the contract void, if the terms of the contract are unacceptable.

Further, Norway has got special Acts limiting the freedom of making contracts. This concerns particularly regulations to protect consumers against unreasonable terms of contract.

The Act concerning the procedure of making contracts is the only general Act regarding contracts. There are some Acts regarding special types of contracts, e.g. the Sale of Goods Act 1988. This Act does not apply to sale and purchase of real estate. However, the Norwegian Parliament (Stortinget) has passed a Sale of Real Estate Act in 1992.

On the other hand, there is no Act concerning construction contracts in Norway.

Even if there is no general Act on contracts or a special construction contracts Act, there are general and special rules for construction and building contracts, not established by the legislator. In principle it is acceptable to deviate from rules not established by statutes. Such an agreement may, however, be disregarded if found to be unreasonable.

An important general rule in Norwegian contract law is that a party is responsible for any kind of breach of contract (delays, defects) if the said party has acted negligently. In some contracts a party is also responsible for breach of contract without having acted negligently.

(Note: The Indemnification Act 1969 applies to liability apart from contract. According to this Act, a person will have to pay damages if he has acted negligently but, as a main rule, only for personal injury or damage to property, not for economic loss only.)

In contract, the parties involved may limit the liability, agree on risk sharing etc. Thus, breach of contract is not necessarily committed even if the usual conditions for this are present. The Act on the procedures for making contracts (1918) will, however, prohibit contracts with unreasonable conditions. The courts will never accept contracts excluding liability if the breach of contract is due to intent or gross negligence.

CONSTRUCTION LAW

The law relating to construction and building contracts is not statutory, but is expressed in standard form contracts such as NS 3430 'General Conditions for Contracts for Construction and Building' (1991) and NS 3430 'General Conditions for Contract for Design and Consultative Work carried out by Architects and Engineers' (1985).

These contracts may apply for residential as well as non-residential buildings. In Norway the same rules do in principle apply for both residential and non-residential buildings. However, reference should also be made to NS 3402 'General Conditions for Contract for Construction of Prefabricated Houses' (1989). This is a design-and-build contract, and the producers' liability for defects or damage is not the same as his liability according to NS 3430 or NS 3403.

NS 3430 is most suitable in the contractual relation between parties who have some professional skill in the execution of commercial contracts, and when the work is of a certain size. The Norwegian Council for Building Standardization, which is responsible for preparing standard form contracts in construction law, has for this reason issued general conditions with a form of agreement for minor construction and building contracts (NS 3408). It has also issued a standard design contract form. This contract is based upon the NS 3430 contract, but with the necessary alterations because of the contractors' obligation to design the work.

'General Conditions' are agreed documents. Most contracts for construction and building as well as contracts for design and consultative work are based upon these documents. This is the case for contracts between private parties and also when the client is an official authority: construction liability as well as post-construction liability are prescribed in the said documents.

However, some clients have got their own general conditions (e.g. Norsk Hydro and other major oil companies). One major aim behind NS 3430 was to make a standard form contract which could be accepted by all participants in the construction industry, without deviations. It is still too early to tell whether this aim is to be fulfilled.

The main rule, however, is that liability is regulated in the contract between the parties, whether standard form conditions are included or not.

According to NS 3403, the producer (architect, consulting engineer) is only liable if he has acted negligently, whereas the contractor according to NS 3430 is under a duty of result (a strict duty). The contractor is liable for any delay (unless he is exculpated by force majeure) and he has a duty to make good, repair or pay damages for any defect, even if he has not acted negligently. If the defect or the damage is caused by someone else, for instance the architect, the consulting engineer, another contractor (not sub-contractor) or the client himself, the contractor is of course not liable.

Build-and-sell contracts are quite common in Norway. A contractor may buy land which he develops as a residential area and then offer the buildings for sale. The Sale of Real Estate Act (1992) does to some extent apply to this kind of contract. According to this Act the seller is, as a main rule, liable for any delay or defect, even if he has not acted negligently. The parties are free to make agreements which deviate from this Act. However, they are not free to deviate from the rules regarding breach of contract for residential buildings when the purchaser is a consumer.

PUBLIC CONTROL BUREAUX

Public control regarding construction and building is administrated by local government through the Building Council (Bygningsrådet).

Before starting the building work, a building licence from the Building Council is required.

A building licence is issued on the basis of drawings and specifications, made according to rules in statutes and regulations.

During the building work, the Building Council has a right, but no obligation, to supervise the work.

The Building Council may be liable if it acts negligently when managing an application for a building licence, e.g. by giving a building licence for a site where the Building Council, but not the client, is aware of a risk for flood damage, landslide etc.

The Building Council may be liable if a member of its staff has given consultative services and during the execution of these services has acted negligently. The question of liability will be judged by the same rules as for private consultants. However, these officers will usually not give such services.

THE LIABILITY OF SUB-CONSULTANTS AND SUB-CONTRACTORS

Sub-consultants and sub-contractors have no contractual relationship with the client. As a main rule they can therefore not be held liable for breach of contract to the client, but they may of course be held liable to the party they are in contract with. On the other hand, they may be sued by the client based on tort (third party liability), and in NS 3430 there is a clause that gives the client a possibility to claim from the sub-contractor if the contractor is bankrupt or insolvent.

See also 1.6.1.

THE LIABILITY OF SUPPLIERS (NOT IN CONTRACT WITH THE CLIENT)

The Sale of Goods Act 1988 regulates liability between the parties. A supplier under contract with a builder or an installer is of course liable to the party he is in contract with. According to the Sale of Goods Act 1988, a supplier may also be liable directly to the client. Unless the client is a consumer, the liability may be renounced in the contract.

THE TENANT'S RIGHTS VIS-À-VIS THE PRODUCERS

The tenant has no contractual relationship to the producers. The producers will therefore not be liable to the tenant in contract. The tenant may, however, be considered a third party, and claim third party liability based on tort.

1.1.1 TERMS AND CONCEPTS

There are no definitions of the term 'building' in the Norwegian legal system, and there is no distinction between building and rebuilding.

In 1989, the Norwegian Parliament (Stortinget) passed a new statute, regarding protection of consumers when engaging professionals to do work on movables or real estate (apart from erecting buildings). This consumer protection Act may therefore be related to repair works on residential buildings and rebuilding.

Fixed installations are automatically included in the 'buildings', but not access roads and paths, parking decks and lamp posts.

Remaining questions under 1.1.1 are dealt with in the previous section.

1.2 During the Guarantee (Maintenance, Defects Liability) Period

1.2.1 DETERMINATION OF THE COMMENCEMENT OF THE PERIOD

The Guarantee Period commences when delivery has taken place.

As a main rule the work is delivered by a formal delivery act: NS 3430, clause 30. The act of delivery must be recorded in writing. The client may refuse to take delivery if any important defects are discovered, i.e. defects that will essentially impede the client's activities.

If the contractor has reported the work to be completed, and the client takes possession without any delivery act, the contractual work shall be regarded as delivered at the end of the fourteenth day from the commencement of use.

1.2.2 SIGNATORIES TO AND NAME OF THE DOCUMENT STARTING THE COMMENCEMENT OF THE PERIOD

The document is called a Certificate of Practical Completion (*Overtagelsesprotokoll*).

The Certificate of Practical Completion is issued after a satisfactory inspection and signed by the parties present. The inspection should be called for with due notice, normally not less than 14 days.

1.2.3 THE PERIOD'S DURATION AND TERMINATION

The client has to report any defect which he has discovered or which he ought to have discovered by normal careful inspection. If he fails to report, he cannot subsequently invoke such a defect after the issuing of the Certificate of Practical Completion.

If the defect is hidden, the period's duration according to NS 3430 is three years.

The end of the period is normally formalized by a guarantee inspection and a maintenance certificate, signed by the parties.

The period is seldom extended. However, defects caused by gross negligence may be claimed after the end of the period.

Work completed after the date of practical completion, for instance completion and repair works, will have a separate period of three years.

1.2.4 TIME ALLOWED FOR DIFFERENT CATEGORIES OF PRODUCERS FOR MAKING GOOD, REPAIRING OR PAYING DAMAGES

Without payment, the contractor is obliged to remedy or repair any fault or defect for which he is responsible under the contract, provided the defect has been invoked in due time (NS 3430, clause 32).

The repair of defects shall be done within 'a reasonable period of time'.

Damages may be claimed for repairs not done within the given time limit.

Unless this leads to major disadvantage for the client, the contractor may postpone the repair until one year after the delivery act. The contractor cannot claim such an extension of his duty to repair when the damage is invoked during the second or third year of the Guarantee Period.

If the contractor does not fulfil his obligation to repair, he has to pay damages, i.e. the costs of the repair. He also has to pay 18 per cent p.a. interest (at the present time) if he does not pay within 30 days after he has received the claim.

If a fault or defect is due to negligence on the part of the architect/ engineer, the consultant will be obliged to pay damages. He has to pay 18 per cent p.a. interest if he does not pay within 30 days after he has received the claim.

1.2.5 KINDS OF DEFECTS AND DAMAGE COVERED BY THE GUARANTEE

The contractor is obliged to repair any defect, whether it has led to damage or not. If he does not comply with this obligation, the client may do the repair at his expense.

If the defect after the delivery has caused damage to other parts of the contractual work, the contractor is also obliged to repair this damage. There is a condition nevertheless that such damage must be close and likely consequences of the defect.

If the repair requires a disproportionate amount of work or costs, the client may have to accept a price reduction. The price reduction shall be at least equal to the savings made by the contractor as a result of his non-contractual performance of the work.

If the defect has caused damage to other parts of the building or con-struction work, e.g. work performed by a sub-contractor, the contractor will be liable if the original defect is due to negligence.

The contractor is not liable for any loss apart from what is described here, unless the defect has been caused intentionally or by gross negligence on the part of the contractor or any person for whom he is responsible, i.e. any of his employees or any of his sub-contractors. For example, as a main rule the contractor is not liable for the client's loss in production, caused by a defect. This is a limitation of liability for consequential losses compared to the ordinary rules of liability in Norwegian contract law.

These rules apply only when the damage is caused by defects. If the contractor causes damage in any other way, then he is liable according to Norwegian tort law – 'i.e. as a main rule he is liable if he has acted negligently.

According to NS 3403 the liability of the architect/engineer is limited to NOK 1 500·000 for each event, and NOK 4 500·000 for the whole assignment.

1.2.6 EXCULPATION OF PRODUCER

THE CONTRACTOR (NS 3430)

In Norwegian construction law the contractor is under a so-called 'strict' duty – a duty of result.

As mentioned above (1.2.3), the client has to report any defect which he has discovered or which he ought to have discovered by normal careful inspection. If he fails to report, he cannot subsequently invoke such a defect after the issuing of the Certificate of Practical Completion. However, this rule does not apply to defects caused intentionally or by gross negligence on the part of the contractor or any person for whom he is responsible. If the client starts to use the building or part thereof without any delivery act, the same rule applies if he has failed to report any defect within 14 days from the commencement of use.

Defects which the client could not reasonably be supposed to have discovered by the time of the issuing of the Certificate of Practical Completion, must be invoked by the client without unreasonable delay after he discovered or should have discovered the defects, at latest three years from the date of delivery.

THE DESIGNER/CONSULTANT (NS 3403)

The designer/consultant is under a duty of reasonable skill and care, according to the main rule on professional liability in Norwegian contract law.

If the architect/engineer is responsible for the defect, the client has to state his claim within a reasonable period following the discovery of the defect. If he does not do this, he cannot subsequently invoke such a defect. Claims for compensation may under no circumstances be filed later than three years after the building has been taken over by the client. This time limit does not apply if the damage has been caused by intent or gross negligence.

If the client cannot make a claim to the producer because the defect is not reported in due time, the person engaged by the client to inspect and/or issue the Certificate of Practical Completion will be liable to the client if he ought to have, but did not, discover or report the defect.

DESIGN-AND-BUILD CONTRACTS

In design-and-build contracts the contractor is under a duty of result for both his obligation to design and build.

THE RISK OF MATERIALS AND WORK

According to NS 3430, clause 15, the contractor bears the risk of materials and work until delivery has taken place. The same rule applies to materials and work which the client has procured and placed in the contractor's possession.

However, the client takes the risk if force majeure occurs, but only if the force majeure event is extreme (e.g. war, riots, natural disaster).

When delivery has taken place, the risk is transferred to the client.

MISCELLANEOUS

If the producer has used experimental techniques, or techniques with a known risk of failure, the producer will be liable if damage occurs, unless the techniques were used at the client's demand or with the client's consent. In this case he will not be liable.

If the user of the building has not followed instructions as to maintenance or operation of the building, the producer will not be held liable.

As stated before, the contractor is obliged without payment to remedy or repair any fault or defect for which he is responsible under the contract, and which has been invoked in due time, even if he has not acted negligently – but there are important limitations as to liability for consequential losses. See 1.2.5 for further details.

1.2.7 CONSEQUENCES OF PRODUCER'S NON-COMPLIANCE WITH UNEQUIVOCAL DUTY TO MAKE GOOD, REPAIR OR PAY DAMAGES

If the contractor does not comply with his duty to repair, whatever the reason, repairs may be done at his expense. According to NS 3430, the contractor shall place in favour of the client a surety bond for the performance of his obligations under the contract (performance bond). The performance bond shall be equivalent to 15 per cent of the contract price, whereas the surety bond for the contractor's liability and obligations during the Guarantee Period shall be equal to 3 per cent of the contract price the first year, 2 per cent in the second year and 1 per cent in the third year.

Until delivery has taken place, the client will also have retention money, 5 per cent of the contract price.

The architect/engineer almost always carries professional indemnity insurance. According to NS 3403 the liability is limited to NOK 1 500 000 for each event and NOK 4 500 000 for the whole assignment. See 1.2.5.

1.5 From expiry of the Guarantee (Maintanence, Defects Liability) Period

1.5.1 DETERMINATION OF THE COMMENCEMENT OF THE (POST-GUARANTEE) PERIOD DURING WHICH A PRODUCER MAY BE HELD LIABLE

The Guarantee Period is three years from the date of delivery.

For work performed after the delivery, i.e. repairing of defects, the guarantee is three years from the date of performance.

Before the end of the Guarantee Period, either party may require a joint guarantee inspection of the work.

The producer (the contractor as well as the consultant) is usually not liable for any fault or defect which appears after the end of the Guarantee Period. Even faults or defects which could not reasonably have been dis-covered earlier will normally not involve any liability for the producer. He will only be liable if the fault or the defect has been caused intentionally or by gross negligence by him or by any person for whom he is responsible (NS 3430, clause 32.10; NS 3403, clause 12.3).

1.5.2 NAME OF THE DOCUMENT STATING THE COMMENCEMENT OF THE POST-GUARANTEE PERIOD; ITS SIGNATORIES

Each of the parties may call for a guarantee inspection. Regulations as recorded under 1.2.2 will apply.

1.5.3 PERIOD(S) OF LIMITATION

As mentioned under 1.5.1, after the expiry of the Guarantee Period the producer will only be liable if damage is caused intentionally or by gross negligence. If that is the case the Limitation Act 1979 will apply: action must be taken within one year from the date of the client's knowledge of the defect, but not later than 13 years from delivery.

1.5.4 TIME ALLOWED (DIFFERENT CATEGORIES OF PRODUCERS) FOR MAKING GOOD, REPAIRING OR PAYING DAMAGES

If a producer is liable in the post-Guarantee Period (because the defects in the work have been caused intentionally or by gross negligence), he is only liable to pay damages. Rules as recorded under 1.2.4 will apply.

1.5.5 KINDS OF DEFECTS AND DAMAGE QUALIFYING FOR DAMAGES

Reference is made to 1.2.5. Liability in the post-guarantee period occurs only if the defects/damage are caused by the producer's intent or gross negligence. In this case all kinds of defects/damage qualify for damages and the only limitation to the producer's liability will be extinction with time according to the Limitation Act 1979. See 1.5.3.

1.5.6 EXCULPATION OF PRODUCER

See 1.5.5.

1.5.7 CONSEQUENCES OF PRODUCER'S NON-COMPLIANCE WITH UNEQUIVOCAL DUTY TO REPAIR

See 1.5.4. After the Guarantee Period the producer will not be obliged to repair, only to pay damages.

1.6 Transfer of ownership and producers' liability

1.6.1 RULES

A successive owner has, like a tenant, no contractual relationship with the producer. He can therefore only be considered as a third party and claim third party liability based on tort. As a rule, outside a contractual relationship a party cannot successfully claim damages for economic loss only, but only for economic loss in connection with personal injury or damage to property. The producer will therefore usually not be liable towards the successive owner even when it can be proved that he has acted negligently.

The successive owner may of course claim damages from the previous owner (the client), who is the other party to his contract (unless the liability is limited in the contract). The client may seek recourse from the producer within the limits stipulated in the contract between these parties.

The client's right to claim damages from the producer can also be transferred to the successive owner in the sales agreement.

The rules that are discussed here are the traditional rules. However, there are some court decisions from the last decades that allow for successive owners to make direct action claims against the producer, based on contract law. The consequences of these decisions are uncertain and disputed, and therefore cannot be outlined here.

[2–3: *Not applicable*]

Residential buildings

Category I: Housing built for sale after completion ('unknown purchaser' or 'speculative building')

1.1   General

In Norway, the legal system regarding liability and insurance for residential and non-residential buildings is basically the same. However, the field of residential buildings may have the advantage of stronger consumer protection. The sales methods developed for the residential market may also introduce some new angles.

Generally, refer to the section on non-residential buildings. Supplementary information follows.

1.1.0   LAW OR CONTRACT OR BOTH

When a house is built for sale after completion, the parties involved will enter into a contract for that specific purpose. Real estate agents and producers often have their own general conditions. The principle of freedom in terms of contract will apply. The background law (mainly the Sale of Real Estate Act 1992) will be applied when necessary. Some articles in the Real Estate Act 1992 cannot be dispensed with by arrangement between the parties when a residential building is for the purchaser's own, personal use (consumer protection). The Act applies only to erected buildings.

1.1.1   TERMS AND CONCEPTS

See 1.1.1 in the section on non-residential buildings.

1.2 During the Guarantee (Maintenance, Defects Liability) Period

1.2.1 DETERMINATION OF THE COMMENCEMENT OF THE PERIOD

The commencement of the period will normally be stated in the contract, and is usually at the time when the purchaser has received the deed of conveyance (title deed) and has taken possession of the building.

1.2.2 SIGNATORIES TO AND NAME OF THE DOCUMENT STATING THE COMMENCEMENT OF THE PERIOD

See 1.2.1.

1.2.3 THE PERIOD'S DURATION AND TERMINATION

If not specified in the contract, the Limitation Act 1979 will apply: as a main rule action has to be taken within three years of commencement of the period. If the purchaser is not aware of the defect by this time, the period may be extended. Then he has to take action within one year from the date of his knowledge of the defect, but not later than 13 years from the commencement of the period. In the Sale of Real Estate Act 1992 the duration and termination is fixed in a special clause. The Guarantee Period is five years.

The vendor will, in particular if he is a producer, be obliged to repair any fault or defect. He will also have the right to do so instead of paying damages. This will usually be regulated in the contract. The repair of defects shall be done within a reasonable period of time. Reference is also made to 1.2.4, non-residential buildings.

A serious fault or defect may allow the purchaser to terminate the contract.

[1.2.4: *Not applicable*]

1.2.5 KINDS OF DEFECTS AND DAMAGE COVERED BY THE GUARANTEE

If the building's condition is not according to the contract, the vendor is obliged to make good, repair or pay damages, unless the contract is terminated. This is the case even if the vendor has not acted negligently, unless he has renounced liability in the contract. Any fault or defect may be invoked, regardless of being caused by the contractor or the architect/consulting engineer.

The Sale of Real Estate Act 1992 has got a special clause for consumer protection: a vendor is not allowed to limit his liability when he is selling a residential building for the purchaser's personal use.

The Act does nevertheless limit his liability and the vendor will not be liable for consequential losses unless he has acted negligently.

1.2.6 EXCULPATION OF PRODUCER

The purchaser has to report faults or defects which he has discovered or which he ought to have discovered within a reasonable time after the discovery. If he fails to do so, he cannot subsequently invoke such a defect. This rule does not apply to defects caused intentionally or by gross negligence.

1.2.7 CONSEQUENCES OF PRODUCER'S NON-COMPLIANCE WITH UNEQUIVOCAL DUTY TO MAKE GOOD, REPAIR OR PAY DAMAGES

If the producer (vendor) does not make good, repair or pay damages, whatever the reason, the purchaser may retain a part of the payment, sufficient to cover the cost of the repair.

If the producer's breach of contract is fundamental, the purchaser may terminate the contract.

1.5 From expiry of the Guarantee (Maintenance, Defects Liability) Period

1.5.1 DETERMINATION OF THE COMMENCEMENT OF THE (POST-GUARANTEE) PERIOD DURING WHICH A PRODUCER MAY BE HELD LIABLE

As explained under 1.2.1, the Guarantee Period will usually be specified in the contract. If that is not the case, the Limitation Act 1979 will apply, see 1.2.3.

The vendor is normally not liable for any fault or defect which appears after the end of the Guarantee Period. He will be liable only if the fault or the defect has been caused intentionally or by gross neglience. The Limitation Act 1979 will then apply – see 1.2.3.

After the Guarantee Period the vendor (producer) will not be obliged to repair, only to pay damages.

[1.5.2–1.5.7: *Not applicable*]

1.6   Transfer of ownership and producers' liability

Reference is made to 1.6 in the section on non-residential buildings. The Sale of Real Estate Act 1992 allows a purchaser to claim damages from the producer.

[2–3: *Not applicable*]

Category II: Housing built for a client known to the producer(s)

1.1   General

1.1.0   LAW OR CONTRACT OR BOTH

There is no special Act prescribing the producer's liability. However, the Ministry of Justice has appointed a committee to prepare a proposal for an Act for the building and sale of new houses with plot of land, as well as contracts covering the building of houses on the plot of the client. This Act will also apply to buildings built for sale after completion ('unknown purchaser'). The committee started its work in 1988, and it is uncertain when the Act will be passed.

There is a standard form contract, NS 3402 'General Conditions for Contract concerning Construction of Prefabricated Houses' (1989), which is quite extensively used. This is a design-and-build contract.

Apart from NS 3402, there are no nationwide standard form contracts for residential buildings.

[1.1.1: *Not applicable*]

1.2   During the Guarantee (Maintenance, Defects Liability) Period

According to NS 3402, the Guarantee Period is five years. The producer is liable for any fault or defect caused by design, construction or material defects. He is also liable for any damage to the contractual work that is caused by a defect. However, he is only obliged to remedy or repair. He is not liable for other consequential losses, unless the defect has been caused by negligence. Even then the liability is limited to 10 per cent of the contract price. If the defect has been caused intentionally or by gross negligence, there is no limitation on his liability.

According to NS 3402, the producer bears the risk of materials and work until delivery has taken place, even when force majeure occurs. There are no qualifications to this rule.

See also the section on non-residential buildings.

1.5 From expiry of the Guarantee (Maintenance, Defects Liability) Period

See the section on non-residential buildings.

1.6 Transfer of ownership and producers' liability

See the section on non-residential buildings.

Category III: Housing built by an owner/occupier from – or partly from – a construction kit delivery by a specialized firm

1.1 General

1.1.0 LAW OR CONTRACT OR BOTH

This is a contract of sale regulated by the Sale of Goods Act 1988.

There is a standard form contract, NS 3404 'General Conditions for Sale and Delivery of Kits for System built Houses' (1989), and the Sale of Goods Act 1988 has got regulations regarding the vendor's liability. These regulations cannot be limited in contracts with consumers.

[1.1.1: *Not applicable*]

1.2 During the Guarantee (Maintenance, Defects Liability) Period

1.2.1 DETERMINATION OF THE COMMENCEMENT OF THE PERIOD

The Guarantee Period starts when the construction kit is delivered.

[1.2.2: *Not applicable*]

1.2.3 THE PERIOD'S DURATION AND TERMINATION

The period's duration is fixed in the Sale of Goods Act 1988. It is five years if the purchaser is a consumer. If not, the Guarantee Period is two years. The period cannot be shortened in contract with a consumer. According to NS 3404 the period's duration is five years.

1.2.4 TIME ALLOWED FOR DIFFERENT CATEGORIES OF PRODUCERS FOR MAKING GOOD, REPAIRING OR PAYING DAMAGES

NS 3404, clause 6.1

If defects are discovered in the materials delivered, and these are discovered before the materials are utilized, the purchaser may demand that the vendor makes a new delivery.

If a new delivery is not promptly effected, the purchaser may terminate the contract for that particular part of the delivery and claim damages for subsequent extra expenses.

This is also in accordance with the Sale of Goods Act 1988.

NS 3404, clause 6.2

If faults are discovered in the finished building, and these are caused by errors in drawings or descriptions or by faults in materials etc. delivered by the vendor, it is the vendor's duty to provide and pay for work and deliveries for the repair of these faults.

If repair is not done within a reasonable period after the faults are made known, the purchaser may instead claim compensation from the vendor.

Reference is also made to the section on non-residential buildings.

1.2.5 KINDS OF DEFECTS AND DAMAGE COVERED BY THE GUARANTEE

According to NS 3404, the producer is obliged to rectify any defect, whether it has led to damage or not. If the producer does not comply with this obligation, the client may do the repair at his cost.

The client cannot claim damages exceeding the cost of correcting the defect (i.e. consequential losses) unless the defect has been caused by negligence.

If the repair requires a disproportionate amount of work or costs, the client may have to accept a price reduction. The price reduction shall be at least equal to the savings made by the contractor as a result of his non-contractual performance of the work.

These rules are approximately the same as in the Sale of Goods Act 1988.

1.2.6   EXCULPATION OF PRODUCER

According to NS 3404, clauses 6.6 and 6.7, the purchaser can only make a claim if the faults are made known within a reasonable time after they should have been discovered. If a fault is not made known within five years from delivery, a claim cannot be made even if the fault was impossible to discover at an earlier date. However, defects caused by fraud or gross negligence may be claimed after the end of this period.

1.2.7   CONSEQUENCES OF PRODUCER'S NON-COMPLIANCE WITH UNEQUIVOCAL DUTY TO MAKE GOOD, REPAIR OR PAY DAMAGES

The purchaser may (wholly or partly) terminate the contract if the building is not completed.

If NS 3404 is part of the contract, the producer has to place a surety bond for liability equivalent to 5 per cent of the contract price.

If the final account is unsettled, the purchaser may retain the necessary amount as security.

1.5   From expiry of the Guarantee (Maintenance, Defects Liability) Period

1.5.1   DETERMINATION OF THE COMMENCEMENT OF THE (POST-GUARANTEE) PERIOD DURING WHICH A PRODUCER MAY BE HELD LIABLE

The Guarantee Period is either two or five years from the date of delivery, see 1.2.3. The vendor (producer) is usually not liable for any fault or defect which appears after the end of the Guarantee Period. The vendor is in this case only liable in the case of fraud or gross negligence.

[1.5.2: *Not applicable*]

1.5.3 PERIOD(S) OF LIMITATION

In case of fraud or gross negligence by the vendor, action must be taken within one year from the date of the purchaser's knowledge of the defect, but not later than 13 years from delivery. This is according to the Limitation Act 1979.

1.5.4 TIME ALLOWED (DIFFERENT CATEGORIES OF PRODUCERS) FOR MAKING GOOD, REPAIRING OR PAYING DAMAGES

If the vendor is liable in the post-Guarantee Period, he will only be obliged to pay damages. Rules as recorded in the section on non-residential buildings, 1.2.4, will apply.

1.5.5 KINDS OF DEFECTS AND DAMAGE QUALIFYING FOR DAMAGES

See the section on non-residential buildings.

1.5.6 EXCULPATION OF PRODUCER

See the section on non-residential buildings.

1.5.7 CONSEQUENCES OF PRODUCER'S NON-COMPLIANCE WITH UNEQUIVOCAL DUTY TO REPAIR

See the section on non-residential buildings.

[2.3: *Not applicable*]

Portugal

J. Ferry Borges and A. Torres Mascarenhas

1 LIABILITY

1.1 General

The general concept of construction works comprises both buildings and other civil engineering works, such as towers, bridges, roads and dams.[1]

The concept of producer includes designer (architect, engineer, consulting firm), builder (contractor, installer etc.) and other participants engaged or commissioned by the client, as indicated in the List of Terms (Appendix B). The distinction between designer and builder is necessary due to different liability rules for these types of producers. In the following, designers' liability is not dealt with in detail.

A distinction is made between private works and state works, the latter being those that are supported totally or in part by central or regional, state or municipal administration. The works carried out by public corporations may also be included in the last category, pending the decision of the competent minister.

1.1.0 LAW OR CONTRACT OR BOTH

There are two specific situations:

- the builder's post-construction liability is prescribed by the Civil Code 1967[2] (*Código Civil*) when applied to private works in general;
- for state works, the contractual system as well as the builder's liability, are covered in specific laws[3] namely Decree-Law no. 235/86 of 18 August

[1] See Interpretative Document, Mechanical Resistance and Stability, EEC, TC1/011, III/3754/90-EN.
[2] Portuguese Civil Code 1967, last revised by Decree-Law no. 496/77, of 25 November.
[3] Programas de Concurso Tipo e Cadernos de Encargos Tipo para as Empreitadas de Obras Públicas e Legislação Complementar. Imprensa Nacional – Casa da Moeda, Lisbon 1988;

which revised Decree-Law no. 44871 of 19 February 1969 (Legal Regime for Public Construction Works – *Regime Jurídico para Empreitadas de Obras Públicas*) and introduced into Portuguese legislation the prescriptions of EC Directives, namely 71/304/CEE and 71/305/CEE. Decree-Law no. 396/90 of 11 December introduced new conditions required by Directive 89/440/CEE.

There are no specific laws defining designers' liability.[4]

1.1.1 TERMS AND CONCEPTS

The terms 'construction works' and 'producer' were dealt with in 1.1.

The concept of construction works includes all civil engineering works. It is normal that the 'builder' who signed the contract with the client is held liable for all the works, though they may be executed by sub-contractors or depend on suppliers.

Generally the 'tenant' has no rights vis-à-vis the producers.

1.2 During the Guarantee (Maintenance, Defects Liability) Period

The guarantee period defined by civil law only applies to builders (contractors) and not to designers.

1.2.1 DETERMINATION OF THE COMMENCEMENT OF THE PERIOD

There are two specific situations:

- for private works in general the guarantee period starts when the building is delivered to the client, who accepts it, without any specific document;
- for state works Decree-Law no. 235/86 specifies that the guarantee period starts with the signature of the Certificate of Provisional Reception (*auto de recepção provisória*).

Regime Jurídico para os Empreiteiros de Obras Públicas, Decree-Law no. 48871, of 19 February 1969, amended by Decree-Law no. 235/86, of 18 August 1986.
[4] Relação das Disposições Legais a Observar pelos Téncicos Responsáveis dos Projectos de Obras e sua Execução. Relatório de Actualização no. 12, LNEC, publicado no Diário da República I Série, no. 109, 12 May 1989; A. Cardoso Disposições Portuguesas Aplicáveis à Construção e Engenharia Civil, Boletim do Gabinete Técnico de Habitação da Câmara Municipal de Lisboa, vol. 6, no. 43, 1982.

1.2.2 SIGNATORIES TO AND NAME OF THE DOCUMENT STATING THE COMMENCEMENT OF THE PERIOD

As we have said, the commencement of the period in the case of private construction works in general does not depend on any document; when dealing with state works the Certificate of Provisional Reception is signed by the client and the builder, or by their representatives. This should be done after inspection of the works.

1.2.3 THE PERIOD'S DURATION AND TERMINATION

There are two specific situations:

- when the contract does not mention the duration of this period, it is prescribed by the Civil Code that the period is five years;
- when dealing with state works Decree-Law no. 235/86 indicates two years as the duration of the period of guarantee, when this period is omitted in contract.

The period of guarantee terminates with the final reception and the signature of the Certificate of Final Reception.

1.2.4 TIME ALLOWED FOR DIFFERENT CATEGORIES OF PRODUCERS FOR MAKING GOOD, REPAIRING OR PAYING DAMAGES

There are two specific situations:

(a) For private works: The client who accepts the construction has made sure it has no apparent defects. However, if he happens to discover any during the guarantee period he must report them to the builder within 30 days of the discovery and have him repair them in a reasonable time;
(b) For public works: According to Decree-Law no. 235/86, if the defect is identified during the inspection indicated in 1.2.2, the client should put his reservation in the Certificate of Provisional Reception and simultaneously notify the builder so that he makes good within a reasonable time fixed by the client. The builder has ten days in which to appeal and the client must then answer this complaint in the following 30 days.

1.2.5 KINDS OF DEFECTS AND DAMAGE COVERED BY THE GUARANTEE

There are two specific situations:

- When dealing with private works in general, and although the Civil Code does not refer in detail to this subject, as does Decree-Law no. 235/86, the client can even have the constructions done again if the defects that he has found cannot be eliminated. This does not exclude the client applying for consequential damages resulting from this situation.
- According to Decree-Law no. 235/86 the builder is liable for all deficiencies and errors in the execution of the project and for those related to the quality, shape and dimensions of the applied materials and components both when the design does not specify the standards to be followed and if the adopted ones are different from those specified.
- However, the builder is not liable if the errors are due to orders given him by the client. The client or the contractor is liable for technical deficiencies or errors in design or in any other element, according to who has generated the erroneous instruction. This implies that the cost for remedying should be met by whoever is responsible.

1.2.6 EXCULPATION OF PRODUCER

This is too complicated to summarize but the general outline is:

- When dealing with private works the builder is exculpated from defects and damage depending on an express agreement with the client.
- When dealing with state works the builder is exculpated from defects and damage due to misleading information supplied by the client, as indicated in 1.2.5.

    Inspectors are liable to the client from whom they receive orders and to whom they are financially subordinated.

    Generally speaking, actions by a third party and other causes are explicitly considered and the damages borne by the client except when they should be covered by an insurance of the producer's, according to the contract.

    Liability of the authors of documents, such as norms, standards, agréments and specifications, is not defined.

1.2.7 CONSEQUENCES OF PRODUCER'S NON-COMPLIANCE WITH UNEQUIVOCAL DUTY TO MAKE GOOD, REPAIR OR PAY DAMAGES

There are two specific situations:

- When dealing with private works in general:
  (a) scenario 1 – the client may take the case to court in order either to force the producer to make good or to receive financial compensation so that he is reimbursed for expenses incurred with another producer with whom he has in the meantime contracted;
  (b) scenario 2 – if bankruptcy occurred he is one of the creditors;
  (c) scenario 3 – if the producer cannot be traced and if those that were responsible for the company affairs cannot be traced either, the client will have to face the consequences.
- When dealing with state works the client:
  (a) scenario 1 – may commission another producer to make good and pay him out of the value of the bid bonds and, if these are not sufficient, out of the producers other assets;
  (b) scenario 2 – (same situation as scenario 1) and if, again, the client must have recourse to the producer's assets he will have a corresponding claim on these;
  (c) scenario 3 – same situation as in building for a private client.
- When a client who faces any of these situations complains, the government body that chose the producer to execute the state works can intervene.

1.5   From expiry of the Guarantee (Maintenance, Defects Liability) Period

1.5.1   DETERMINATION OF THE COMMENCEMENT OF THE (POST-GUARANTEE) PERIOD, DURING WHICH A PRODUCER MAY BE HELD LIABLE

As soon as the guarantee period is over (see 1.2.3) all contractual bonds are also over and no post-guarantee period begins.

1.5.2   NAME OF THE DOCUMENT STATING THE COMMENCEMENT OF THE POST-GUARANTEE PERIOD; ITS SIGNATORIES

When dealing with state works the document that ends all contractual bonds is the Certificate of Final Reception (*auto de recepção definitiva*) which is signed both by the client and the producer or by their representatives.

[1.5.3–1.5.7: *Not applicable*]

1.6 Transfer of ownership and producers' liability

Portuguese law does not consider this case explicitly. Attention has been called to the need of improving legislation to cover this case.

2 PROPERTY (OR MATERIAL DAMAGE) INSURANCE

Insurance for covering defects and unspecified damage in the post-construction period is not usual in Portugal. Usual insurance only covers damage caused by fire or natural hazards and sometimes earthquake damage. Contracts of All Risks insurance at the building stage may be extended to cover the guarantee period, however they should not be considered as property insurance.[5]

3 LIABILITY AND PROFESSIONAL INDEMNITY INSURANCE

3.1 General

This section covers defects and damage occurring after the client has taken over the construction.

3.1.0 LAW OR CONTRACT OR BOTH

Until 1991 there was no specific law covering professional indemnity insurance. Liability and professional indemnity insurance were not compulsory and up to that time seldom required by clients.

From 1991 Decree-Law no. 445/91 on licensing of private construction works and Decree-Law no. 11/92 impose obligatory liability insurance for designers and contractors involved in private construction works, covering a post-construction period of five years.

Decree-Law no. 48871 refers to insurance in case of natural hazard (article 170).

---

[5] A. Carranca, E. Vieira, F. Valadas Fernandes and M. Madeira, Responsabilidades dos Intervenientes no Acto de Construir e o Seguro. Congresso 81, Ordem dos Engnheiros, Tema 2 Comunicação 8, Lisbon, December 1981; F. B. Oliveira, Seguro de Obras e Projectos, 1o Encontro Nacional de Geotecnia, Sociedade Portuguesa de Geotecnia, Lisbon, November 1985; J. Ferry Borges, Qualidade na Construção, Curso 167, Laboratório Nacional de Engenharia Civil, January, 1988; Encontro Nacional sobre Qualidade na Construção. Laboratório Nacional de Engenharia Civil, Lisbon, June 1986; 2o Encontro Nacional sobre Qualidade na Construção. Laboratório Nacional de Engenharia Civil, Lisbon, June 1990.

3.1.1 TERMS AND CONCEPTS

Nothing to add.

3.1.2 TECHNOLOGY AND INSURANCE TERMS AND CONDITIONS

Insurance conditions are established in each specific case according to market conditions. They are not much influenced by technology conditions.

Due to the new legislation referred to above a more active intervention by the Insurance Institute of Portugal will be required.

3.1.3 INSURER'S TECHNICAL CONTROL

Up to the present insurers seldom oversaw design and site work, nor did they consult clients or supervision bureaux. This situation is, however, currently changing.

3.1.4 INSURERS' POSSIBLE INFLUENCE UPON BUILDING TECHNOLOGY

There is an official rating of construction firms by the Contractors Classifying Board (Comissão de Alvarás de Obras Públicas e Particulares) which belongs to the Market Council for Public and Private Construction Works (Conselho de Mercados de Obras Públicas e Particulares), created at the Ministry of Public Works, Transportation and Communications by Decree-Law no. 99/88 of 23 March. This classification is based on the evaluation of technical, economic and financial capacities, as well as ethical behaviour, and is also influenced by the quality of previous works. Insurers do not use this classification for fixing rates.

3.1.5 NEW TECHNOLOGIES

The agrément system is used in Portugal. However, insurers seldom utilize this information.

3.1.6 THE INSURANCE MARKET, GENERAL IMPRESSION

The new legislation prescribes a generalization of construction insurance, which up to the present was covering only a fraction of the works.

3.1.7 COLLECTIVE INSURANCE FOR PROFESSIONAL OR OCCUPATIONAL GROUPINGS. GRADATION OF INSURANCE CONTRIBUTIONS

Collective professional insurance is not used.

3.2 During the Guarantee (Maintenance, Defects Liability) Period

See 3.1.

3.5 After the Guarantee (Maintenance, Defects Liability) Period

See under 3.1.0.

Quebec, Canada[*]

Daniel Alain Dagenais

AUTHOR'S NOTE

This chapter contains only a summary of the subjects. For a more thorough analysis, several factors should be added and exceptions included.

Any remarks made with respect to the new Civil Code of Quebec (see 1.1) must be carefully considered. Amended several times since its draft bill and subject to further amendments, the new Code has yet to be tested by the courts and its application is uncertain.

1 LIABILITY

1.1 General

1.1.0 LAW OR CONTRACT OR BOTH

The forms of liability applicable to a producer differ depending on whether he constructs the building for a third party or for himself with the intention of selling it.

In the first case, the producer is subject to three forms of liability:

- a statutory guarantee of five years which is a requirement of public order (Civil Code, article 1688);
- ordinary contractual liability which may be excluded, restricted or extended by the contract (Civil Code, article 1065);
- delictual liability in the event that a fault occurs outside the preceding framework (Civil Code, article 1053).

Practice dictates a period of one year after the work is completed at which time the contractual holdback (generally 10 per cent of the cost of the work) is released. This period runs simultaneously with a contractual guarantee

---

[*] Original text in French.

(the new Code includes an equivalent statutory guarantee). But the different forms of liability supplement one another.

The producer who sells the building is not subject to the guarantee of article 1688. However, he is obliged by law to warrant against latent defects (Civil Code, article 1522). In some cases, normal contractual and delictual forms of liability may be added.

For identification purposes, I will refer to each of the four forms of liability by the article number of the Civil Code which deals with them: 1688, 1065, 1053 and 1522.

The Quebec Construction Federation (Fédération de la construction du Québec), the Quebec Provincial Housing Association (Association provinciale des constructeurs d'habitation du Québec) and the Quebec Construction Association (Association de la construction du Québec), organizations of construction companies, each implemented a 'guarantee plan for new houses' by which they act as the contractor's guarantor to complete certain works or repair them when necessary.

Lastly, on 18 December 1991, after 30 years of work, the Quebec Government passed a law amending the Civil Code of Quebec. The new Civil Code of Quebec (CCQ) as we now know it, is expected to take effect in 1993 at an undetermined date. With respect to the topics at hand, the new Code does not make any revolutionary changes. But some amendments should be pointed out and I will refer to them occasionally. The articles of the present Civil Code (CC) and their equivalents in the new Code are as follows:

- CC, article 1688 becomes CCQ, article 2118
- CC, article 1065 becomes CCQ, article 1590
- CC, article 1053 becomes CCQ, article 1457
- CC, article 1522 becomes CCQ, article 1726

Excerpts from the Civil Code

'Article 1688
If a building perish in whole or in part within five years, from a defect in construction, or even from the unfavourable nature of the ground, the architect superintending the work, and the builder are jointly and severally liable for the loss.

Article 1688
Si l'édifice périt en tout ou en partie dans les cinq ans, par le vice de la construction ou même par le vice du sol, l'architecte qui surveille l'ouvrage et l'entrepreneur sont responsables de la perte conjointement et solidairement.

Article 1689
If, in the case stated in the last preceding article, the architect do not superintend the work, he is

Article 1689
Si, dans de cas de l'article précédent, l'architecte ne surveille pas l'ouvrage, il n'est responsable

liable for the loss only which is occasioned by defect or error in the plan furnished by him.

Article 1696
Masons, carpenters, and other workmen, who undertake work by contract, for a fixed price, are subject to the rules prescribed in this section. They are regarded as contractors with respect to such work.

Article 1065
Every obligation renders the debtor liable in damages in case of a breach of it on his part. The creditor may, in cases which admit of it, demand also a specific performance of the obligation, and that he be authorized to execute it at the debtor's expense, or that the contract from which the obligation arises be set aside; subject to the special provisions contained in this Code, and without prejudice, in either case, to his claim for damages.

Article 1053
Every person capable of discerning right from wrong is responsible for the damage caused by his fault to another, whether by positive act, imprudence, neglect or want of skill.

Article 1522
The seller is obliged by law to warrant the buyer against such latent defects in the thing sold, and its accessories, as render it unfit for the use for which it was intended, or so diminish its use-

que de la perte occasionnée par les défauts ou erreurs du plan qu'il a fourni.

Article 1696
Les maçons, charpentiers et autres ouvriers qui se chargent de quelque ouvrage par marché pour un prix fixe sont soumis aux règles contenues dans cette section. Ils sont considérés comme entrepreneurs relativement à ces ouvrages.

Article 1065
Toute obligation rend le débiteur passible de dommages en cas de contravention de sa part; dans les cas qui le permettent, le créancier peut aussi demander l'exécution de l'obligation même, et l'autorisation de la faire exécuter aux dépens du débiteur, ou la résolution du contrat d'où nait l'obligation; sauf les exceptions contenues dans ce Code et sans préjudice à son recours pour les dommages-intérêts dans tous les cas.

Article 1053
Toute personne capable de discerner le bien du mal, est responsable du dommage causé par sa faute à autrui, soit par son fait, soit par imprudence, négligence ou inhabileté.

Article 1522
Le vendeur est tenu de garantir l'acheteur à raison des défauts cachés de la chose vendue et de ses accessoires, qui la rendent impropre à l'usage auquel on la destine, ou qui diminuent telle-

fulness that the buyer would not have bought it, or would not have given so large a price, if he had known them.

ment son utilité que l'acquéreur ne l'aurait pas achetée, ou n'en aurait pas donné si haut prix, s'il les avait connus.

Article 1530

The redhibitory action, resulting from the obligation of warranty against latent defects, must be brought with reasonable diligence, according to the nature of the defect and the usage of the place where the sale is made.

In the case an animal stricken with tuberculosis, the redhibitory action shall be deemed to be brought within a reasonable delay if so brought within ninety days of the delivery, and, in such case, the burden of proof that the animal was not so stricken at the time of delivery shall lie on the vendor.

Article 1530

L'action rédhibitoire résultant de l'obligation de garantie à raison des vices cachés, doit être intentée avec diligence raisonnable, suivant la nature du vice et suivant l'usage du lieu où la vente s'est faite.

S'il s'agit d'un animal atteint de tuberculose, l'action redhibitoire est considérée intentée dans un délai raisonnable si elle l'est dans les quatre-vingt-dix jours de la livraison et, dans ce cas, la preuve que l'animal n'en était pas atteint au moment de la vente incombe au vendeur.'

1.1.1 TERMS AND CONCEPTS

'BUILDING'

The producer's liability should be analyzed in the same way whether it pertains to new construction or rebuilding and whether it pertains to a residential or non-residential building. The entire building and each of its sections or 'fixed installations' are covered equally, subject to the proviso that the statutory guarantee under CC, article 1688 only applies to total or partial perishing or the threat of perishing and not to mere defects of poor workmanship which do not endanger the soundness of the building or prevent it from being used for the purpose for which it was intended. Defects of poor workmanship continue to be covered by other forms of liability. The new Code is more or less to the same effect. CCQ, article 2188 only applies to the loss of the work, which should correspond to 'perishing' under present CC, article 1688.

'PRODUCER'

According to CC, article 1688 and CCQ, new article 2118 this means:

- the architect, *if he supervises* the work;
- the engineer, *if he supervises* the work;
- the general contractor;
- under certain conditions, masons, carpenters and other workmen or sub-contractors.

Article 2124 of the new Code adds the real estate promoter to this list.

The liability of architects and engineers who do not supervise the work is restricted to the loss resulting from a defect or error in their plan (present CC, article 1689 and CCQ, new article 2121).

According to CC, article 1522 or CCQ, new article 1726, the liable party is the vendor (see 1.2.7).

According to other liability forms this is whoever is involved in the design, execution or supervision of the work. The public inspection board is not liable for these claims unless it actively intervenes pursuant to provisions set forth in certain laws (Crown Liability Act, or others).

The contractual or delictual form of liability applies depending on whether or not the damage occurred during the execution of a contract. Thus, a sub-contractor or a supplier of materials bound by contract only to the general contractor is nonetheless delictually liable toward the client.

'TENANT'

The tenant benefits from delictual recourses against the producer.

1.2 During the Guarantee (Maintenance, Defects Liability) Period

In this section, I will deal exclusively with two forms of legal liability which apply to the producer (1688 or 1522).

Ordinary forms of contractual and delictual liability will be analyzed under the following heading. As they are forms of liability for proven fault, they do not strictly speaking constitute a guarantee even though they generate liability.

This provision should help to clarify the system and simplify the presentation of answers. However, it should be noted that each of the four forms of liability apply as of the acceptance; the statutory forms of liability end with the expiry of the guarantee period, while the other two may be extended.

References to CC, article 2120 – which will introduce a guarantee of one year against poor workmanship – are made in the following text as needed.

1.2.1 DETERMINATION OF THE COMMENCEMENT OF THE PERIOD

ARTICLE 1688

The guarantee starts at the 'end of the work'. Consensus seems to be that this is the date of the provisional acceptance, that is, when the architect acknowledges that the work is finished – subject to minor defects – and when the one-year period starts at the end of which the holdback is released.

The new Civil Code also refers to the 'end of the work' both in articles 2118 and 2120. In article 2120 it is defined as taking place when the work is executed and in a condition to be used in accordance with the use for which it is intended.

ARTICLE 1522 OR NEW ARTICLE 1726

Date of the sale.

1.2.2 SIGNATORIES TO AND NAME OF THE DOCUMENT STATING THE COMMENCEMENT OF THE PERIOD

ARTICLE 1688

The document is usually signed by the architect after consultation with the engineers regarding their respective fields. If not, it will be signed by another professional who acted as site supervisor or director.

The document, which does not have any official title, may be called:

- Certificate of Provisional Acceptance;
- Completion of Work Certificate;
- Certificate of Substantial Completion of the Work;
- Certificate of Provisional Acceptance.

A document is not absolutely necessary but is common practice. The delivery date may also be considered.

One year later, the architect will issue a Final Certificate.

The new Civil Code is not any clearer. It does, however, describe 'acceptance of the work, as the act by which the client declares that he accepts it, with or without reservations' (CCQ, article 2110). It takes place at the 'end of the work'.

ARTICLE 1522

The date of the sale is recorded by notarial act. The same applies under the new Code.

1.2.3 THE PERIOD'S DURATION AND TERMINATION

ARTICLE 1688

According to the combined provisions of articles 1688 and 2259, the damage must appear within five years following the 'end of the work'. From that time, the client has five years to file his claim.

If the damage appears gradually, it is deemed to have occurred at the end of the first five-year period and the client then has ten years after the 'end of the work' to file his claim.

No formalities mark the end of the term, which cannot be extended.

Under article 2118 of the new Civil Code of Quebec, the damage must occur within five years following the end of the work for the guarantee to come into play. Under article 2120, which pertains to poor workmanship, the guarantee lasts one year after the end of the work. The prescription period for filing the claim will normally be three years after the loss first appears (CCQ, article 2925) even if the appearance is gradual (CCQ, article 2926). No formality marks the end of these periods.

ARTICLE 1522

The duration of the period is not fixed by statute or by any document. The defect which was hidden at the time of the sale does not necessarily have to appear within a given period. However, as soon as it does appear, the client should notify the producer and pursuant to CC, article 1530 initiate proceedings 'within a reasonable delay'. Insofar as real property transactions are concerned, this delay may vary from a few months to several years, depending on the circumstances of the case and, especially, depending on the length of the negotiations between the parties.

The new Civil Code is more or less to the same effect but points out that when the defect appears gradually, the delay runs from the day on which the seriousness of it was suspected. In the event of a latent defect known or presumed to be known by the vendor, the delay is extended (CCQ, article 1739).

1.2.4 TIME ALLOWED FOR DIFFERENT CATEGORIES OF PRODUCERS FOR MAKING GOOD, REPAIRING OR PAYING DAMAGES

ARTICLES 1668 AND 1522

The period of time extended to the producer is entirely a matter for negotiation between the parties. Strictly speaking, however, there is no fixed period and it is possible for the client to show that, as a result of the discovery of the defect, he has lost confidence in the producer and prefers to have repairs carried out by a third party.

Usually, however, the client will formally notify the producer to verify and identify the damage and indicate the remedial measures that he intends to undertake. To do this, the client will give him a few weeks and will expect the repairs to be undertaken shortly afterwards. This formal notice is sometimes mandatory.

Standard form contracts often require that the producer be given five working days to carry out the repair or provide an acceptable timetable.

The new Code is similar to the present Code, except that it indicates that the formal notice must give the producer 'sufficient time for performance' (CCQ, article 1594 ff.).

1.2.5 KINDS OF DEFECTS AND DAMAGE COVERED BY THE GUARANTEE

ARTICLE 1688

The statutory guarantee under article 1688 does not apply to ordinary poor workmanship or minor defects which, as unpleasant and inconvenient as they may be, do not affect the solidity or the intended use of the building. Any defect likely to lead to major damage may be covered by the guarantee. Although the Code uses the word 'perishing', case law tends to interpret it more and more widely.

Examples:

(a) Cases where 1688 was applied:
  - seepage of water endangering the foundations;
  - sagging of overhanging slabs;
  - loss of roof insulation.
(b) Cases where the application of 1688 was refused:
  - floors whose tiles lifted;
  - plumbing joints that burst;
  - freezing pipes.
(c) Doubtful cases:
  - seepage of water through the roof.

There is every reason to believe that the loss of the work pursuant to CCQ, article 2118 is equivalent to 'perishing'.

New Article 2120 covers poor workmanship which is excluded from article 2118. I believe that it can only pertain to poor workmanship which was hidden at the time of delivery.

ARTICLE 1522

All defects are covered by the guarantee. In reality, the definition of 'defect' and 'damage' in our terminology corresponds very closely to the meaning of 'defect' under article 1522 or CCQ, new article 1726.

The only proviso is that the defect must have been latent at the time of the sale. There are both subjective and objective criteria in this respect but the normal rule is that the defect must have been hidden from a diligent purchaser or even an expert who made a brief survey of the building. However, insofar as new buildings are concerned, admittedly, a purchaser is not required to have the building surveyed by another expert. There is a presumption, so to speak, that a producer who fails to tell the purchaser about the defect did not see it and, therefore, neither would another expert.

The new Code clarifies the law by stating that a defect is apparent when it can be noticed by a prudent and diligent purchaser without the need for expert assistance (CCQ, article 1726).

The guarantee applies even if the vendor was unaware of the defect. The bonafide vendor may stipulate that he is not bound by any guarantee. There is a presumption that the producer is aware of the defects of the thing which he sells.

In each and every case, a contractual guarantee may be added to these statutory guarantees.

1.2.6 EXCULPATION OF PRODUCER

ARTICLE 1688

The different categories of producers subject to CC, article 1688 are jointly and severally liable toward each other insofar as the owner is concerned, an error by any of them being the responsibility of the others. Only at a later stage, after the client has been indemnified, will it be possible to establish which of them must reimburse the others. It will be the one who committed the basic error or omission: an error in design, if professionals are concerned, or an error in execution, if the contractor is involved. The latter is liable for faults committed by his sub-contractors but, in turn, has a right of recourse against them.

None of the producers can claim exoneration in relation to the client on the grounds that he acted in accordance with trade practices or that one of the others was at fault. The producers have a strict liability to perform which extends to defects in construction and defects in the ground. Their only defence is to prove an Act of God, a fortuitous event, fault of a third party (other than one of the producers) or fault of the client. The last-mentioned arises when the building is not used for its intended purpose or is not properly maintained, if this is causal, and only if the client was properly informed in cases where special maintenance is required.

The client's acceptance of an incident which eventually leads to damage covered by CC, article 1668 is not grounds for defence because this article contains a requirement of public order and is not restricted to latent defects. However, the client must not allow a situation of which he is aware to deteriorate; he has an obligation to mitigate the damage as much as possible.

It is important to note that this form of liability tends to protect not only the client who allegedly is unable to judge the quality of the producer's work but also the public. If a client who is familiar with building techniques imposes his designs on the producer he may, by doing so, reduce or completely avert the latter's liability.

A fortuitous event or an Act of God is an unforeseeable and unavoidable occurrence, such as rainfall which is twice the maximum recorded over the past one hundred years and not just rain which is unusually heavy but foreseeable. In Canada, a major earthquake would no doubt be an Act of God although small earthquakes are frequent and therefore foreseeable. This is subject to the proviso that the producer is liable for defects in the ground which existed at the time of construction.

Impact of a vehicle as well as vandalism or sabotage would also be valid grounds of defence but only when committed by a third party.

The provisions of the new Code give producers significant new means of exoneration. Joint and several liability is initially presumed but each one may be exempted if he is able to prove that he is not responsible for the breach of duty or that the defect results from decisions imposed by the client (CCQ, article 2119).

The one-year poor workmanship guarantee set forth in new article 2120 will not make such grounds of exoneration available. However, it is not certain whether it will be imposed by legislation; if it is not, it could be restricted, even excluded, by contract. Moreover, liability is not joint but several.

ARTICLE 1522

Only the producer who sold the building is subject to this guarantee which does not apply to defects which were latent at the time of the sale. He may exonerate himself by proving that:

- there is no defect, just a normal characteristic of the building and it is of minor importance;
- the defect did not exist at the time of the sale but appeared subsequently;
- if the defect was present at that time, it was visible;
- the damage is due to a fortuitous event, an Act of God, the fault of a third party or the owner's fault or any cause other than a defect present at the time of the sale.

Ignorance of the defect is not a valid ground for defence but his knowledge of it makes him more blameworthy and therefore liable not only to refund all or part of the sale price (in which case in return for the building) but also to pay damages. As previously stated, by law the producer is usually presumed to know the defects of the building. In this respect, his obligation is similar to a strict obligation to perform.

Basically, the new Code contains these rules.

1.2.7 CONSEQUENCES OF PRODUCER'S NON-COMPLIANCE WITH UNEQUIVOCAL DUTY TO MAKE GOOD, REPAIR OR PAY DAMAGES

ARTICLE 1688

A client who finds a defect within the specified periods may require any of the producers involved to repair it. He may also opt to have the repairs carried out by a third party and claim compensation from one of the original producers or from all of them, subject to the provisions of the Code with respect to demand letters. A producer cannot exonerate himself by pleading the fault of another. Thus, for instance, the architect who supervised the work must indemnify the client for direct damages resulting from the plumber's fault. Or, the general contractor must assume the costs of a defect resulting from an error in the plans even if he followed them to the letter. However, the party who indemnified the client will have a right of recourse or subrogation against the party responsible for the original error or omission.

In addition, the client also has direct recourse against the producers' liability insurers.

If none of the producers can be identified, or their liability insurers located, the client has no recourse except against his own property insurers although little or no direct insurance exists which specifically covers defects and damage.

The same situation applies under CCQ, new article 2120.

Under CCQ, new article 2118 we saw that producers have additional grounds for exoneration.

ARTICLE 1522

'Article 1526	**Article 1526**
The buyer has the option of returning the thing and recovering the price of it, or of keeping the thing and recovering a part of the price according to an estimation of its value.	L'acheteur a le choix de rendre la chose et de se faire restituer le prix, ou de garder la chose et de se faire rendre une partie du prix suivant évaluation.'

The buyer can only resort to the guarantee against his vendor or the vendor's liability insurer.

However, he may also initiate an ordinary claim against another party who committed a causal fault. This claim would be in tort as it has no contractual basis. It may pertain to any party intervening in the construction of the building.

Recourses against the vendor are similar in the new Code. CCQ, article 1730 adds that under certain conditions the manufacturer, distributor and supplier of materials must also abide by the guarantee. How will this provision apply to buildings? If it pertains to a component of the building, the buyer will no doubt have a direct latent defect recourse against the parties. But will he be able to exercise this guarantee against a contractor or sub-contractor who is not the vendor but who may be compared to the manufacturer? Case law will have to decide.

1.5 From the expiry of the Guarantee (Maintenance, Defects Liability) Period

1.5.1 DETERMINATION OF THE COMMENCEMENT OF THE (POST-GUARANTEE) PERIOD DURING WHICH A PRODUCER MAY BE HELD LIABLE

When the statutory guarantee under article 1688 no longer applies because the time period has elapsed since the construction, the producer is still not protected from possible lawsuits. The ordinary forms of liability under articles 1065 and 1053 continue to apply for 30 years following the completion of the construction.

However, remember that for our purposes these forms of liability may be invoked after acceptance, provided the defect exists – the presence of the defect usually constituting damage.

Let us now consider these ordinary forms.

There is no specific mention in the new Code that these contractural forms are added to or supersede the aforementioned forms of guarantees. We will have to wait and see what the courts will decide to do. The following discussion assumes that the future state of the law will reflect present law in these matters.

1.5.2 NAME OF THE DOCUMENT STATING THE COMMENCEMENT OF THE POST-GUARANTEE PERIOD; ITS SIGNATORIES

No document exists fixing the end of the guarantee period or the beginning of the post-guarantee period.

1.5.3 PERIOD(S) OF LIMITATION

PRESENT LAW

For a non-residential building, the relevant contractual prescription will be either the commercial prescription of five years or, for some public buildings, the 30 years prescription.

In delictual matters, the prescription will be two years for damage to property and one year for personal injury effective the date of the discovery of the defect.

The delays are the same for every category of producer.

NEW LAW

In almost every case, the prescription will be three years (CCQ, articles 2925 and 2926).

1.5.4 TIME ALLOWED (DIFFERENT CATEGORIES OF PRODUCERS) FOR MAKING GOOD, REPAIRING OR PAYING DAMAGES

ARTICLES 1053 AND 1065

Same situation as in 1.2.4, also applies for the new Code.

1.5.5 KINDS OF DEFECTS AND DAMAGE QUALIFYING FOR DAMAGES

Beyond the guaranteed defects and damage, all damage justifies compensation if fault on the part of the producer or producers can be proven.

In theory, a defect justifies compensation as soon as it is discovered, if it can be shown by experts that in all likelihood it will subsequently develop into damage. In practice, such cases are extremely rare.

1.5.6 EXCULPATION OF PRODUCER

Contractual or delictual forms of liability are for proven breach of duty and, unlike CC, article 1688 or CCQ, article 2118 there is no joint and several liability among the producers, unless they undertook the same thing.

Therefore, each one of them may exonerate himself if the client does not prove his error or omission. Some customary grounds of exoneration are as follows: trade practices were respected, plans and specifications were reasonably followed, the client accepted the risk which he knew about, the breach of duty of a third party or the owner or an Act of God.

1.5.7 CONSEQUENCES OF PRODUCER'S NON-COMPLIANCE WITH UNEQUIVOCAL DUTY TO REPAIR

Only the producers who committed a breach of duty are required to indemnify the client, the successive owner or the third party who sustained a loss. If the parties are bound by contract, the producer's liability will be easier to prove than if the recourse is only delictual. But once this obstacle is overcome, the amount of the compensation is about the same: all damage directly related to the defect will be compensated. In contract, however, compensation is limited to damage which was foreseeable when the contract was entered into.

The client may have the damage repaired by a third party and then sue the producer.

If the producer no longer exists or has become insolvent, the client bears the loss unless he has recourse to property or liability insurance policies.

1.6 Transfer of ownership and producers' liability

See 1.2.7 and 1.5.7.

2 PROPERTY (OR MATERIAL DAMAGE) INSURANCE

I am unaware of any direct insurance which covers, in Quebec, the loss sustained by the client or his successor further to a defect in construction, defect in the ground or latent defect after construction for either residential or non-residential buildings. If such policies exist, they are so few in number that they are not worth mentioning.

It may happen, for instance, that an insurer has to indemnify an owner or a tenant for loss resulting from a fire caused by a construction defect. However, this is not within the realm of this chapter.

Normally, the following items are specifically excluded, even from All Risk policies:

- normal wear and tear, gradual deterioration, latent defects or inherent defects;
- costs related to the proper execution of the work and which are necessitated by defects in materials, their use or selection, manpower, plans or specifications.

Builders' risk policies also exist but they have little to do with the post-construction insurance which concerns us.

3 LIABILITY AND PROFESSIONAL INDEMNITY INSURANCE[1]

Professionals, such as architects, engineers, soil experts or other works inspectors and supervisors are able to insure their liability quite easily. Some professional associations negotiate standard form policies with some insurers and then recommend them to their members.

On the other hand, contractors cannot insure their liability for error and omission. I only know of one policy which offers this coverage but am not familiar with any contractor who has subscribed to it. The cost would be too high. Construction companies in my country are not sufficiently regulated and their skills are not monitored enough for an insurer to assume the risk of covering them once they have left the job site.

This section only refers to the insurance available to professionals.

3.1 General

3.1.0 LAW OR CONTRACT OR BOTH

Liability insurance is not imposed by law. Some efforts in this respect have been made through a professional code but, to date, no regulations have been enacted for architects and engineers to compel construction professionals, or some of them, to insure their liability.

Insurance is therefore voluntary. Rarely, however, do you come across a professional who is not insured for a minimum.

The client or his creditor may require that professionals insure their liability. This, at the very least, is prudent but does not appear to be a fixed rule.

3.1.1 TERMS AND CONCEPTS

'*BUILDING*'

The producer's liability insurance is treated in much the same way whether it involves a new building or reconstruction, or a residential or non-residential building. The entire building and each of its sections or 'fixed installations' are similarly covered.

'*INSURANCE POLICY*'

The following excerpt is an example found in a policy covering professionals:

---

[1] The new Civil Code will not significantly change the contents of this section. Consequently, I will not refer to it.

'The insurer will pay on behalf of the insured all sums which the latter shall be legally obligated to pay as damages if legal liability arises out of the performance by the insured in his capacity as an architect or an engineer, of professional services for others, and such legal liability is caused by an error, omission or negligent act.'

This policy covers work carried out or, rather, claims made against the professional, on a continual basis, rather than by job site and is renewable every year. It covers the professional for all past work.

In some special cases, policies or endorsements are issued to cover specific projects or specific types of work.

3.1.2 TECHNOLOGY AND INSURANCE TERMS AND CONDITIONS

The following are answers to the Questionnaire, section 3.1.2, questions (a)–(f): see Appendix A.

(a) not usually, unless policies applicable to special projects are concerned;
(b) same answer;
(c) same answer;
(d) no;
(e) no;
(f) yes.

Terms and conditions are usually in relation to:

- the insured's experience;
- previous claims;
- his field of work;
- the fees he charges;
- the franchise and excess he bears.

3.1.3 INSURER'S TECHNICAL CONTROL

3.1.4 INSURERS' POSSIBLE INFLUENCE UPON BUILDING TECHNOLOGY

With regard to both these issues, this seldom happens. Some insurers or large brokers may have a technical office or may call upon consultants to assist them in evaluating a risk which they are asked to cover and which would be excluded from their standard policy. However, this is not common practice.

3.1.5 NEW TECHNOLOGIES

As professional liability insurance alone is concerned, it must be understood that all work carried out by professionals is covered, whether or not it involves new technology.

As we will see further on, some projects such as bridges or dams may be excluded. In such cases, insurance is usually available provided the insured pays an additional premium. When negotiating this premium, technology may be taken into account, however, the insurer will tend to rely on the professional's past experience which he more or less trusts rather than the risks inherent in the technology. One thing is certain, this insurance is usually available.

3.1.6 THE INSURANCE MARKET, GENERAL IMPRESSION

Presently, the market is virtually closed. The degree of specialization expected from the insurer and his team (re-insurers, brokers, experts, investigators and lawyers), the considerable increase in the amount of indemnities awarded by the courts, the scope of liability assumed by producers, the costs of defence and the increasing number of law suits limit the advent of new insurers on the market and lead existing ones either to raise their premiums and restrict insurance coverage or purely and simply withdraw from the market.

3.1.7 COLLECTIVE INSURANCE FOR PROFESSIONAL OR OCCUPATIONAL GROUPINGS. GRADATION OF INSURANCE CONDITIONS

Professional corporations do not formally negotiate with insurers. Drawing up a standard form contract is left to other organizations like the Association of Consulting Engineers of Quebec. More favourable insurance conditions may possibly be negotiated through training courses or other similar arrangements.

3.2 During the Guarantee (Maintenance, Defects Liability) Period

3.2.0 EXISTENCE OF LIABILITY AND PROFESSIONAL INDEMNITY INSURANCE

See the comments made under 3. Insurance usually covers all of the insured's liability regardless of when the claim is made.

3.2.7 RIGHT OF RECOURSE AGAINST A PRODUCER

Irrelevant.

3.5 After the Guarantee (Maintenance, Defects Liability) Period

3.5.1 NORMAL DURATION OF COVER; CANCELLATION

Usually, coverage is on an annual basis.

Non-payment of the premium usually results in the cancellation of the contract subject to written notice of 15 days.

'Article 2567	**Article 2567**
Except for transport insurance, the insurer or the insured may resile the contract by written notice.	L'assureur ou l'assuré peut, sauf le cas de l'assurance de transport, résilier le contract moyennant un avis écrit.
The notice takes effect upon receipt if it proceeds from the insured, and fifteen days after receipt at the last known address if it proceeds from the insurer.	L'avis prend effet dès réception s'il émane de l'assuré et quinze jours après réception à la dernière adresse connue s'il émane de l'assureur.'

Insurance is usually applied to first-time claims made during the term of the policy even if the errors, omissions or negligence took place in the past.

3.5.2 LEVEL OF PREMIUMS, EXCESS AND FRANCHISE

No obstacles stand in the way of reducing the excess to zero, but it does not happen frequently. Usually, the insured bears at least the first $5000 of the claim but the insurer must all the same bear the costs of defence. Some producers have much higher excesses or franchises ($1 000 000 or more).

The amount of the premiums may vary enormously for any category of professional insured.

3.5.3 TIME LAPSES BETWEEN OBSERVATION OF DAMAGE AND INVESTIGATION

3.5.4 TIME ALLOWED BETWEEN THE INSURER'S RECOGNITION OF A CLAIMS EVENT AND HIS PAYING OUT

With regard to both the above, the period between the time the client discovered the damage and filed his claim for compensation is covered by provisions pertaining to the prescription period for recourses (see 1.2.3 and 1.5.3).

As soon as a claim is made, the insured must notify his insurer whose first step will be to initiate an investigation. There is no fixed period for this or for initiating an expert analysis or its conclusions. In the case of a large claim, the insurer usually needs at least six months before he is able to decide on the indemnity.

There is no fixed period for the insurer to make the payment. However, the client may demand interest at the current rate of 10 per cent per annum.

3.5.5 KINDS OF DEFECTS AND DAMAGE WHICH CONSTITUTE CLAIMS EVENTS; ONUS OF PROOF

See 1.5.5.

Under article 1688, the plaintiff is not required to identify the producer at fault. Under article 1522, he only has recourse against the vendor pursuant to the guarantee. Under articles 1065 and 1053, he must identify the party at fault and prove the said fault.

3.5.6 EXCLUSIONS

Under article 2563 of the Civil Code, the insured's intentional fault cannot be covered. Otherwise, civil liability may be insured (Civil Code, article 2600), even for negligence.

Insurers, by contract, generally exclude some activities except in return for payment of an additional premium:

- execution of work not usually carried out by the producer concerned;
- work on fairs or exhibitions;
- surveying, soil investigation;
- work related to tunnels, bridges or dams;
- joint ventures;
- turnkey projects;
- work related to nuclear energy;
- testing material.

By contract, they will also exclude:

- infringement on copyrights, patents or trademarks;
- insolvency or bankruptcy of the insured;
- failure to carry the work out on time, except as a result of an error or an omission which is covered;
- third party liability assumed by contract;
- cost estimates;
- libel, defamation.

Scotland

Alan Wightman

In consultation with Peter Anderson, George Burnet,
Robert Blair and Eric Purves

1 LIABILITY

1.1 General

1.1.0 LAW OR CONTRACT OR BOTH

It should be noted that it is not possible to describe the law and practice in the UK without recognizing that Scotland has its own national identity and distinctive institutions within the framework of the UK. It has its own legal system, its own educational system, building standards regulations and traditional practices which are significantly different from other parts of the UK.

The common law of Scotland in relation to obligations is mainly derived from Roman law and is based more on principles rather than precedents. In respect of actions for negligence, known in Scotland as 'delict', two cases in Scottish legal history illustrate the two fundamental 'tests':

(a) breach of the duty of the 'professional man of ordinary skill acting with ordinary care': the test of the standard of care (*Hunter v Hanley*);[1]
(b) the duty of care owed by the defender to the pursuer – the test of proximity (*Donoghue v Stevenson*).[2]

Liability for negligence can arise both in contract and in delict. Where there is a contractual relationship, the pursuer will usually raise an action in contract since this generally represents an easier route towards obtaining redress in the courts. Very often there is no real difference between the obligations of an architect in negligence and his obligations under his

[1] 1955 SLT 213.
[2] 1932 SC(HL) 31.

contract with the client, because criticism of him usually turns upon breach of an implied condition to use the duties of ordinary skill and care in design and/or inspection.

1.1.1 TERMS AND CONCEPTS

For RIAS definitions of terms, see Appendix to this chapter.

1.2 During the Guarantee (Maintenance, Defects Liability) Period

1.2.1 DETERMINATION OF THE COMMENCEMENT OF THE PERIOD

The Guarantee Period (usually referred to in the UK by the quaint term 'Defects Liability Period'), begins on the date when the architect issues a Certificate of Practical Completion or when the building owner accepts delivery of the building. If standard forms of contract are not used, or if an architect is not employed, a court would probably hold that the Guarantee Period begins when the building owner accepts delivery of the completed work, or takes occupation.

At the point of delivery it is said that the owner takes beneficial occupation or use, and thus acquires legal possession. The significance of delivery can be summarized as follows:

(a) The owner assumes legal responsibility for the security and insurance cover of the building or construction.
(b) The contractor having surrendered possession, he may enter the site or building only with the owner's permission, to carry out remedial, or other work.
(c) The 'limitation clock' can be started with certainty, i.e. the date can fix the point from which the period of liability is reckoned.

[1.2.2: *Not applicable*]

1.2.3 THE PERIOD'S DURATION AND TERMINATION

For most building contracts, the Guarantee Period is usually 12 months, but for small contracts, it is often six months.

The contractor remains liable for defective work which is not apparent at the time of the architect's inspection. The duration of such liability for latent damage is under review by the Scottish Law Commission; as it stands at present the law provides for two prescriptive periods:

(a) the short period is five years from the time when damage is discovered or ought, with reasonable diligence, to have been discovered;
(b) the long period is 20 years from the date of the damage and is intended to be an absolute 'cut-off'. There are, however, doubts as to how this rule should be interpreted.

Proposals by the Scottish Construction industry (professionals and contractors) have been submitted to the Scottish Law Commission for the reform of the law concerning latent damage. Our three main recommendations are as follows:

(a) The short period should be reduced to three years in the interests of early settlement of disputes.
(b) The long period should be reduced to ten years in line with the Consumer Protection Act and for the avoidance of stale claims.
(c) There should only be one start date for both parties to the contract, i.e. the date on which delivery or substantial completion occurs (the limitation clock then starts ticking with recognizable certainty).

1.2.4 TIME ALLOWED FOR DIFFERENT CATEGORIES OF PRODUCERS FOR MAKING GOOD, REPAIRING OR PAYING DAMAGES

The time allowed for 'making good' etc. depends on the type of producer who is alleged to have failed in his/her obligations to the client:

(a) Architect or engineer: The obligations and the time needed to fulfil them will be stated in the Conditions of Engagement.
(b) Statutory authority exercising a control function: The time allowed for discharging its duties is normally laid down by the relevant government department.
(c) A contractor: His obligations will be set out in his contract. It is a general rule that a contractor shall remedy defects expeditiously. Where a dispute arises regarding an alleged breach of contract the arbiter or court will decide on matters of time for acting upon any award.

If the client should refuse access to the contractor to carry out remedial work, then the courts would, depending on circumstances, probably regard such refusal as in effect frustrating the contract – in such a case, the contractor would be relieved of his obligation to do the remedial work.

1.2.5 KINDS OF DEFECTS AND DAMAGE COVERED BY THE GUARANTEE

The kind of defects and damage which the contractor is responsible for depends on the terms of his contract and, in particular, on the requirements of the drawings and specification.

On completion of the project at the end of the Guarantee Period, when the architect issues the Final Certificate, it is generally believed that he is not expressing a conclusive view as to quality of materials or standards of workmanship, except in respect of those matters where the specification states that certain items 'are to be to the reasonable satisfaction of the Architect'. In one of the more common forms of contract (the JCT Standard Form) there is now some doubt as to this interpretation of the architect's obligations in these respects. There appears to be a trend in certain cases for the courts to hold that in certifying completed work, quality and standards of workmanship generally are to be included in the architect's judgement. It is of critical importance that when issuing the Certificate of Practical Completion and the Final Certificate the architect must act impartially as between the parties to the contract.

In the case of contractual liability, the liability of the contractor for defects is written into the contract between the parties. The contractor's liability to 'make good' defects during the Guarantee Period covers any matter which is at variance with the drawings and/or specification and any failure to observe good practice in the execution and supervision of work. The decision as to what matters are at variance rests with the architect.

1.2.6 EXCULPATION OF PRODUCER

An architect or engineer will be liable for defective design work for which he is responsible and which is subsequently proved to have been done negligently, resulting in a loss. In order to prove negligence, it must be established that the actions of the defender are ones which no professional man of ordinary skill would have taken if he had been acting with ordinary care (*Hunter v Hanley*).[3] Thus, for example, if an architect can satisfy the court that he adopted a course of action which an ordinary competent architect (not a genius) would follow, then he will be found not to have been negligent.

An architect or engineer would probably not be liable for defects resulting in a loss in the following circumstances:

(a) fraud on the part of the contractor, e.g. deliberate concealment of important items of work;
(b) failure in respect of some experimental technique which was adopted with the client's consent;
(c) failure in respect of a product specified by the client which the architect or engineer had no reason to question;
(d) bad or inadequate maintenance by the client provided that he was made aware of special requirements;
(e) damnum fatale (see definition at the end of this chapter).

[3] 1955 SLT 213.

The extent to which each of these circumstances would provide an adequate defence cannot be stated with certainty. Much depends, for example in respect of (b), (c) and (d), on the state of knowledge of the client and the reliance placed on the architect's expertise.

An area of continuing difficulty exists with respect to inspection and supervision. Where defective work occurs without detection and subsequent damage occurs, there is dispute as to who is responsible. In the 'administration' of a building contract, the architect's duties include certification of interim payments to the contractor, issuing Certificates of Practical Completion or Final Completion, monitoring of progress etc. In order to fulfil these duties, the architect needs to make periodic inspections of work on site. The architect is not responsible for supervision – that remains the responsibility of the contractor. This being so, it must be said that the definition of the duty of inspection is far from clear. To quote Peter Anderson:[4]

'The courts tend to interpret widely the duty of periodic inspection and to say that architects are responsible for being on site at certain critical stages and may therefore be responsible if a problem arises later which can be said to have been capable of detection or observation at one of those critical stages. In other words, the inspection duty is so wide that it can sometimes get fairly close to "supervision".'

In recent years the courts have expanded the number of situations in which a duty of care can arise and these developments in the law are significant for the professional designer. It is established that the duty of care of a professional man to his client can arise not only in delict but also under a contract between them. However, such duty may also be owed to people other than the client based on the foreseeability of injury to a third party; this is known as 'the neighbourhood principle'.

LOCAL BUILDING AUTHORITY

A local authority which is given statutory powers to administer the Scottish Building Standards Regulations has certain duties to inspect work in progress and to issue a Completion Certificate when it is satisfied that requirements of the Regulations are fulfilled. Hitherto local authorities have not normally carried PI (Professional Indemnity) insurance cover: this situation is likely to change as new obligations are placed upon local authorities to act in more accountable ways.

A local authority in fulfilment of its statutory powers owes a duty of care in respect of matters concerning a threat to the health and safety of the occupants of the building. Note, that this duty of care extends only to the occupants or users of the building and arises in delict in respect of negligent

[4] Legal Adviser to RIAS Insurance Services and to the RIAS.

inspection or failure to inspect (*Anns v London Borough of Merton*[5]). Such duty does not extend to developers (*Peabody Donation Fund v Sir Lindsay Parkinson & Co*).[6] Both these cases quoted were decided in the English courts, but their judgements have been reviewed in subsequent cases (*Murphy v Brentwood District Council*[7]) where the principle of an imminent threat to health and safety had been clarified. The overriding principle which at least the Sottish courts will apply, is that a local authority will normally be considered liable if it breaches its duty of reasonable care in fulfilment of its statutory duties, i.e. duties laid down by an Act of Parliament. Liability will, however, only arise if the court considers that Parliament has imposed a duty which may give rise to civil liability rather than simply a duty to fulfil certain tasks of public interest and administration.

1.2.7 CONSEQUENCES OF PRODUCER'S NON-COMPLIANCE WITH UNEQUIVOCAL DUTY TO MAKE GOOD, REPAIR OR PAY DAMAGES

The consequences of any party failing to meet his obligation or to make good, repair, or pay damages, depend on the particular circumstances but, in general, the aggrieved party can raise an action in the courts to obtain redress. If the party whom it is alleged has failed in his obligations has gone bankrupt, there is nothing to be gained by proceeding with the action.

1.5 From expiry of the Guarantee (Maintenance, Defects Liability) Period

[1.5.1–1.5.5: *Not applicable*]

1.5.6 EXCULPATION OF PRODUCER

During and after the Guarantee Period, questions of exculpation of the producer depend on whether the duty of care or breach of contract arises out of a statutory duty, a delictual liability or a contractual obligation.

Statutory duty

A manufacturer, for example, is bound to supply or deliver a product which is reasonably fit for its intended purpose. This obligation arises under the Consumer Protection Act 1987 whereby the supplier of a product is under a

---

[5] [1978] AC 28, [1978] 5 BLR 1, HL.
[6] [1985] AC 210, [1985] 3 All ER 417, HL.
[7] [1990] 3 WLR 414, CA.

strict liability for defect which causes damage. Section 5 of the Act extended strict liability to property damage provided that the property was of a type ordinarily intended and mainly used as a private dwelling and provided the damage resulted in a loss of at least £275. A defender can succeed in his/her defence if it can be shown that:

(a) Damage did not result from the alleged defect.
(b) There had been unreasonable usage, e.g. by the claimant ignoring warnings or instructions on the product or by an unreasonable use which is not foreseeable.
(c) Damage was of such a trivial nature as not to justify legal action.

Duty of care

A designer, for example an architect, an engineer or a contractor under a 'design-and-build' contract, may be liable for a failure to exercise a professional duty of care under the common law of delict. A defender can succeed in his/her defence if the court can be satisfied that:

(a) Damage did not result from the alleged defect.
(b) There had been unreasonable usage, and/or the building or construction had suffered because of inadequate or improper care and maintenance.
(c) The design error can be directly related to the 'state of the art' when the error was made.
(d) An error of judgement was made in circumstances where a reasonable and proper amount of care and skill had been exercised, such as would be used by another ordinary architect of repute in similar circumstances.

It is generally believed that an architect is not under an obligation to supply the owner with detailed guidance concerning the care and maintenance of his/her building. The RIAS in the publication of its Practice Guide makes the following recommendation to its members:

'The client having been informed as to maintenance commitments arising from design proposals and budget limitations, the architect should compile the relevant information concerning care and maintenance in an Owner's manual. This should include manufacturers' instructions to the user, expected durability and requirements for regular inspection. This obligation is not only in the client's interest, but also in the architect's interest, especially where future defects can be traced to a lack of care on the part of the building owner.'

To ignore this advice is to run the risk that the defence of unreasonable usage and lack of careful maintenance (see 'Duty of care', para (b), above) will not be available to the defender.

Breach of contract

A contractor is normally responsible for defective work during the guarantee period and in the post-guarantee period for latent damage arising from a defect in the contract work. His defence depends on whether he can satisfy the court that:

(a) The defect is directly due to circumstances outwith his control.
(b) He complied with his contractual obligation and thus the defect was due to a design and/or specification error.
(c) The risk of an unpredictable event resulting in damage/loss was apparent to both parties to the contract and the contractor had done all he could to minimise the risk.

[1.5.7: *Not applicable*]

1.6　Transfer of ownership and producers' liability

1.6.1　RULES

A successive owner, who has no contractual relationship with the producer, would have to raise an action in delict for alleged defects resulting in damage. To be able to raise such an action, the owner would have to prove a duty of care existed which met the test of 'proximity' (Lord Atkin's dictum in *Donoghue v Stevenson*).[8] Even if such duty of care can be shown to exist, if the loss is 'economic loss', the courts will regard such loss as irrecoverable.

The growing practice in England, and now in Scotland, whereby developers require tenants to accept full responsibility for the care and maintenance of the building throughout the period of the lease, places very onerous obligations on the tenant or tenants. In the event of an alleged defect resulting in economic loss, tenant(s) have sought to provide a route for possible redress; this is made possible by establishing a contractual relationship, known as a Duty of Care Agreement, between the tenant(s) and the professional designers. Such agreements extend the risks carried by the architect and consulting engineers; in effect, they create rights to claim for losses which could not exist in delict.

In view of the power of developers to insist on such agreements, a standard form of agreement is now published which seeks to place a limit on the additional obligations imposed on the professionals.

The appearance of these Duty of Care Agreements is a strange phenomenon. They have been described as providing mythical comfort for the developer and his tenant(s). The potential value of such agreements depends on insurance backing. If, for example, the insurance industry find

---

[8] 1932 SC(HL) 31.

that a growing number of claims pursued under these agreements against design teams are successful, then they will simply withdraw the additional cover granted on renewal of the professional indemnity cover. Thus the illusory value of Duty of Care Agreements is such that they provide no real comfort to anyone, but much potential expense and fashery.

2 PROPERTY (OR MATERIAL DAMAGE) INSURANCE

2.1 General

2.1.0 LAW OR CONTRACT OR BOTH

Property insurance is not demanded by law, but is normally required by institutions who lend money under a mortgage policy.

The policy holder is the legal purchaser of the property.

[2.1.1–2.1.6: *Not applicable*]

2.2 During the Guarantee (Maintenance, Defects Liability) Period

BUILDING CONTRACTS GENERALLY

At the point in time when the Certificate of Practical Completion is issued and accordingly the owner takes possession of the building, the contractor is no longer responsible for the insurance of the property.

NHBC

The National House Building Council operate a protection scheme for new houses and flats which are built and sold by NHBC registered Members. It is intended that the scheme, known as 'Buildmark', will be extended at a future date to cover conversion and renewal work in the housing field. This type of work is, however, covered by a different version of the NHBC protection scheme which provides a six-year warranty. The 'Buildmark' scheme provides a ten-year warranty against damage which occurs because of faulty building. The basis of the scheme is essentially one of quality assurance in three respects:

(a) that the builder is considered to be competent before he is granted registration as a member of the scheme;
(b) that the NHBC published Technical Requirements are mandatory;
(c) that the operations on site are subject to unannounced spot-check inspections by NHBC inspectors.

The 'Buildmark' scheme provides the following protection:

(a) Up to £10 000, or 10 per cent of the agreed purchase price, whichever is greater, against builder insolvency or fraud where, after exchange of missives, a homebuyer cannot recover either monies paid to a Member or additional sums needed to complete the home.

The £10 000 or 10 per cent limit only applies before NHBC certify the home as complete. After that, i.e. during the first two years of the ten-year period, the limit of cover in the insolvency situation is the same as three to ten-year cover.

(b) The builder is required to remedy, at his own expense, any defective work or damage which arises as a result of his not conforming to NHBC Standards for design, materials or workmanship current at the time of construction. This liability applies for a period of two years from the date on which the Council certifies the dwelling as being satisfactorily complete. If the builder becomes insolvent NHBC take over this liability.

(c) If the homebuyer and builder are in dispute the NHBC can conciliate and issue recommendations. If the builder is required to carry out work and fails to do so, the NHBC will do it on the builder's behalf.

(d) From the beginning of the third year to the end of the tenth year, there is cover against major damage arising from structural defects resulting from a failure to comply with NHBC's Standards. The cover includes the drainage system, external harling/render and other forms of cladding.

(e) The limit of cover for an individual house or flat is the original purchase price increased by 12 per cent per annum compound.

(f) The benefits of the scheme automatically transfer from one owner to another throughout the ten-year period.

The 'Buildmark' scheme can thus be seen as providing an essential link between insurance cover and technical performance. Theoretically the Council reserves the right of recourse against a registered builder or professional consultant but in practice the need for such action has never occurred in Scotland because of the small number of serious defects. The claims made to date have not been the consequence of palpable negligence.

ALL RISKS POLICIES

There is a variety of 'All Risks' policies which an owner can obtain, covering such risks as fire, storm, frost, settlement, damage etc. These policies are normally renewable annually.

[2.5: *Not applicable*]

2.6 Transfer of ownership and insurance cover

All Risks insurance cover follows the owner. On disposal of the property, the policy is cancelled. The successive owner takes out whatever type of cover he wishes, or mortgage demands.

3 LIABILITY AND PROFESSIONAL INDEMNITY INSURANCE

3.1 General

3.1.0 LAW OR CONTRACT OR BOTH

PI insurance cover is not required by law, but many clients in the public sector require such cover as a condition of appointment of the architect and other consultants. PI cover is normally available for all those who exercise a design responsibility, including contractors. The intention of such a policy is to protect the insured party in respect of claims for damages arising out of a breach of a duty of care.

See also 3.1.6.

3.1.1 TERMS AND CONCEPTS

Three features of PI insurance policies should be noted:

(a) They are 'legal liability policies', which means that they indemnify the insured person if he/she is legally liable to pay damages. These policies also cover the cost of defending an action by a pursuer alleging negligence.
(b) Since most claims are settled out of court either before or after litigation is actually commenced, the larger proportion of sums paid out by underwriters is to deal with these circumstances rather than making payment of damages awarded by a court.
(c) In the rare event of a dispute arising between the insured and the insurance company as to the defence of a threatened action, most policies provide for the matter to be referred to Queen's Counsel who then act as arbiter. (This is known as the QC clause.)

The traditional PI insurance policy covers an architect or engineer for their liability for damages arising from a breach of professional duty of care in the conduct of their business. Such duty is governed by the common law of delict. As previously noted, if an architect or engineer signs a Duty of Care Agreement with a developer, this would in effect extend his duty of care from one of delict, to one of contract and would extend the risks carried by the architect or engineer.

Traditional PI policies are underwritten by insurance companies and also by Lloyds Underwriters. A policy needs to be renewed on a year-by-year basis, which means that there can be no certainty that cover will be available in future years at a cost which is predictable and affordable.

A PI policy differs from other types of legal policy in one very important way. The policy is known as a 'claims made policy'. This means that the policy which is in force when the claim is made, and not the one existing when the fault occurred, is the one which is applicable. In other words, the terms of the policy which provide the basis for settling a claim are those in force when:

(a) a claim is first made; or
(b) the circumstances which may give rise to a claim, are first notified.

[3.1.2–3.1.5: *Not applicable*]

3.1.6   THE INSURANCE MARKET, GENERAL IMPRESSION

There is general support for the view that compulsory PI insurance is necessary, provided that it is adopted on a uniform basis by all who have a design function, whether on the professional or the contracting side of the construction industry. It has proved difficult to instigate a joint agreement with the other design professionals on this matter. Thus the RIAS has decided to adopt a policy of compulsory PI insurance for its members in order to protect a client's interests in the event of a successful claim for negligence. Members are, however, advised to ensure that, when negotiating terms of engagement, the client agrees that other consultants whom he/she proposes to engage on the project will be required to have similar PI insurance cover.

3.1.7   COLLECTIVE INSURANCE FOR PROFESSIONAL OR OCCUPATIONAL GROUPINGS. GRADATION OF INSURANCE CONDITIONS

WREN INSURANCE

An alternative type of policy has been devised by about 40 large architectural practices in the UK who have set up a company known as Wren Insurance. The twin objectives of the Wren scheme are to provide reasonable stability to facilitate sensible business planning and reasonable certainty in the continuity of future cover. The scheme is a non-profit-making arrangement whereby the members cover their potential liabilities by subscribing to a 'mutual' fund. Thus the Wren Scheme promotes a long-term view among its members whose mutual interest is to build up adequate funds and to benefit from each other's experience. The members appoint a committee of directors which has the responsibility of managing the funds and handling claims. The directors also have the duty to devise risk managing procedures to be adopted by member firms; the objective here is to ensure that all members are operating at about the same level of competence in the control of their professional responsibilities.

RIAS INSURANCE SERVICES

RIAS Insurance Services is a joint venture between the RIAS and Bowring UK Ltd intended to provide architects with an insurance service which is linked to the Practice Information Service of the RIAS. The objective is to create a loss prevention service and to provide a balance between commercial interests of the insurance company and the need for the RIAS to ensure that standards of fairness and competence are maintained. This characteristic of the scheme, namely the mutual interest of the company and RIAS members, has provided a close link between the claims experience of the company and the effectiveness of the RIAS Practice Information Service.

This mutual interest is reflected in the composition of the Board of RIAS Insurance Services. The Board consists of 50 per cent of RIAS members and 50 per cent of representatives of the insurance brokers, Bowrings, with an RIAS member as chairman.

The standard policy indemnifies the insured against any claim or claims made against them by reason of a breach of professional duty. In fixing the premium, each risk is assessed separately, according to various factors concerning the particular firm of architects and the nature and location of its work. When the annual revision of rating is undertaken only the claims experience of members of the RIAS is taken into account. In a review covering five years of claims experience, as at November 1989, by RIAS Insurance Services, claims have been analysed under various headings: Who are the claimants? At what stage of the project? Circumstances of the claims? Class of buildings in which claims have arisen?

These statistics reveal significant information concerning the nature of claims over the five-year period as follows:

- 85 per cent of circumstances notified or claims made arise from events in the pre-contract and contract stages, or in the period up to five years from building completion. This proportion has remained steady over the five-year period.
- 54 per cent of claims relate to residential buildings, i.e. houses, cottages, flats and tenements. This proportion has also remained fairly steady over the five-year period.
- In 1987/88, 84 per cent of circumstances notified, or claims made, relate

to design and specification work, i.e. pre-contract work. This compares with a 30 per cent average over the five-year period. In the same year circumstances/claims in respect of costs, over-runs, maladministration of contract and administrative failures in respect of the usual approval procedures, accounted for 46 per cent. This compares with a 30 per cent average over the five-year period.

[**3.2**: *Not applicable*]

3.5   After the Guarantee (Maintenance, Defects Liability) Period

3.5.1   NORMAL DURATION OF COVER; CANCELLATION

If an architectural practice fails to renew its PI policy, there is no cover in force for any new claim/circumstance which may arise although there may have been a policy in force when the work which gave rise to the subsequent problem was executed.

Failure to renew a PI policy can arise by reason of insurance cover being no longer available at an affordable cost, or the claims experience of the practice is so bad that insurance is unobtainable.

[3.5.2–3.5.6: *Not applicable*]

APPENDIX I   ROYAL INCORPORATION OF ARCHITECTS IN SCOTLAND: CONSTRUCTION LAW TERMS AND DEFINITIONS

GENERAL

Client
The individual or corporate body who engages the Architect or other Designer, and who specifies the requirements for the project.

Brief
A co-ordinated statement of requirements for a project issued by the Client which describes the functional requirements and the priorities in respect of speed of building, performance, quality and cost control.

Designer
One who carries responsibility for some specified aspects of the design process.

Producer
A collective term to describe those who contribute to the realization of a completed project, whether as Architect, Engineer, specialist Designer, Contractor, sub-contractor or supplier and including those exercising a control function, whether employed by a Statutory Authority (e.g. Health & Safety Executive) or by local government (e.g. a building inspector). Where a Client makes stipulations regarding design or choice of materials or components, or carries out inspection procedures, he is then included in the concept of 'Producer' to the extent of his actual intervention.

Consumer
A person who purchases goods or professional services not in the course of business.

Statutory Authority
A public or private body with statutory powers in respect of utility services, or of control of pollution, safety, public health or environmental standards.

Quality Assurance
Predetermined and systematic planning which aims to ensure that a product or service will satisfy stated criteria of quality in ways which can be continuously audited.

LEGAL

Pursuer
The person suing (bringing/instigating) an action in the courts.

Probative Contract
A probative contract is a document which, because of certain features which give confidence, affords prima facie proof of its own contents. This means that it can be produced in court without further evidence of authenticity. The principal classes are contracts signed before two witnesses, who also subscribe their signatures. Where the signatory has written the words 'adopted as holograph' above his signature in his own hand, the effect of this is as if the whole document was written in the signatory's own hand.

Non-Probative Contract
Is any other contract which can be oral or in writing but not signed before witnesses or adopted as holograph. It cannot therefore be produced as

evidence in court without further proof unless its authenticity is conceded by the other party.

Prescription
Prescription is the passing of a period of time which confers positive rights (e.g. property rights) or which, regarded from another viewpoint, cuts off rights, i.e. negative prescription. In building contracts the most common form is negative prescription by which a contractual or delictual obligation, and its correlative rights, are extinguished. Where such an obligation has been in existence for a period of five years without interruption by any relevant claim or acknowledgement, the correlative right to make a claim is extinguished – this is known as the period of the short negative prescription.

When a contract is in probative form the prescriptive period is 20 years – the long negative prescription. The long negative prescription is important also in respect to latent defects. The short negative prescription runs from the date when the damage was discoverable. The long negative prescriptive period runs from the date when the damage occurred.

Relevant Claim
A claim for loss and/or expense directly related to a duty or obligation, presented within the relevant prescriptive period. The effect is to terminate the running of the prescriptive period and to 'wind the clock back to the beginning'. If the claim is subsequently abandoned, a new prescriptive period would then begin.

Limitation
The effect of lapse of time to make the pursuer's rights unenforceable by court action. Limitation is procedural whereas prescription is substantive. It only applies to actions concerning personal injury or death.

Delict
The original meaning of this word is 'a wrong' but it is normally applied to mean a deliberate or negligent act by which a person's rights are infringed which gives rise to an obligation to make reparation (i.e. to make good, so far as possible in terms of money, loss caused by a legal wrong).

Duty of Care
The obligation to avoid acts which are reasonably foreseeable as causing harm to persons closely affected by such acts. (Lord Atkin's dictum in *Donoghue v Stevenson* 1932 SC(HL) 31.)

Duty of Care Agreement
A written agreement entered into by an Architect, Engineer or other Designer in which he/she accepts a duty of care towards a successive Owner

or Tenant of the property in respect of the performance of his/her services to the Client. The resulting contractual relationship enables a successive Owner or Tenant to seek redress for an alleged failure in the exercise of a duty of care directly with the Architect or Engineer responsible.

Negligence

This covers the situation where a person's rights are infringed by failure to exercise the degree of care required by law in the particular circumstances, which also gives rise to an obligation to make reparation.

Professional Negligence

A failure to exercise a duty of care as a result of which harm has been caused to the pursuer. The test of negligence in respect of a professional person 'is the standard of the ordinary skilled man exercising and professing to have that special skill. A man need not possess the highest expert skill – it is sufficient if he exercises the ordinary skill of an ordinary competent man exercising that particular art' (Mr Justice McNair). In other words, the test is to match an individual, e.g. an architect, with an 'ordinary competent architect' and not against a genius.

Strict Liability

An imposition of liability on the defender for his actions regardless of fault, or of intention, or of negligence (see *Lord Advocate v Reo Stakis* 1982 SLT 140). The most common examples of this are the liability of an employer for his employee's wrongs, and secondly, producers of goods who are strictly liable for injuries caused by defects in their products under the Consumer Protection Act 1987.

Risk

Exposure to possible injury, damage, or economic loss/gain arising from involvement in a project or enterprise.

Error of Judgement

A mistake in circumstances where a reasonable and proper degree of care and skill has been exercised such as would be used by another ordinary Architect of repute in similar circumstances.

Damage (Physical)

A failure to meet the standard of performance and/or correctness of construction, which makes the completed work unfit for the purpose described or intended, and thus liable to repair or replacement.

Damages (Monetary)

Damages are a monetary recompense to put an injured party who has suffered a loss in a position he would have been, had there not been an injury resulting in such loss, so far as money can so do.

Damnum Fatale

Also called an 'act of God'. Some accident which has resulted directly and solely from natural causes without human intervention and which could not have been prevented by any amount of foresight and care (e.g. an earthquake).

Patent Defect

An item of work or component supplied which is readily identified as not complying with the requirements of the contract as described in the drawings or specification or reasonably inferred therefrom.

Latent Defect

An item of incorrect work which does not become readily apparent until some time after the defective work was carried out or a Final Certificate has been issued when it may manifest itself in some kind of damage.

Minor Defects

Small items of incomplete or unsatisfactory work which are relatively easily remedied and do not have a significant effect on the beneficial occupation of the building or works.

Date of Accrual

The moment in time when a contractual or delictual obligation becomes enforceable and a prescriptive period begins. Generally in claims for compensation the time begins for both short and long prescriptive periods when damage occurs. However, in respect of the short negative prescriptive period as it relates to latent defects, this does not begin until the defect is discovered or ought with reasonable diligence to have been discovered. The date of accrual is sometimes known as the 'starting point' from which time begins to run for either prescriptive period.

PRACTICE

Inspection

Periodic assessment of the progress and quality of the Works to determine that they are being executed generally in accordance with the contract documents.

Detailed Inspection

The day to day examination by the Site Architect of the Contractor's progress and quality of the Works to verify conformity with the Specification and to check that the Contractor is maintaining proper control over all his operations.

Supervision
The detailed planning and control by the Contractor of all operations in the discharge of his contractual obligations to manage the project correctly and on time.

Warranty
A guarantee that work has been done in accordance with the contract or that a product complies with a stated standard and was manufactured from stated materials.

Building Warrant
The formal approval of the Local Authority to plans and specifications submitted as required by the Scottish Building Regulations.

Practical Completion
Completion for the practical purposes of allowing the building Owner to take possession of the building or works and to use them as intended irrespective of minor defects and unimportant items of incomplete work within the contract. The test of 'practical' is whether the Owner can have beneficial occupation at the time when the Certificate of Practical Completion is issued, or the Owner accepts that beneficial occupation is possible.

Substantial Completion
When a construction, or a separately defined section of it, is complete, or is ready to be handed over to the Owner, despite minor defects or unimportant items of incomplete work within the contract.

Delivery
The formal handing over on practical or substantial completion of the building or construction or professional service or a separately defined part thereof by the Producer to the Owner or person who commissioned the work.

Post-Construction Guarantee Period
The period following Delivery (Defects Liability Period) in which the Contractor remains under an obligation to make good defects in the completed Works.

Certificate of Making Good Defects
A certificate issued by the Architect or Consulting Engineer that the Contractor has made good defects in workmanship, materials and components in satisfaction of his contractual obligations.

Final Certificate
The certificate signed by the architect which provides conclusive evidence

that where the quality of materials or standard of workmanship are stated in the bills or specification to be to the reasonable satisfaction of the architect, they are to that standard; and that the provisions of the contract requiring adjustment of the contract sum have been complied with, apart from any arithmetical errors.

Certificate of Completion
A certificate issued by the Local Authority when it is satisfied that all aspects of the Scottish Building Regulations, and any particular conditions attached to the Building Warrant, have been complied with.

Maintenance
The routine inspection and care during the lifetime of a building by the Owner or, alternatively, by a Tenant who has taken responsibility for the maintenance of the property under a full-repairing lease for the duration of the lease.

Singapore

Liability
Anthony Lavers

Insurance
George Thomas

1 LIABILITY

1.1 General

The legal system in Singapore is a common law system and its derivation from the British model is readily apparent. There are four ways (relevant to this paper) in which the Singaporean and UK systems are connected:

(a) The Judicial Committee of the Privy Council is still Singapore's ultimate appellate court, although neighbouring Malaysia has now severed this link.

(b) Singaporean courts draw heavily in many areas of law upon British (and to a lesser extent other Commonwealth) authorities for the purposes of precedent. This is partly because of (a), partly a reflection of the pre-independence traditions which have survived and the British training of a high proportion of the judiciary and the legal profession and partly because of (c).

(c) By section 5(1) of Singapore's Civil Law Act, English law (specifically) is continuously received insofar as it affects a whole range of specified matters which can be subsumed under the general heading of commercial or mercantile law. This is particularly of relevance to this chapter because this includes the reception of the law relating both to contract and to insurance. This reception applies both to case law and to statutes. The qualification should be noted, however, that the reception does not occur when there is existing Singaporean provision on the point in question.

(d) Many of Singapore's statutes are based to a greater or lesser extent on UK statutes. This has important consequences in such areas as limitation of actions. Where there is close similarity, the relevant British case law may be used for purposes of interpretation.

However, this does not mean that the law relating to post-construction liability is identical with that of the UK and certain critical differences will be noted in this chapter.

1.1.0 LAW OR CONTRACT OR BOTH

Producers' post-construction liability is prescribed by law and is also provided for to some extent in contracts.

BY LAW

Under the Housing Developers Rules 1985 (made under the Housing Developers Control and Licensing Act[1]) and under the Sale of Commercial Properties Rules 1985 (made under the Sale of Commercial Properties Act 1979) a standard pro forma agreement must be used for the sale of units within building projects. The pro forma agreements include the following basic provision (with certain variations according to the type of property):

'Any defect, shrinkage or other fault in the building which shall become apparent within a period of twelve months after the date of delivery of vacant possession to the Purchaser or after the date of issue of the ... Certificate of Fitness [granted by the building control authority – the BCD: author's note], for the occupation of the building (whichever shall be the earlier) and which shall be due to defective workmanship or materials or to the building not having been constructed in accordance with the said specifications and plans (amended or unamended as the case may be) shall be made good by the Vendor at its own cost and expense within one (1) month of its having received written notice thereof from the Purchaser and if the said defects, shrinkage or other faults in the building have not been made good by the Vendor, the Purchaser shall be entitled to recover from the Vendor the costs of making good the same and the Purchaser may deduct such costs from any sum which has been held by the Purchaser's solicitors as stakeholder for the Vendor.'[2]

The rules of tort relating to negligence on the part of contractors, architects, engineers and other consultants are generally the same in

---

[1] (Revision) Cap. 250 of 1970.
[2] Example taken is from clause 18 of Form D in the Schedule to the Housing Developers Rules: see rule 12(1).

Singapore as in Britain, although it should be noted that this position is not invariable. Because tort is not part of mercantile law there is no obligation to receive English law. Statutes such as the Defective Premises Act 1972 and the Occupiers Liability Acts 1957 and 1984, for example, are of no application in Singapore. In *Mathew Bohman v Jurong Town Corporation*[3] Singapore's Court of Appeal declined to hold a landlord liable in tort for alleged defects in industrial premises – the point being that modern English law on the issue was ignored.

The first point to note is the statutory identification of producers other than builders and developers, by reason of responsibility. The producers identified by the Building Control Act 1989 who could incur liability for failure to discharge their statutory responsibilities under that Act are: the 'qualified person', who must be either registered as an architect under the Architects Act or as a professional engineer under the Professional Engineers Act, who must prepare and submit plans for approval; the registered accredited checker (a new concept under the 1989 Act) who provides an independent check on the 'key structural elements' of the building plan; the 'qualified person' who may (but need not) be the same qualified person as the designer who supervises the building works; and the site supervisor who must give immediate supervision to the structural elements of certain classes of building. Inspecting engineers are not strictly producers but potentially liable in respect of their duty in carrying out their periodical inspections being every five years for commercial buildings and every ten years for multi-storey residential buildings.

The Architects Act 1991 was introduced on 2 August 1991. It replaces the previous Act but still provides for mandatory registration of architects and for a Code of Professional Conduct and Ethics (by regulations). The main new departure in the 1991 Act is that corporations may be licensed by the Board of Architects to supply architectural services, subject to requirements set out in detail in Part VI of the Act which ensure that they will be under the control of registered architects and will be insured against professional liability. The attitude of the law to practice by unregistered consultants can be seen from *Raymond Banham v Consolidated Hotels*[4] where the (non-resident) plaintiff engineers failed to recover unpaid fees. The contract for professional services was illegal because the engineer was unregistered. Similarly, in *BEP Akitek v Pontiac Land*,[5] architects failed in an action because, as a corporation, they could not supply architectural services. Note that this case was decided before the new Architects Act came into force, which permits the licensing of corporations.

The second point to note is that the building control authority (the BCD) cannot be regarded as a producer for purposes of legal liability. Although

---

[3] [1981] MLJ 134.
[4] (1976) I MLJ 5.
[5] (1991), unreported.

plans, drawings, specifications and other documents are submitted to the authority (under the Building Control Act 1989) and inspections are made during the construction process and for the purposes of granting Temporary Occupation Licences (TOL) and Certificates of Fitness (for occupation) (COF), no legal liability attaches to the building authority. A typical provision is that:[6]

> 'No liability shall lie against the Government or any public officer by reason of the fact that any building works are carried out in accordance with the provisions of this Act or that such building works or plans of the building works are subject to inspection or approval by the Building Authority or the public officer, nor shall anything in this Act make it obligatory for the Building Authority to inspect any building or building works or the site of any proposed building to ascertain whether the provisions of this Act are complied with or whether any plans, certificates, notices or other documents submitted to him are accurate.
>
> No matter or thing done by the Building Authority or by any public officer shall if it were done bona fide for the purpose of carrying out the provisions of this Act subject him or such person personally to any action, liability, claim or demand whatsoever.'

It is worthy of note as an indicator of where the main burden of responsibility is borne under the law of Singapore that in the professional Codes of Conduct of both the architecture and engineering professions, it is mandatory to exercise due diligence to ensure that there is no contravention of or failure to comply with 'any written law' (i.e. the Building Regulations) by any person and to report any such contravention or failure (i.e. to the building authority (see the Professional Engineers (Disciplinary Committee and Code of Professional Conduct and Ethics) Rules 1972 and 1981 amendment, which inserted the above provision, and also the Architects Rules 1982).

UNDER CONTRACT

In general, the liability of the producer under the contract is similar to that in Britain. In a contract for professional services the usual implied term of reasonable care and skill applies, and the English Supply of Goods and Services Act, section 13 is imported to Singapore by the Civil Law Act, section 5(1). Thus in a contract for the supply of a service where the supplier is acting in the course of a business, there is an implied term that the supplier will carry out the service with reasonable care and skill (section 18 states that 'business' includes profession, so architects, engineers and quantity surveyors are clearly covered by this provision).

The standard form of building contract itself was originally based upon

---

[6] Building Control Act 1989, Part VI.

an English model. The Singapore Institute of Architects Form of Contract 1970 edition (SIA 70) is based upon the old RIBA/JCT form. Clauses 15(2) and 15(3) provide for making good of defects appearing within the Defects Liability Period at the contractor's own expense. SIA 70 made the Defects Liability Period six months usually if no other period was stated. This would now be stated as 12 months (SIA 70 is still used).

Under clause 30(7) of SIA 70, the contractor is to some extent relieved of liability for defects when the Final Certificate is issued, because as between the contracting parties it represents 'conclusive evidence . . . that the Works have been properly carried out and completed in accordance with the terms of . . . Contract' subject to exceptions for:

(a) fraud, dishonesty or fraudulent concealment relating to the works;
(b) defects which reasonable inspection or examination at any reasonable time during the carrying out of the works or before the issue of the said certificate would not have disclosed;
(c) computational errors.

Clause 30(7) does not, of course, exclude general rights of action in law against the contractor, even if the case does not fall within one of the exceptions. In any event, the doctrine of privity of contract means that any right of a third party in tort against the contractor is unaffected by the certification provision in the contract.

The new form of contract is the 1987 edition of the SIA Form (SIA 87).

SIA 87 represents a material change in contractual provision in that the main contractor is made responsible for any defects in design and workmanship by sub-contractors, as well as by himself, although it can be argued that the same result can be achieved by implication of terms.

1.1.1 TERMS AND CONCEPTS

'BUILDING'

There are statutory definitions, notably the Building Control Act 1989, section 2:

'"building" means any permanent or temporary building and includes any structure or erection of whatever kind or nature (whether permanent or temporary) and in particular

(a) a wall (including a retaining wall), partition, gate, fence, paling, platform, staging, post, pillar, shoring, hoarding or frame;
(b) a slip, dock, wharf, pier, jetty, landing stage or landing bridge;
(c) a culvert, crossing, bridge, underpass or tunnel; or
(d) a sewage treatment plant, sewer, drain, swimming pool or tank for the storage of any solid, liquid or gaseous matter.'

This is not, however, a definition of general application and could not be applied exclusively.

The only distinction to be observed between the construction of a new building and rebuilding (i.e. partial reconstruction) is that the former would be covered by the Housing Developers Rules 1985 (see 1.1.0) as a unit in a housing project, whereas the latter would normally not be, and the requirement of the statutory Defects Liability Period in the pro forma contract would thus not apply in the case of rebuilding.

'PRODUCERS'

Developer

The developer may be the same party as the contractor but can be a separate 'producer' as a licensed developer undertaking the project (see 1.1.0).

Contractor

Especially under the new standard form SIA 87, the contractor is often the most likely to be liable after the taking over of the building. His liability in contract is not dependent upon 'fault', and he will bear (at least initially) liability for defects attributable to the work of sub-contractors and suppliers, for design, workmanship and materials.

He may be liable both in tort and (to the client) for breach of contract.

Professionals

Architects and engineers in particular (but also other consultants), bear primary legal responsibility in respect of design failure or failure due to inadequate supervision under the building control legislation (see 1.1.0).

Sub-contractors and suppliers

They will normally owe an indemnity to the main contractor in respect of their breaches of contract which occasion him loss but, under certain circumstances, may be sued in tort under *Junior Books Ltd v Veitchi Co*,[7] presuming that the case would be followed in Singapore. Since *D and F Estates v Church of England Commissioners*[8] this appears a less likely possibility, although *Junior Books* is apparently still good law on its own facts.

A supplier normally supplies materials as a supplier of the contractor, whose remedy would be under the contract of supply, since the contractor

---

[7] [1982] 2 All ER 201.
[8] [1988] 2 All ER 992, HL.

would be responsible to the client. A supplier could be sued in tort under ordinary negligence principles.

Building authority

Despite their functions of approval of plans and inspection, they cannot be under any legal liability (see building control legislation in 1.1.0). The authority may be a 'producer' within the definition in that it exercises a function within the process of construction, but it is a function to which legal liability does not attach.

SUMMARY˙

To summarise: it is not exclusively those parties with whom the client has a contractual relationship who are 'producers': sub-contractors, sub-consultants and suppliers may all bear legal responsibility, at least to the client and potentially to other persons in tort.

The building control authority may be regarded as a 'producer', but not if that term connotes legal responsibility for the role played.

'TENANT'

As in English law, a tenant can in theory sue producers in tort if sufficient proximity can be shown to establish a duty of care. The law of negligence in Singapore is basically similar. However, note the departure in this general area in *Mathew Bohman v Jurong Town Corporation* (see 1.1.0).

1.2 During the Guarantee (Maintenance, Defects Liability) Period

1.2.1 DETERMINATION OF THE COMMENCEMENT OF THE PERIOD

The Defects Liability Period (DLP) commences with the date specified in the Certificate of Practical Completion of the Works (see Appendix to SIA 70) and with issue of the Completion Certificate under SIA 87 (where it is called the Maintenance Period). As under the English system the existence of the DLP does not prevent rights of action in tort by strangers to the contract to whom a duty of care in negligence is owed, nor does it preclude general legal rights of recovery by the client (subject to the explanation in 1.1.0).

1.2.2 SIGNATORIES TO AND NAME OF THE DOCUMENT STATING THE COMMENCEMENT OF THE PERIOD

The Certificate of Practical Completion would normally be signed by the architect or by an engineer. The project must be supervised (at least nominally) by a registered architect, or professional engineer, or another person designated as a 'qualified person'.

1.2.3 THE PERIOD'S DURATION AND TERMINATION

Fixed by contract as between contractor and client, but between vendor/developer and purchaser (i.e. 'consumer') legislation now requires the use of a pro forma specifying one year duration (formerly six months). The Housing Developers Rules 1985 and Sale of Commercial Properties Rules 1985 (as amended) cover the sale of new property and the effect is that the vendor/developer has to make good 'Any defect, shrinkage or other fault in the building' due to defective workmanship or materials or non-compliance with plans and specifications, which is observed within 12 months after the date of delivery of vacant possession to the purchaser.

Consequently, vendors/developers normally include a 12-month period in the construction agreement with the contractor so that they can off-load their liability to the purchaser onto the contractor. The period could, however, be increased by contractual provision. A certificate of execution of the making good of defects would be issued when it was completed.

1.2.4 TIME ALLOWED FOR DIFFERENT CATEGORIES OF PRODUCERS FOR MAKING GOOD, REPAIRING OR PAYING DAMAGES

The contractor must, under SIA 70, remedy 'within a reasonable time' after receipt of the Schedule which shall be delivered not later than 14 days after the expiry of the DLP. Under SIA 87 the contractor must 'forthwith' repair and make good upon the receipt of such a Schedule. Note that under SIA 87 the architect can direct that a defect be not remedied but that there is instead a reduction in the contract sum (clause 27(4)).

1.2.5 KINDS OF DEFECTS AND DAMAGE COVERED BY THE GUARANTEE

SIA 70, clause 15(1): 'Any defects, shrinkages or other faults . . . which are due to materials or workmanship not in accordance with this contract.'

SIA 87, clause 27(1): 'any defects, omissions or other faults . . . if the cause of the same is due or found to be due to any breach by the Contractor of any of his obligations express or implied under the Contract or those of any sub-contractor or supplier, direct or indirect, and whether Designated, Nominated or otherwise.'

Otherwise, the kinds of damage which are compensatable in damages are dependent on the normal rules of general law, i.e. must flow naturally from the breach. If the claim were in tort, the rules of foreseeability/remoteness apply. The position is basically the same as in English law. No information is available on defects not leading to damage qualifying for damages.

1.2.6 EXCULPATION OF PRODUCER

CONTRACT

If the contractor complies with the requirements of the architect as to remedying of defects occurring within the DLP (or if there are none), he is partially exempted from liability for defects subsequently appearing, in that he can rely upon the architect's certificate of satisfactory execution of work in any proceedings between himself and the client. Even this partial exemption is, however, subject to the important exceptions (set out in 1.1.0) contained in clause 30(7) of SIA 70, i.e. latent defects and fraudulent concealments, liability for which continues after the expiry of the DLP.

In any event, the expiry of the DLP does not remove a client's general rights against a contractor and, being contractual, it has no effect in exculpating the contractor vis-à-vis third parties who are not subject to it.

SUPPLY OF SERVICES AND EXCLUSION CLAUSES

Under section 13 of the Supply of Goods and Services Act a contract for the supply of services in the course of a business (which by section 18 includes a profession) contains an implied term that the supplier will carry out the service with reasonable care and skill. This can be varied by express agreement under section 16, but that section, as will be the case generally, is subject to the controls imposed by the Unfair Contract Terms Act 1977, the applicability of which is the subject of some jurisprudential dispute in Singapore. In this author's view it probably applies.

LIMITATION

The law relating to limitation may limit the duration of the liability of producers: see 1.5.3.

SUMMARY

The general position regarding the liability of producers is the same as in English law. Professional producers such as architects and engineers will be judged by the standard of competence and skill current in their profession. Lai Kew Chai J, a member of the High Court of Singapore, in the 5th Sreenivasan Oration in 1985 reaffirmed the belief of the judiciary in the *'Bolam'* principle,[9] i.e. the ordinary competent practitioner standard. There is then a duty of care, and not of result. The contractual duty under a construction contract, however, can be viewed as a duty of result.

Thus a professional who follows accepted practices is exculpated, irrespective of result. A contractor who fails to produce what he contracted for is not exculpated, even if his failure was reasonable.

1.2.7 CONSEQUENCES OF PRODUCER'S NON-COMPLIANCE WITH UNEQUIVOCAL DUTY TO MAKE GOOD, REPAIR OR PAY DAMAGES

The position is similar to English law. If a producer refuses to comply, he can be sued. If he is unavailable, for whatever reason, only the existence of insurance gives a remedy. Other producers may be sued if a particular producer is unavailable/unable to pay, e.g. if the contractor is in liquidation, the architect might be sued for negligent supervision of the contractor's work.

1.5 After the Guarantee (Maintenance, Defects Liability) Period

1.5.1 DETERMINATION OF THE COMMENCEMENT OF THE (POST-GUARANTEE) PERIOD DURING WHICH A PRODUCER MAY BE HELD LIABLE

CONTRACT

Date on which breach of contract occurs

Since this may be difficult to determine with accuracy in the course of a construction project, the date of completion of the building is normally used as the starting-point. This is appropriate because breaches may be and are remedied by the contractor up to that point. It is delivery of the defective building which in effect constitutes the breach.

---

[9] *Bolam v Hospital Management Committee* [1957] 1 WLR 582.

TORT

Pirelli General v Oscar Faber and Partners[10] established that the starting-point in tort is when the cause of action accrues which is when the damage occurs, and not, as was previously thought at various times, when the damage is discovered, or discoverable, or when the duty of care is breached (i.e. the same time as contract). This has been adopted in Singapore in *United Indo-Singapura v Foo See Juan*,[11] although the correctness of that decision is doubted by some.

Note, that if the defect is 'fraudulently concealed', which has been construed widely in such construction cases as *Archer v Moss*,[12] the starting-point of liability is when the damage was discovered or discoverable, whichever is the earlier.[13]

1.5.2 NAME OF THE DOCUMENT STATING THE COMMENCEMENT OF THE POST-GUARANTEE PERIOD; ITS SIGNATORIES

There is no such document in Singapore.

1.5.3 PERIOD(S) OF LIMITATION

Singapore's Limitation Act is similar but not identical to the Limitation Act 1980 in Britain. Section 6(1)(a) provides that 'actions founded on a contract or on tort . . . shall not be brought after the expiration of six years from the date on which the cause of action accrued.'

Note that there is no distinction between simple contracts and contracts under seal and no 12-year period for the latter as under English law.

(The period is three years in cases involving personal injury.)

Note also that the English Latent Damage Act 1986 is of no application in Singapore – there is no 15-year long-stop.

The limitation periods are the same for all categories of producers.

1.5.4 TIME ALLOWED (DIFFERENT CATEGORIES OF PRODUCERS) FOR MAKING GOOD, REPAIRING OR PAYING DAMAGES

No specific rules – as to contractual provision, see 1.2.4.

---

[10] [1983] 1 All ER 65.
[11] (1985) 2 MLJ 11.
[12] [1971] 1 All ER 747.
[13] Singapore's Limitation Act, section 29.

1.5.5 KINDS OF DEFECTS AND DAMAGE QUALIFYING FOR DAMAGES

No information is available on defects not having led to damage qualifying for damages.

The general rules as to the nature of damage qualifying for damages are as in 1.2.5.

1.5.6 EXCULPATION OF PRODUCER

The general rules applied after the expiry of the guarantee period are not different from those valid during that period (see 1.2.6).

1.5.7 CONSEQUENCES OF PRODUCER'S NON-COMPLIANCE WITH UNEQUIVOCAL DUTY TO REPAIR

The general rules are those outlined in 1.2.7.

1.6 Transfer of ownership and producers' liability

1.6.1 RULES

Successive owners may have rights against producers in tort but not, of course, in contract.

2 PROPERTY (OR MATERIAL DAMAGE) INSURANCE

Property insurance exists in Singapore under that name, but the cover offered usually only extends to those aspects which are stated to be outside the scope of CIB 87's interest. Therefore, only a very brief outline is given in section 2. The only company to offer a kind of defects policy is the (American) Federal Insurance Co. This insurance has not made a large impact on the market.

2.1 General

2.1.0 LAW OR CONTRACT OR BOTH

Property insurance is not required by law. It is often required by mortgagees, especially banks, and this trend seems to be increasing.

2.1.1. TERMS AND CONCEPTS

The 'policy holder' will normally be the owner, although the tenant may be required to insure under the terms of the lease.

2.1.2 TECHNOLOGY AND INSURANCE TERMS AND CONDITIONS

Do terms and conditions depend upon:

(a) Insurers collecting technical information about project?
No.
(b) Geographical location of building, height etc.?
To some extent – concentration of risk is avoided.
(c) Actual technology used?
Yes – especially materials.
(d) Levels of competence of producers?
No.
(e) Whether insurer uses own control organization?
Sometimes.

2.1.3 INSURER'S TECHNICAL CONTROL

Does the insurer:
– Have his own organization for controls?
Sometimes.
– Consult the client's control bureau?
Never.
– Consult the public control bureau?
Never.

There is no influence on terms and conditions.

2.1.4 INSURERS' POSSIBLE INFLUENCE UPON BUILDING TECHNOLOGY

Do the insurers:

• Pool technological experiences?
To some extent.
• Convey technological experience to producers?
No.

- To clients?
  No – only through premiums.

  Is their experience taken into account:

- By authorities issuing codes and norms?
  To some extent.
- By standardization organizations?
  No.
- By agrément organizations?
  No.
- By authors of textbooks/handbooks?
  No.

2.1.5 NEW TECHNOLOGY

Assuming access to all relevant information it would be theoretically possible but practically very difficult indeed to insure an innovative building against defects (except for the 'normal' risks of fire, flood etc.). Terms and conditions would have to be negotiated through a broker with international connections, since a local insurer would be unlikely to deal with the matter.

2.1.6 THE INSURANCE MARKET, GENERAL IMPRESSION

The market is very competitive but mainly on a national rather than an international basis, with a tendency towards oligopoly.

2.2 During the Guarantee (Maintenance, Defects Liability) Period

2.2.0 EXISTENCE OF THIS KIND OF INSURANCE

No. Property insurance does not cover losses during the Guarantee Period due to damage.

[2.2.7: *Not applicable*]

2.5 After the Guarantee (Maintenance, Defects Liability) Period

2.5.1 NORMAL DURATION OF COVER; CANCELLATION

Cover is always annual. Premium and terms normally remain the same but not necessarily.

There is a standard cancellation clause available to either party on given notice. It is rarely invoked by the insurer except for non-payment of premium, but it could be, without reason. The insured can cancel, although there is a penalty in proportion to the premium returnable.

2.5.2 LEVELS OF PREMIUMS; EXCESS; FRANCHISE

Note: there is cover only for *excluded* causes such as fire etc. No data are available on defects cover premiums.

Office complex	0.03 per cent of cost
Condominium	0.02 per cent of cost
Hotel	0.05 per cent of cost

Typical excess: S \$200 000.
It is impossible to give typical figures for franchise.

2.5.4 TIME ALLOWED FOR NOTICE OF CLAIM AND FOR PAYMENT OF DAMAGES

There are no time limits for notice of claim or payment of damage.

2.5.5 KINDS OF DEFECTS AND DAMAGE WHICH CONSTITUTE CLAIMS EVENTS

No information is available on defects not having led to damage constituting claims events. The only fixed rules on kind of damage are in the policies – there is no limit on extent of damage – \$1 can constitute a claims event.

2.5.6 EXCLUSIONS

This question is not really relevant, since it is the unexpected natural occurrences such as storms, earthquakes etc. and fire against which property insurance in Singapore does cover.

The insurer may be entitled to avoid the contract or reduce the quantum if the neglect to observe proper maintenance by the insured causes or contributes to the damage.

2.6   Transfer of ownership and insurance cover

2.6.1   RULES

The insurance cover can follow the building by simple transfer, although each contract is made with the owner. There is no new inspection and procedures involved in transfer are minimal.

3   LIABILITY AND PROFESSIONAL INDEMNITY INSURANCE

3.1   General

3.1.0   LAW OR CONTRACT OR BOTH

Liability/professional indemnity insurance is not imposed by law upon producers of buildings. It is sometimes required by clients or by financiers or mortgagees. American clients/financiers are renowned for requiring such cover.

3.1.1   TERMS AND CONCEPTS

'BUILDING'

See 1.1.1.

'INSURANCE POLICY'

Apart from the above, the only 'project insurance' is carried by contractors during the running of the project and it terminates with practical completion.

3.1.2   TECHNOLOGY AND INSURANCE TERMS AND CONDITIONS

Do terms and conditions depend upon:

(a) Insurers collecting technical information about projects?
    To some extent.
(b) Geographical location of building, height etc.?
    To some extent.
(c) Actual technology used?
    To some extent.

(d) Levels of competence of producers?
Yes – competence and skill.
(e) Whether insurer uses own control organization?
Sometimes.
(f) Claims record of insured party?
Yes, very much so.

3.1.3 INSURER'S TECHNICAL CONTROL

Does the insurer:

- Follow design and site work through own organization?
Seldom.
- Consult client's control bureau?
Never.
- Consult public control bureau?
Never.
- Manner of organizing control has no real influence on terms and conditions.

3.1.4 INSURERS' POSSIBLE INFLUENCE UPON BUILDING TECHNOLOGY

Insurers pool technological experiences to some extent. They do not convey their information to producers or clients. Their experience is not taken into account except by the public authorities through a semi-official body – the Insurance Commissioner.

Insurers do categorize, classify and rate the producers as proposers. These facts are not published.

3.1.5 NEW TECHNOLOGIES

See 2.1.5 – same answer. Any proposals involving new technology would be likely to go outside Singapore, probably back to London.

3.1.6 THE INSURANCE MARKET, GENERAL IMPRESSION

The market is very conservative about indemnity insurance, which is relatively new and could not be described as competitive. Brokers refer anything new or very large to London.

The Association of Consulting Engineers of Singapore has recently made good progress in arranging a professional indemnity scheme with brokers.

3.1.7 COLLECTIVE INSURANCE FOR PROFESSIONAL OR OCCUPATIONAL GROUPINGS. GRADATION OF INSURANCE CONDITIONS

The only model professional indemnity contracts in existence are for doctors (a scheme for lawyers has just started). No such schemes exist for producers of buildings, although consulting engineers have negotiated a scheme (see 3.1.6) and architects are known to be interested.

No more favourable terms can be obtained at present by commitment to continuing education, but insurers expressed interest in the idea.

Note: Mandatory continuing education is just taking its first steps. Currently, architects, surveyors and engineers are discussing profession-wide insurance schemes, although no action has been taken. Some firms, especially those with international connections, carry individual insurance.

3.2 During the Guarantee (Maintenance, Defects Liability) Period

3.2.0 EXISTENCE OF LIABILITY AND PROFESSIONAL INDEMNITY INSURANCE

It is possible to obtain cover for the Defects Liability/Maintenance Period.

3.2.7 RIGHT OF RECOURSE AGAINST A PRODUCER

The insurer cannot claim against the client's retention monies.

3.5 After the Guarantee (Maintenance, Defects Liability) Period

3.5.1 NORMAL DURATION OF COVER; CANCELLATION

Cover is always annual. Non-payment of premium would result in cancellation of the policy by the insurer. There is a Standard Cancellation Clause (see 2.5.1).

3.5.2 LEVEL OF PREMIUMS, EXCESS AND FRANCHISE

It is impossible to generalise about levels of premiums. It is not possible to reduce the excess to zero – it is a strong element of practice here to use excess as an incentive to careful operation. The factors which determine premiums and franchise are basically:

(a) length of time in practice;
(b) type of work undertaken;
(c) claims record;
(d) gross fees, i.e. exposure;
(e) partnership/sole proprietorship/limited company;
(f) qualifications.

As stated, attitudes towards professional indemnity insurance are very conservative. Many professionals operate without any cover at all because the cost is prohibitive or at least unacceptable. The market has so far failed to change this disturbing state of affairs to any great extent.

3.5.3 TIME LAPSES BETWEEN OBSERVATION OF DAMAGE AND INVESTIGATION

The system of limitation of actions has been described in 1.2.6. Beyond this there is no specific provision covering any of these periods of time.

3.5.4 TIME ALLOWED BETWEEN THE INSURER'S RECOGNITION OF A CLAIMS EVENT AND HIS PAYING OUT

There is no fixed penalty for delay, but the High Court of Singapore has awarded interest for delay in payment by insurers in a workmen's compensation case.

3.5.5 KINDS OF DEFECTS AND DAMAGE WHICH CONSTITUTE CLAIMS EVENTS; ONUS OF PROOF

No information is available on defects not having led to damage constituting a claims event.

The only rules on kind of damage are in the insurance policy. Beyond the general law, any sum is claimable. There is no de minimis rule.

Res ipsa loquitur itself would not normally be applied but the plaintiff's task is made easier by the provisions of the building control legislation (see 1.1.0) which identifies the responsibilities of producers. Otherwise the burden of proof rests with the plaintiff.

3.5.6 EXCLUSIONS

The cover would not exclude any forms of breach of duty which could lead to damage to a building. The standard exclusions in professional indemnity policies cover such matters as defamation, loss of documents and malicious damage.

Spain

Alfredo Cámara Manso

INTRODUCTION

In writing the present report we have aimed to follow the standards and guidelines established for the overviews. This means that in order to analyse some points in depth others could only be summarily presented.

1 LIABILITY

1.1 General

Spanish legislation does not distinguish between residential and non-residential construction.

Although the 'developer' is well established in the construction process that is not so in law. This does not mean that the developer bears no responsibility to his clients.

1.1.0 LAW OR CONTRACT OR BOTH

Producers' post-reception liability is established by law, article 1591 of the Civil Code:

> 'The contractor of a building which is ruined due to faulty construction [*vicios de la construcción*] is liable for damages if the ruin [*la ruina*] occurs within a period of ten years from the time of completion; equal liability and for the same period of time applies to the architect directing the works if the ruin is due to faulty foundations [*vicio del suelo*] or a fault in the direction.'

All producers are subject to the same legal rules concerning liability within the scope of their participation.

1.1.1 TERMS AND CONCEPTS

'BUILDING'

Legislation does not provide a definition of the term 'building', nor does it distinguish between new construction and rebuilding: in both cases producers are bound by the same type of liability.

Utilities from which each residence, premises or portion of the building derives its supply are considered an integral part of the building.

'PRODUCER'

The term 'producer' covers several parties participating in the construction process, all of them liable before the law.

The Civil Code identifies as producers only the contractor and the architect. This is explained by the fact that in the nineteenth century, when the Civil Code was enacted, the only producers were the aforementioned.

Since then court rulings have extended liability to professions that emerged to cope with the increasing diversification of the construction process: the developer, the site supervisor (*arquitecto técnico*) and the engineer.

All producers, whether private or public (see below), have the same legal liability.

The tenant does not have recourse to direct legal action against producers but only against the developer or client.

1.2 During the Guarantee (Maintenance, Defects Liability) Period

Current legislation does not define this period as the guarantee period. Article 1591 of the Civil Code states that producers, by reason of their participation in the building process, are liable for damage which derives from their professional activity over a period of ten years.

The concept defined by this article can be termed 'guarantee period'.

1.2.1 DETERMINATION OF THE COMMENCEMENT OF THE PERIOD

The date of completion, recorded and certified, is the date of the commencement of the guarantee period.

1.2.2 SIGNATORIES TO AND NAME OF THE DOCUMENT STATING THE COMMENCEMENT OF THE PERIOD

According to the model approved by the public administration the document which establishes the beginning of the guarantee period is the

Certificate of Practical Completion (*Certificado Final de la Dirección de la Obra*).

This document certifies the date of completion of the works under the supervision of the participating professionals, who are the signatories of the document.

This certificate is mandatory for residential buildings but not for other types for which it may or may not be used.

1.2.3 THE PERIOD'S DURATION AND TERMINATION

The duration of the guarantee period is established by article 1591 of the Civil Code as ten years from the date of completion of the works. The period may be extended by contractual agreement, but the practice is rare.

1.2.4 TIME ALLOWED FOR DIFFERENT CATEGORIES OF PRODUCERS FOR MAKING GOOD, REPAIRING OR PAYING DAMAGES

Law does not fix a term for repairs or payments. Should the producer fail to make the required payments or repairs, the client's only recourse is to legal action.

1.2.5 KINDS OF DEFECTS AND DAMAGE COVERED BY THE GUARANTEE

When the defect does not cause damage ('ruin') there is no obligation to repair the defect. The concept of 'ruin' must, however, be understood in the widest sense, i.e. as any occurrence of damage threatening the loss of the building or rendering it unsuitable for its intended purpose.

No fixed rules exist to determine the type and extent of the damage. General *de jure* principles determine under which circumstances compensation may be claimed to repair the damage in order to restore stability and safety or avoid discomfort and lack of functionality.

1.2.6 EXCULPATION OF PRODUCER

Exoneration exists under the following circumstances:

(a) when the damage is caused by the action of a third party or by an interruption in the normal provision of services to the property, unless the latter results from a defect;
(b) in case of force majeure.

When the cause of damage is misuse of the building's facilities or lack of maintenance, then the producer may be exonerated.

The utilization of experimental techniques is not a cause for exoneration: the producer, due to his specialized knowledge, is expected to show a particular diligence in his performance.

Errors and omissions which cause damage leading to ruin entail liability. In principle, in Spain the producers are under a duty of care but this has progressively been changing towards a duty of result. Thus, at present, if a producer follows accepted or prescribed practices which later prove to be erroneous and cause damage the courts may find him liable for negligence. New legislation currently under preparation is expected explicitly to adopt the principle of duty of result.

The principle of liability is subjective, that is to say that it implies a negligent act of omission contrary to professional rules.

1.2.7 CONSEQUENCES OF PRODUCER'S NON-COMPLIANCE WITH UNEQUIVOCAL DUTY TO MAKE GOOD, REPAIR OR PAY DAMAGES

If two or more producers are found guilty of negligence and one of them is insolvent, then the other(s) must pay the damages. In practice, since the contractor is not under an obligation to be insured for the duration of the guarantee period, and under the period he may become insolvent, it is often the architect or the site supervisor's insurer who has to pay damages.

If only one party is liable and this is insolvent, no payment of damages will take place.

1.5 From expiry of the Guarantee (Maintenance, Defects Liability) Period

The post-guarantee period begins at the expiration of the guarantee period which, as indicated above, is ten years. There is no document establishing the beginning of the post-guarantee period.

Claims may be made in the post-guarantee period only in the case of fraud or bad faith in the professional activity of the producer and within a maximum period of thirty years.

1.6 Transfer of ownership and producers' liability

1.6.1 RULES

The successive owner is subrogated in the same position as the client and has the same rights to claim damages from a producer.

2 PROPERTY (OR MATERIAL DAMAGE) INSURANCE

This kind of insurance is neither defined nor required by Spanish law, although it is now beginning to be marketed for building within the private sector. Unfortunately, as conditions vary widely it is not possible to provide precise information in this respect.

3 LIABILITY AND PROFESSIONAL INDEMNITY INSURANCE

3.1 General

3.1.0 LAW OR CONTRACT OR BOTH

Civil liability insurance is not required by law, but voluntary.

3.1.1 TERMS AND CONCEPTS

'BUILDING'

There is no legal definition of this term.

'INSURANCE POLICY'

Producers are covered by different kinds of insurance. The contractor subscribes to a 'project policy' (*póliza única de obra*) insurance, that is, he subscribes to a policy insurance for each project. The participating professionals insure their liability though a type of 'continuous' or 'running' insurance.

'POLICY HOLDER', 'INSURED PARTY'

The beneficiary and the subscriber are considered to be the same person.

3.1.2 TECHNOLOGY AND INSURANCE TERMS AND CONDITIONS

In the case of 'running' insurance the terms and conditions determined by the insurer, such as premiums, franchises, minimum damage and exclusions, are not based on information connected with the nature of the project. Neither the characteristics of the building (geographic location, closures, height and technology used) nor that of the participating professionals (organization of the design team, professional competence, responsibility of the suppliers) are of any relevance to the terms and conditions of the insurance.

This type of insurance is also independent of the supervision office established for the execution of the works. The insurer does not usually have his own supervision office; private and public laboratories are used only in relation to 'project' policies.

In the case of 'project insurance' the insurer has access to, and may use, technical information relating to the project, as well as other characteristics of the projected building (geographical, design-related etc.), to determine terms and conditions.

3.1.3   INSURER'S TECHNICAL CONTROL

In the case of 'running insurance' the insurer does not oversee either the planning or the execution by means of his own supervision facilities, as he does not possess these facilities.

The insurer oversees the execution of the works only in the case of 'project insurance'; this supervision is performed by private and public supervision offices.

[3.1.4: *Not applicable*]

3.1.5   NEW TECHNOLOGIES

The system of insurance for buildings using new technologies is not different from the system used for more traditional buildings, although premiums and special conditions of the insurance policy are determined in each case.

3.1.6   THE INSURANCE MARKET, GENERAL IMPRESSION

It is not possible to define the insurance market in relation to new technologies as there is not a sufficient number of buildings of this kind in existence to justify a market for them.

3.1.7 COLLECTIVE INSURANCE FOR PROFESSIONAL OR OCCUPATIONAL GROUPINGS. GRADATION OF INSURANCE CONDITIONS

Contracts are issued collectively on a provincial basis or as individual policies. No preferential premiums are established for professionals engaging in continuous technical training.

3.2 During the Guarantee (Maintenance, Defects Liability) Period

The contracts have a duration of one year and claims for events or interventions during or previous to the period insured are guaranteed. Professional indemnity insurance is not usually issued to cover liability for a particular project but for any projects undertaken in the exercise of the profession.

3.2.0 EXISTENCE OF LIABILITY AND PROFESSIONAL INDEMNITY INSURANCE

Contracts covering the complete guarantee period do not exist although this period may be covered by renewable one-year contracts.

[3.2.7: *Not applicable*]

3.5 After the Guarantee (Maintenance, Defects Liability) Period

After the ten-year guarantee period liability terminates, neither Spanish law nor the insurance market refers to the post-guarantee period.

Sweden

Håkan Albrecht, Kjell Jutehammar, Anders G. Kleberg,
Bo Linander and Christer Skagerberg

1 LIABILITY

1.1 General

The following description outlines the legal relationship in Sweden between producer and client. The purchase of land with a house which is ready to move into is not dealt with here. The purchase of land is regulated by the Code of Land Laws 1972 (Jordabalken). The rules concerning purchase of property are to a large extent not compulsory, i.e. the parties concerned have the option of negotiating outside the regulations.

1.1.0 LAW OR CONTRACT OR BOTH

There is a fundamental difference between a client who is a professional, e.g. an owner of a block of flats or a tenant-owner society, and one who is a private individual, a 'consumer'. The legal position between professional clients and producers is the same for both residential and non-residential sectors. When the client is a professional the legal position is governed solely by contract, but when he is a private individual (a 'consumer') the legal position is sometimes governed by contract and laws relating to consumer protection.

PUBLIC CONTROL

Public control relating to construction and building is administrated by the local authorities. The function of public control is regulated in the Planning and Building Act 1987 (Plan- och Bygglagen).

Local authority approval is required for new buildings and for extensive rebuilding of existing buildings. The local authority checks the drawings and specifications and then grants or rejects the application. The building inspector employed by the local authority inspects the works at specific stages of construction to ensure that the work has been carried out in accordance with the requirements of the building regulations. In both these capacities the local authority may act wrongly or negligently to the detriment of the client. In such a case the local authority can be sued in tort and the client, if successful, will be able to recover his loss.

The question of liability is determined in accordance with the Damages Act 1972 (Skadeståndslagen). Damages are due if the local authority has, with regard to the nature and purpose of the activity, neglected demands that can be reasonably made on it. In recent years the Supreme Court has, in some cases, decided against the local authority and awarded damages.

Decisions of the Supreme Court stipulate that the local authority is liable principally when it, due to particular knowledge about local conditions or its special resources, has a better opportunity than the client to realize the risk of damage and to judge what measures should be taken. In this connection the client's qualifications must also be taken into account. However, when there is a risk of any considerable degree, the local authority must not neglect to see to it that preventative measures are taken, even if both the producer and the client have a great deal of experience. Another aspect to consider is how important an inspection is in relation to the building regulations, e.g. if the inspection is compulsory, or if spot checks are carried out, or if there is an inspection for some particular reason. The Supreme Court has also ruled that a series of small mistakes and omissions by the local authority might mean that a reasonable standard has not been maintained.

THE PROFESSIONAL CLIENT

The main rules governing the commercial relationship between client and producer are laid down in a set of General Conditions of Contract for Building, Civil Engineering and Installation work – AB 92 (Allmänna bestämmelser för byggnads- anläggnings- och installationsentreprenader) – which are agreed to by the main bodies representing clients and contractors in the building and construction industry. In April 1992 a modified version of the General Conditions was agreed to, in which among other things the contractor's liability is enlarged in what concerns the post-guarantee period.

There is a similar agreement concerning the relationship between client and designer with Conditions of Contract for Architects and Consulting Engineers – ABK 87 (Allmänna bestämmelser för konsultuppdrag inom arkitekt- och ingenjörsverksamhet) agreed to by the main bodies representing clients and designers.

There are other general conditions of contract, jointly agreed by the industry, dealing with, e.g., the design-and-build concept – ABT 74 (Allmänna bestämmelser för totalentreprenader avsedda för byggnads-, anläggnings- och installationsentreprenader). There are, furthermore, agreed conditions of contract for sub-contracting – AFU 83 (Allmänna bestämmelser för underentreprenader) – and for suppliers of various kinds of building components on site. ABT 74 and AFU 83 are subject to modification and new versions are expected during autumn 1992.

These general conditions of contract, and others regarding the building industry, effectively constitute the prevailing law for resolving legal matters within the construction and building industry between commercial parties.

The rules mentioned in this part are used for both new construction and for rebuilding.

WHEN THE CLIENT IS A PRIVATE INDIVIDUAL (CONSUMER)

When the client is a private individual there is, in the first place, a difference between work carried out under contract and the purchase of material and fittings used in the building.

For contracts in general the main rules are laid down in AB 92. These rules have, in the first place, been worked out for commercial relationships. It has become apparent that AB 72, preceding AB 92, was less suitable for contracts between 'consumers' (see above) and professionals or craftsmen. In a case between the Consumer Ombudsman and a producer, the Court of Market declared that some of the terms in AB 72 were unreasonable in contracts between consumers and professionals or craftsmen.

Therefore a contract known as the General Conditions of Contract for Building Single Family Houses for Consumers – ABS 80 (Allmänna bestämmelser för småhusentreprenader, där enskild konsument är köpare) with associated appendices has been agreed by authorities and organizations representing consumers and organizations representing producers. ABS 80 with its appendices has been adjusted to the situation whereby only one producer is liable to the consumer for the supply of the complete building. By an Act of Parliament, compliance with ABS 80 and appendices has become a condition for state subsidy for new building (but not for rebuilding). ABS 80 as well is subject to modification. The new version will be published in 1993. In the new General Conditions an enlarged liability for the contractor in accordance with AB 92 (see 1.1.0) can be expected.

The consumer's purchase of material and fittings for the house is regulated by law in the Consumer Sales Act 1990 (Konsumentköplagen). This Act, which replaced the Consumer Sales Act 1973, came into force in 1991. It gives the consumer protection on some important points.

The purchase of prefabricated single family houses is quite common in Sweden. The Consumer Sales Act applies to this kind of purchase, unless it is covered by a ten-year warranty fulfilling the conditions for government loans (see 2.1.0). In that case, the provisions in the Act are non-compulsory.

A set of general conditions of contract called General Conditions of Delivery – AL 84 (Allmänna leveransbestämmelser) then normally governs the purchase.

ABS 80 with appendices is not normally used when the client has engaged more than one producer. Thus nothing has been agreed on what is to be used when two or more contractors are engaged by the client. Therefore with regard to new building, only ABS 80 with appendices will be dealt with in the following text.

When consumers engage architects and consulting engineers ABK 87 normally applies.

The Consumer Services Act 1985 (Konsumenttjänstlagen) came into force in 1986. This Act governs agreements between consumers and professionals and tradesmen on different types of work and services that professionals and tradesmen undertake for consumers. Among other things the Act governs work in connection with rebuilding and repair work to property. Since 1991, the Act applies to new buildings as well. However, the provisions of the Act are non-compulsory in the case of single family houses with a ten-year warranty fulfilling the conditions for government loans. See 2.1.0.

1.1.1 TERMS AND CONCEPTS

'BUILDING'

The concept of 'building' is not defined by law but is dealt with in the theme of one of the Acts mentioned above. In the travaux préparatoires to the Planning and Building Act 1986 there is a short report on the legal usage of this concept in the building regulations. The legal usage corresponds, on the whole, with the definition given by the Swedish Building Standard Institution (Byggstandardiseringen, BST) of a building as something that contains one or several rooms delimited by floor, walls and roof and is mainly situated above ground level.

Section 1.1 described different regulations for new construction and rebuilding. The concept of 'rebuilding' is not defined in law or by BST. The Planning and Building Act defines the changes to an existing building that require local authority permission. The governmental body that issues building regulations, 'Boverket', has begun to distinguish between 'new building' and 'rebuilding' for the simple reason that some regulations can be relaxed when it comes to rebuilding where the requirements applying to new building would make a rebuilding scheme completely impractical and utterly uneconomic.

BST defines installation as 'a fixed internal system of provision in the building' and regards it as being included in the 'building'. Access roads, paths etc. are, on the other hand, never included in the concept of 'building'.

'PRODUCER'

According to the definition in the List of Terms (Appendix B), the local authority, giving building permission and making inspections, is also a producer.

Producers with whom the client has a contract are liable for work done by sub-contractors and others engaged by the producer to execute the contract.

It is laid down in the Damages Act which parties are responsible for damages in non-contractual situations. This statute stipulates that someone who has suffered damage has a right to compensation for personal injury and damage to his property from the one who has caused the damage by negligence. If the damage is caused by an authority, the one who has suffered also has the right to damages for pure economic loss under the conditions mentioned in 1.1.

Thus, in the case of a sub-contractor or a supplier who supplies and installs his components on site, the client can only sue the sub-contractor/supplier in tort for causing the client personal injury or physical damage to his property, but not for the pure economic loss that the client might have incurred as a result of the sub-contractor's/supplier's negligent behaviour.

'TENANT'

A tenant would in any normal situation sue the landlord if he has suffered damage or been caused economic loss due to defects in the building. However, the tenant can, as an alternative, sue the producer in tort and, if the producer has been negligent, be awarded damages for personal injury and damage to his property, but not for pure economic loss.

1.2 During the Guarantee (Maintenance, Defects Liability) Period

Both AB 92 and ABS 80 contain a Guarantee Period. The Consumer Sales Act and the Consumer Services Act contain regulations concerning a guarantee. A guarantee according to these Acts implies a presumption that the product is defective if it diverges from what has been promised.

The General Conditions of Contract for Architects and Consulting Engineers (ABK 87) does not contain a Guarantee Period. The consultant is, as stipulated in ABK 87, liable for damage that is discovered during the contractor's Guarantee Period if it is caused by error or omission on the part of the consultant. This liability of the consultant is, however, limited to seven years after the consultant work was completed. After the termination of the contractor's Guarantee Period the consultant is liable for damage only when it is caused through gross negligence.

1.2.1 DETERMINATION OF THE COMMENCEMENT OF THE PERIOD

AB 92 declares the commencement of the Guarantee Period to be the day the works have been approved of. The works can only be approved of by way of having a final inspection (slutbesiktning) whereby the works are checked and gone through by an inspector (besiktningsförrättare): cf. 1.2.2. If final inspection is not carried out within the time prescribed due to failure on the part of the client, the Guarantee Period will commence on the day the inspection should have taken place.

ABS 80 stipulates the commencement of the Guarantee Period to be when the building is completed and the client is permitted to move in. A final inspection is held by an inspector who decides if the building is completed.

It is stipulated in AB 92 that the inspector is required to give his approval notwithstanding that a limited number of minor defects still remain. However, the contractor has to remedy these remaining defects without delay.

In ABS 80 it is stipulated that the inspector should decide on the final inspection whether the building is in a completed state for moving in or not. External work, that is not essential for the functioning of the house, does not have to be completed. However, remaining work must be carried out within a reasonable time from the takeover by the client.

If a guarantee is offered in accordance with the Consumer Sales Act or the Consumer Services Act, the commencement of the Guarantee Period is when the property is handed over to the client, or when the contractor completed his work. There are no rules about final inspections in these cases.

1.2.2 SIGNATORIES TO AND NAME OF THE DOCUMENT STATING THE COMMENCEMENT OF THE PERIOD

The title of the certificate used as laid down in AB 92 and ABS 80 is the Certificate of Practical Completion (Slutbesiktningsutlåtande).

The only signatory to the Certificate of Practical Completion is the inspector. AB 92 stipulates that the inspection shall be carried out by a competent person appointed by the client. According to ABS 80 the inspector is to be appointed jointly by the producer and the client. AB 92 as well as ABS 80 states that the parties are to be notified to attend the inspection.

1.2.3 THE PERIOD'S DURATION AND TERMINATION

AB 92 stipulates a Guarantee Period of two years, unless otherwise is agreed upon in other contract documents.

Just before the Guarantee Period has elapsed, the client arranges for the guarantee inspection (Garantibesiktning). As a result of that inspection, the inspector either approves the works or finds fault with them. After approving the works, the inspector will issue a Certificate of Maintenance (Garantibesiktningsutlåtande).

The inspector can under certain circumstances order an extension of the Guarantee Period up to a maximum of double the original period relating to such work that is necessary for remedying defects that have only become apparent during the Guarantee Period.

ABS 80 does not give a specific length of the Guarantee Period. ABS 80 only states that the Guarantee Period runs during the specific time agreed upon in the individual contract. As mentioned under 1.1, the use of ABS 80 with contract is, in certain cases, a condition for state subsidies. Another condition prescribes a minimum Guarantee Period of two years in these cases. The rules of guarantee inspection, as described above, are essentially the same in ABS 80.

The Consumer Sales Act stipulates that the time allowed for claims is two years unless the parties have agreed on a longer period.

1.2.4 TIME ALLOWED FOR DIFFERENT CATEGORIES OF PRODUCERS FOR MAKING GOOD, REPAIRING OR PAYING DAMAGES

In AB 92 it is stated that the contractor has to make good without delay and, at the latest, two months after the contractor has received the Certificate of Practical Completion and the Certificate of Maintenance respectively. Any defect which does not seriously affect the existence or appearance of the contract work, or the use of it for its intended purpose, the remedying of which would entail unreasonable costs for the contractor, must be repaired only to such a degree as in view for the prevailing circumstances may be deemed reasonable. Instead the inspector estimates a reduction in payment. There is no time limit given in AB 92 for paying damages.

ABS 80 stipulates that the producer has to make good within a reasonable time. Making good of less serious defects or deficiencies can wait until after the guarantee inspection. There is no time limit in ABS 80 for paying damages.

The Consumer Sales Act and the Consumer Services Act state that the producer has to make good within reasonable time.

1.2.5 KINDS OF DEFECTS AND DAMAGE COVERED BY THE GUARANTEE

In AB 92 and ABS 80 the producer is liable for defects which become apparent during the Guarantee Period.

According to AB 92 and ABS 80 both parties are entitled to request a special inspection during the Guarantee Period, relating to defects which have appeared after the final inspection. During the inspection defects existing in the inspected part of the contract work at the time of the inspection are noted. As a principial rule, action may not be brought concerning defects other than those thus noted. As an exceptional rule, AB 92 states that the client may bring action concerning defects within three months from the day the works have been approved of. ABS 80 states that the client may bring action concerning defects that have not been noted in a Certificate of Completion, a Certificate of Maintenance or a Certificate of Special Inspection if the inspector has not noticed them but should have or if the client draws the producer's attention to them within two months of receiving the certificate.

A guarantee, according to the Consumer Sales Act or the Consumer Services Act, implies a presumption that the product or service is defective if it diverges from what has been promised.

1.2.6 EXCULPATION OF PRODUCER

During the Guarantee Period the producer is liable for defects appearing during the period. The inspector is liable – if nothing else has been agreed – in the case of negligence leading to personal injury, damage to property and to pure economic loss.

The producer is not liable for deterioration and imperfections due to action by a third party, or Acts of God.

In AB 92, defect is defined as being when part of the contract work has not been executed, or has not been executed in accordance with the requirements of the contractual agreement. What defect is depends on the contents of the agreement. If liability for design, function or quality is included as a whole or by a certain reference, the question of defect is judged from another starting-point than if such liability is not included in the agreement. Thus, according to AB 92, during the Guarantee Period the producer is consequently under a 'duty of result'. When the contractor has carried out the work in accordance with the documents supplied by the client, there is no defect even if the design, function or quality of the building is unsatisfactory.

Also ABS 80 with appendices lays down a duty of result which includes the design of the building.

The Consumer Sales Act and the Consumer Service Act also include a duty of result during the Guarantee Period.

ABK 87 lays down that the consultant has a 'duty of care': the consultant must manifest care and execute his work with the skill that is held to be common within the profession. If the consultant breaks this standard he will have to pay damages.

The inspector from the local authority is also under a duty of care, see 1.1.0.

1.2.7 CONSEQUENCES OF PRODUCER'S NON-COMPLIANCE WITH UNEQUIVOCAL DUTY TO MAKE GOOD, REPAIR OR PAY DAMAGES

AB 92 and ABS 80 stipulate that defects to which attention is drawn in the inspection report and for which the producer is liable must be remedied by the producer, otherwise they may be remedied by the client at the expense of the producer. There would be no difference in the alternative situation mentioned in the Questionnaire.

In professional relationships the client often has a performance bond which can be used to finance the cost of rectifying the defects by employing another contractor to do the work. (Normally this bond gives cover of up to 5 per cent of the value of the contract sum for eventualities of this type during the Guarantee Period.)

For single family houses there is a ten-year warranty protecting the client in case the producer goes bankrupt and for that reason does not fulfil his contractual obligations (see 2.1.0).

1.5 From expiry of the Guarantee (Maintenance, Defects Liability) Period

1.5.1 DETERMINATION OF THE COMMENCEMENT OF THE (POST-GUARANTEE) PERIOD DURING WHICH A PRODUCER MAY BE HELD LIABLE

In AB 92 and ABS 80 the commencement of the post-guarantee period may be postponed in the case of a defect having occurred and been repaired during the guarantee period. Also see 1.2.4.

In AB 92 it is stated that the duration of the Guarantee Period is two years unless otherwise is agreed upon in other contract documents, starting from the day when the contracted work, or part of it, has been approved. As mentioned, if the final inspection is not carried out within the time laid down, due to failure on the part of the client, the Guarantee Period shall start on the day when the inspection should rightfully have been carried out.

ABS 80 states that the duration of the Guarantee Period shall be laid down in the individual contract, starting from the day the house is completed and the client is permitted to move in.

If the guarantee is offered under the Consumer Sales Act or the Consumer Services Act the post-guarantee period starts when the client receives the bought product or when the producer completes his work.

1.5.2 NAME OF THE DOCUMENT STATING THE COMMENCEMENT OF THE POST-GUARANTEE PERIOD; ITS SIGNATORIES

There is no document or formalities giving the commencement of the post-guarantee period. See 1.5.1. In AB 92 it is stated that guarantee inspection must be carried out before the termination of the Guarantee Period, if the parties have not agreed on something else.

ABS 80 stipulates that there is to be a guarantee inspection in connection with the termination of the Guarantee Period.

The only signatory to the inspection report is the inspector.

Neither in the Consumer Sales Act nor in the Consumer Services Act is any formal procedure laid down for the commencement of the post-guarantee period.

1.5.3 PERIOD(S) OF LIMITATION

AB 92 lays down that the contractor is responsible during the post-guarantee period for defects which were not noticed or could not reasonably have been noticed before the termination of the Guarantee Period, in so far as such defects are essential and due to negligence on the part of the contractor. This liability starts from the commencement of the Guarantee Period and ends with the expiry of the period of limitation according to law. The period of limitation is laid down in the Limitation Act 1981 (Preskriptionslagen) and is ten years.

ABS 80 states that the contractor is responsible during the post-guarantee period only for defects due to gross negligence of the contractor. As mentioned above, the contractor's liability will be enlarged in the modified General Conditions which will be published in 1993.

ABK 87 stipulates that the designer is liable for defects due to faulty design which have been discovered before the expiry of the Guarantee Period agreed upon in the contract between the client and the producer. If such a Guarantee Period is not specified in the contract, the designer's liability ceases two years after the issuing of the Certificate of Practical Completion, except in case of gross negligence, where the normal ten-year limitation period applies. In case the works do not start until some years after the assignment has been completed, the designer's liability ceases to exist when seven years have elapsed from the time of fulfilling his assignment. After the termination of the contractor's Guarantee Period the consultant is liable for damage only when caused by gross negligence. The designer will, however, under no circumstances be liable for a longer period than the Limitation Act stipulates.

According to the Consumer Sales Act the period of limitation is two years, unless the parties have agreed on a longer period. In the Consumer Services Act the period for new building and work on buildings is ten years from the date the work was completed.

In tort, the actions must be taken before ten years have elapsed from the time that the negligent act was committed.

1.5.4 TIME ALLOWED (DIFFERENT CATEGORIES OF PRODUCERS) FOR MAKING GOOD, REPAIRING OR PAYING DAMAGES

See 1.2.4.

In tort, the legal principles in the law of damages will be applicable and that means that a tortfeasor normally cannot be forced to make good defective work: the plaintiff will have to be satisfied with obtaining damages from the tortfeasor.

1.5.5 KINDS OF DEFECTS AND DAMAGE QUALIFYING FOR DAMAGES

In AB 92 the producer is responsible for essential defects which were not noticed or could not reasonably have been noticed before the termination of the Guarantee Period, in so far as such defects are due to negligence on the part of the producer. ABS 80 states that the producer is responsible only for defects due to gross negligence. ABK 87 stipulates that the liability of the consultant after the expiry of the Guarantee Period is limited to damage caused by gross negligence.

Both in the Consumer Sales Act and the Consumer Services Act the producer is liable for defects existing when the client received the goods or when the work was completed.

1.5.6 EXCULPATION OF PRODUCER

In AB 92 it is laid down that the liability of the producer is limited to defects caused by negligence. ABS 80 states that the producer is liable only for defects caused by gross negligence. This means that the 'duty of result' at the expiry of the Guarantee Period changes to a 'duty of result' combined with a 'duty of care'. Liability for negligence means that the producer normally is responsible if the contract work is not in accordance with the requirements of the client or with professional execution of work. The restriction to gross negligence implies a relatively low standard of care – after the expiry of the producer's Guarantee Period. Furthermore ABK 87 stipulates that the liability of the consultant is also limited to damage caused by gross negligence.

According to para. 8 of Chapter 5 in AB 72, preceding AB 92, the producer is liable only for defects caused by gross negligence from the producer's side. The meaning of the term 'gross negligence' has been dealt with in two cases, briefly described below.

In an arbitration from 1988, gross neglience in the sense of AB 72 is described as 'serious neglect of normal care combined with the knowledge of an existing risk for a damage to occur'. Further, the arbitration states that 'on judging the degree of risk demanded, regard must be paid to what could have been assumed about the proportion of the eventual damage at the point when the damage was caused'. The arbitration board adds that it is possible that as gross negligence should also be considered, cases of serious negligence where the parties, though not aware of any particular technical risk, should anticipate the likelihood of serious harm in case of failure.

In a judgement from 1992, the Supreme Court has, for the first time, put para. 8 into practice. The facts were as follows. A contractor had erected a building containing a school dining-hall with a wooden panel ceiling. Five years after completion, when the Guarantee Period had expired, the complete ceiling collapsed and fell down on the floor. Fortunately, the dining-hall was empty of people on that occasion and the collapse did not bring about personal injury. The cause of the damage was that the contractor, in contravention of official regulations, had nailed the panel insufficiently. It could not be presumed that the contractor had realized that the attachement was insufficent or involved a risk of collapse.

The Supreme Court was of the opinion that the contractor – himself or with the help of an expert – should have made sure that the attachment was sufficient and safe. Especially considering the risk of personal injury, the court was of the opinion that the contractor's omission was gross negligence.

No precedent has yet been set which clarifies the meaning of gross negligence according to ABS 80, but there are indications that gross negligence may have been committed when the house lacks qualities which are necessary for its functional existence.

Both the Consumer Sales Act and the Consumer Services Act stipulate that the client has the onus of proof.

See also 1.2.6.

1.5.7 CONSEQUENCES OF PRODUCER'S NON-COMPLIANCE WITH UNEQUIVOCAL DUTY TO REPAIR

See 1.2.7.

1.6 Transfer of ownership and producers' liability

A successive owner can only claim damages in an action in tort toward any producer, which means that he has to prove that the producer has been negligent, which is not necessary in an action in contract.

Outside a contractual relationship, as stated above, a party cannot successfully claim damages for pure economic loss, only for personal injury and damage to his property. As mentioned, the Damages Act stipulates that only a public authority is liable to pay damages also for pure economic loss.

By way of exercising his rights under the Code of Land Laws, the successive owner is also able to recover his losses from the previous owner. It is then up to the previous owner to seek recourse from the producer within the economic and time limits that are stipulated in the contract.

2 PROPERTY (OR MATERIAL DAMAGE) INSURANCE

2.1 General

2.1.0 LAW OR CONTRACT OR BOTH

For *non-residential* buildings and *residential, multi-dwelling* buildings there is no insurance in Sweden covering defects in the building during the post-construction period.

Parliament has passed a Building Guarantee Act, which lays down a compulsory warranty for new construction and major reconstruction of blocks of flats. The warranty shall be signed by the client with an authorized warranty organization and cover defects and damages occurring up to ten years after the final inspection. The coming into force of the Act has been postponed several times, for one thing because of the negotiations concerning AB 92. For the present, it is not probable that the Act will come into force. However, an insurance policy providing approximately the same cover as the Building Guarantee can be expected to become compulsory for new construction and major construction of blocks of flats.

When it comes to *single-family houses* on housing estates, a ten-year warranty for buyers is a condition for government subsidies of interest, as well as for single-family houses in tenant-owner co-operatives. This condition is stipulated by law (Nybyggnadsförordningen) and applies to new building. The ten-year warranty scheme provides protection for the client in case of the producer's bankruptcy (Production Warranty). It covers the producer's commitment up to the expiry of the Guarantee Period (including remedying of deficiencies and defects during this period). The scheme also covers damage caused by latent defects up to ten years after completion of the works (Long-term Warranty).

Today, the ten-year warranty in Sweden is offered by AB Bostadsgaranti, a company owned by the government and the Associated General Contractors and House Builders of Sweden. Bostadsgaranti's Long-term Warranty is insured by a consortium formed by the insurance companies Skandia and Trygg-Hansa. The ten-year warranty is also issued by the co-operative insurance company Folksam and by a warranty company initiated by the National Association of Swedish Wooden House Manufacturers.

In the co-operative housing sector there is a Production Warranty and a Long-term Warranty for single-family houses with about the same terms and conditions as AB Bostadsgaranti has. The insurance company Folksam issues the ten-year warranty in this sector. The national federation of the co-operative housing organizations (HSB) is, however, itself responsible for production warranties when an associated organization is producer.

In the case of multi-dwelling houses with tenant-owner co-operatives, AB Bostadsgaranti's undertaking comprises a production warranty and a security for the deposits made by the tenant-owners.

2.1.1 TERMS AND CONCEPTS

'BUILDING'

The concept 'building' is not used to describe the undertakings in the Production Warranty or Long-term Warranty. Instead, 'works according to the contract' is referred to in the certificate of warranty.

'POLICY HOLDER'

For owner-occupation of single-family houses both the client and successive owners are policy holders. For single-family houses and multi-dwelling houses with tenant-owner leases only the client in practice is the policy holder as the tenant-owners association does not sell its property.

2.1.2 TECHNOLOGY AND INSURANCE TERMS AND CONDITIONS

Below, the principles for dealing with warranty applications in AB Bostadsgaranti are described. When it comes to the other organizations providing warranties, the principles are mainly the same. However, the restriction to design-and-build concepts mentioned below does only apply to AB Bostadsgaranti.

A producer who, for the first time, applies for a Production Warranty or a Production Warranty combined with a Long-term Warranty at AB Bostadsgaranti, has to be approved by the warranty company. The inquiry into the affairs of the producer includes his economic, technical and administrative ability to accomplish a project of the kind that the warranty is intended for. The result of the inquiry might be a demand for a performance bond or a rejection of the producer on the grounds of lack of the necessary technical or administrative ability. The producer's affairs are continuously followed up through the accounts etc.

As AB Bostadsgaranti can only give a warranty when a design-and-build concept is agreed on by the parties, there is no reason to inquire into the affairs of sub-contractors or suppliers.

Production and Long-term Warranties can only be given for houses with reliable and well tried materials and technical solutions. When AB Bostadsgaranti judges a project it has already been designed. If the project in some technical respect is not acceptable, AB Bostadsgaranti can refuse a warranty.

2.1.3 INSURER'S TECHNICAL CONTROL

For the ten-year warranty AB Bostadsgaranti uses consultants for technical control. The consultants are chosen by the organization of owners of single-family houses. The quality of the technical control, the final inspection and the guarantee inspection are of direct importance to the terms and conditions of the warranties. The technical control is carried out according to detailed rules with compulsory control of important and critical points and spot tests on other points.

The organization of owners of single-family houses, AB Bostadsgaranti and the insurance consortium all take part in deciding on the rules of the technical control. The consultants who undertake the technical control have to be free from influence from producers and other interests that might influence their neutrality in the task.

2.1.4 INSURERS' POSSIBLE INFLUENCE UPON BUILDING TECHNOLOGY

The ten-year warranty is still too new to be the basis of any feedback to producers in any systematic manner.

However, one experience is that single-family houses with Production and Long-term Warranties have proportionally few deficiencies and defects.

The experience has not yet been taken into account by those mentioned in the last part of the Questionnaire. The experience of the system has, however, led to an extension of the government loan condition concerning ten-year warranties to all new single-family houses (see above).

2.1.5 NEW TECHNOLOGY

Theoretically it is not possible to insure all projects with known risks.

In practice, however, it is possible to insure innovative building in the present system with a Production Warranty and a Long-term Warranty.

2.1.6 THE INSURANCE MARKET, GENERAL IMPRESSION

The general impression is that the insurance market is interested in solutions with known risks and with low administrative costs. There are, of course, differences between insurance companies depending upon the company's present field activities and resources available to develop new activities.

For this type of warranty the competition is national.

2.2 During the Guarantee (Maintenance, Defects Liability) Period

2.2.0 EXISTENCE OF THIS KIND OF INSURANCE

The producer is liable for deficiencies or defects which become apparent during the Guarantee Period. The Production Warranty, i.e. the insurer, guarantees the completion of the contract agreement including responsibilities of the producer during the Guarantee Period.

2.2.7 RIGHT OF RECOURSE AGAINST THE PRODUCER

The warranty will be triggered only if the producer fails to make good, repair or pay damages. The issuer of the warranty (i.e. the insurer) has a right of recourse against the producer for expenses and other costs that the warranty has caused the issuer.

2.5 After the Guarantee (Maintenance, Defects Liability) Period

2.5.1 NORMAL DURATION OF COVER; CANCELLATION

The Long-term Warranty covers the period from the expiry of the Guarantee Period until ten years after the date of practical completion (normally eight years) – see 1.2.3.

The premium is paid in advance and covers the whole period.
The Long-term Warranty cannot be cancelled.

2.5.2   LEVELS OF PREMIUMS; EXCESS, FRANCHISE

The premium for the Long-term Warranty is at present SEK 6400.

For each deficiency or defect in a single-family house, the owner of the house has to assume a certain excess. The amount is indexbound and at present is SEK 17 200.

2.5.4   TIME ALLOWED FOR NOTICE OF CLAIM AND FOR PAYMENT OF DAMAGES

Damage covered by Long-term Warranty should be reported, without undue delay, to the consortium mentioned under 2.1.0 formed by the Trygg Hansa and Skandia insurance companies, or to the Folksam insurance company. If the owner of the house fails to make such a report, despite the fact he has detected or should have detected the damage, he loses his rights to have the defects remedied or alternatively to get economic compensation.

The Long-term Warranty does not specify the time allowed for repair or payment of damages. That probably implies that it must be done without delay.

2.5.5   KINDS OF DEFECTS AND DAMAGE WHICH CONSTITUTE CLAIMS EVENTS

The Long-term Warranty covers – with some exceptions – remedying of, or economic compensation for, essential damage to the house including a garage with a separate approach to it, even if the garage is not situated on the premises, provided the damage consists of, or is the result of, defects or deficiencies in design, workmanship or materials. The warranty also covers compensation for reasonable extra expenditure for accommodation for the house owner and his family for a maximum period of six months, if damage results in essential parts of the house becoming unfit for use.

The total undertaking, according to the warranty, is limited to an amount corresponding to the amount in the building contract, exclusive of the cost of the land.

As mentioned above, damage covered by the Long-term Warranty should be reported to the consortium formed by Trygg Hansa and Skandia, or to Folksam. The consortium/Folksam decides if the damage is covered by the warranty. Disputes arising from the warranty should, in the first place, be referred to a claims committee (separate committees for the consortium and Folksam). If a party is not satisfied with the decision of the committee, the dispute can be settled in an ordinary court of law.

2.5.6　EXCLUSIONS

The warranty does not cover damage that has been detected or should have been detected during the Guarantee Period, or damage that is the result of such damage not having been sufficiently well remedied if the owner has realized, or should have realized this.

The warranty does not cover damage to refrigeration and freezing equipment, washing machines and similar domestic appliances. (For this kind of equipment there is normally a special three-year guarantee provided by the supplier of such goods.)

Damage due to natural occurrences is also excluded. The system does not have a definition of what constitutes Act of God.

The warranty does not cover damage resulting from age and use, want of proper care or the like.

2.6　Transfer of ownership and insurance cover

2.6.1　RULES

Both the Production Warranty and the Long-term Warranty are transferred to a new owner of the house.

3　LIABILITY AND PROFESSIONAL INDEMNITY INSURANCE

3.1　General

3.1.0　LAW OR CONTRACT OR BOTH

The professional indemnity insurance and liability insurance are not imposed by Swedish law. In ABK 87, however, it is stated in para. 13 that the consulting engineer/architect (the designer) should have such an insurance with an adequate cover unless the parties decide otherwise. In practice, this sort of insurance cover is always required by the client.

According to AB 92 the producer must take out a so-called 'Contractor's All Risk' insurance (CAR) which also includes a liability cover. The insurance policy shall be in force during the construction period and a two-year Guarantee Period.

3.1.1 TERMS AND CONCEPTS

'BUILDING'

See answer given under 1.1.1.

'INSURANCE POLICY'

The most common insurance policy covering the responsibilities that the designer carries according to ABK 87 is a running insurance on a yearly basis. The amount of coverage is usually three times SEK 1 m, which means that the insurance company limits its responsibility to pay up to a maximum SEK 1 m a year for all the damage that can occur within one assignment and, with several assignments within a year, not more than three times SEK 1 m.

The CAR insurance policy is also a running insurance on a yearly basis, both for designers and contractors. The level of premium for such a policy will depend upon the type of assignment or contract to be performed and the premium will be paid by the insured party on top of the premium for his yearly cover which, though, at the end of the insurance policy year will be reduced somewhat, due to the special project insurance cover that has been paid for separately. Understandably, there is no standard cover for project insurance, as is the case with the running insurance cover.

The running insurance follows the insured producer and the project insurance follows the project.

3.1.2 TECHNOLOGY AND INSURANCE TERMS AND CONDITIONS

Professional indemnity for designers and contractors on a running basis (see the Questionnaire, Appendix A):

- (a)–(e): These circumstances are not normally taken into account in a determining way when deciding the terms and level of premium.
- (f): After a while when the insurance company has been able to judge the insured party and his activities through a period of time, the claims record will affect the terms and the level of premium.

Special comment regarding (e): The insurance companies in Sweden do not have standing control bureaux that take an active part when the insurer evaluates the risks involved in issuing a policy for a designer or a contractor, nor do they, other than very exceptionally, visit sites and inspect ongoing works.

3.1.3 INSURER'S TECHNICAL CONTROL

The insurer never does what is suggested under 3.1.2, first paragraph.

The insurer has no control bureau as such of its own in the field of professional indemnity.

3.1.4 INSURERS' POSSIBLE INFLUENCE UPON BUILDING TECHNOLOGY

There is, generally speaking, no organized feed-back from the insurance companies to the building industry. Also, strangely enough, their experience is taken into account to a very limited extent, if at all.

Once the claims record of the insured party becomes worrying from the insurer's point of view, a probable action from the insurer is that the insured party is asked to cease using a technical process or a building component that has become the cause of many claims. Apart from that sort of ad hoc action towards a certain insurance party, there is no systematic way of conveying the insurer's technological experience to producers or clients.

Insurers do, however, categorize and rate their proposals in accordance with the producer's particular activities. The level of premium, whether it is based on the amount of wages paid to the employees during the year or based on the yearly turnover, will vary according to the amount and size of likely claims within the respective field of activity. For instance, architects involved in designing only detached houses have a much lower premium than a consulting engineering firm that designs chemical plants.

Categories etc. are not published.

3.1.5 NEW TECHNOLOGIES

The example under 3.1.5 in the Questionnaire (Appendix A) applies best to a project insurance situation where a risk analysis has to be performed by the insurer before deciding the terms and conditions of the policy. The question of the new technology for the project being used will be one of the elements of the insurer's analysis. Depending upon the nature of the new technology and its estimated risks, the premium will vary. In principle, however, it is perfectly possible to insure innovative building.

3.1.6 THE INSURANCE MARKET, GENERAL IMPRESSION

The competition is fierce. The premiums have for some time therefore been very low. They are now increasing somewhat, due mainly to a sharp increase in claims. There are several insurance companies that issue professional indemnity insurance for producers but the two biggest companies, Skandia and Trygg Hansa, have together the dominant share of the market.

There are also several foreign insurance companies who have put up their subsidiaries in Sweden and who provide insurance for professional indemnity within the building industry.

Foreign insurance companies without a presence in Sweden and concession to operate in Sweden are not allowed to provide insurance in Sweden.

3.1.7 COLLECTIVE INSURANCE FOR PROFESSIONAL OR OCCUPATIONAL GROUPINGS. GRADATION OF INSURANCE CONDITIONS

There are no model professional indemnity insurance contracts for certain groups of professionals in the building industry in Sweden. The closest we get to that situation is probably the cover for architectural work that the Association of Practising Architects (SPA) has negotiated on behalf of their members with a particular insurance company. It is quite a voluntary scheme and many architect firms decide to insure for their professional indemnity elsewhere. On the other hand, the Professional Indemnity Policies for designers are tailormade to suit their activities and the policy is drafted to cover the liability that the designer carries in ABK 87. The same thing goes for contractors and their particular needs. The policy is then based on AB 92.

There is no clear policy within the insurance companies to give reduction on the premium due to the insured party's willingness to follow professional upgrading courses. On the other hand, it might well, along with other elements, influence the premium before issuing a professional indemnity policy on a project basis.

3.2 During the Guarantee (Maintenance, Defects Liability) Period

3.2.0 EXISTENCE OF LIABILITY AND PROFESSIONAL INDEMNITY INSURANCE

Yes.

3.2.7 RIGHT OF RECOURSE AGAINST A PRODUCER

No money is withheld once the Final Inspection is done and the works are approved of (AB 92, Chapter 6, para. 10; the exception to this principle in AB 92, Chapter 6, para. 15 can be disregarded in this context). However, common practice is that the performance bond of 10 per cent of the contract sum is reduced to 5 per cent during the Guarantee Period to be used by the client in the event of the contractor not fulfilling his obligations during the Guarantee Period.

Another matter is the insurer's right of recourse against the liable party, such as the insured party's sub-contractor, or in certain very limited cases the insured party's employee.

3.5 After the Guarantee (Maintenance, Defects Liability) Period

3.5.1 NORMAL DURATION OF COVER; CANCELLATION

The cover for a professional indemnity insurance of the running type is in Sweden almost without exception on an annual basis.

It is extended year by year as long as the insured party pays the premium, which in itself might vary year by year in accordance with variation in wages or turnover regarding the company concerned.

A project insurance cover will naturally be tailormade to the requirements of the particular customer.

A non-payment of this premium does not immediately have the effect of the insurance contract being cancelled. The liability of the insurer commences at the beginning of the period of insurance, even if the premium has not yet been paid. This applies only if the premium is paid within 14 days from the date when the insurer sent out the premium notice.

If the premium for a renewed contract is paid too late, the contract shall nevertheless be renewed if the premium is paid within two months from the date when the insurer sent out the premium notice.

The insurer is entitled to give notice of cancellation of the contract if the insured party is late in paying his premium. Both the insurer and the insured party are entitled to give notice of cancellation of the contract when damage has occurred. The insured is entitled to give notice of cancellation of the contract if the need for insurance cover ceases wholly or to a significant degree.

The insurance policy covers all damage that is discovered during the insurance year on the condition, of course, that the insurance premium is duly paid.

3.5.2 LEVEL OF PREMIUM, EXCESS AND FRANCHISE

A professional indemnity insurance premium for designers will vary according to their activities, e.g. an architect firm designing solely detached houses will pay roughly 0.8 per cent of the total sum for wages per year, whereas a consultant engineering firm which concentrates on much more risky activities, e.g. on chemical technical projects, will pay roughly 3.5 per cent of the total sum of wages paid per year. Equivalent differences will be found with the Contractor's All Risks cover.

The duration of cover does not vary in relation to the level of premium. The prevalent time of duration is one year.

It is possible for the designer to reduce the excess to zero, but in practice this is not done. The normal level of excess is 10 per cent of the claimed sum on each and every damage. The excess must, however, never fall below one basic amount[1] and not exceed 10 basic amounts.

The tendency has been for premiums to go up in view of a significant rise in claims over the last few years. The level of excess does not seem to change much, at least when it comes to CAR insurance, and this is probably, to a certain extent at least, due to the fact that the major companies have an inclination to decentralize into autonomous divisions making a high excess heavy to carry for the individual division. Franchise arrangements are not commonly used by Swedish insurers.

The premiums are likely to increase considerably in years to come unless the increasing number of claims seen in this section of insurance is significantly curbed.

3.5.3 TIME LAPSES BETWEEN OBSERVATION OF DAMAGE AND INVESTIGATION

(a) In ABK 87 it is stated that if the client has the intention to claim for damages, he must, after getting to know of the actual damage, inform the designer of his intention without delay.

(b) In the event of the damage occurring during execution of the works, the client can, according to AB 92, claim for damages within three months of the issuing of the Certificate of Practical Completion, whereas if the damage occurs during the Guarantee Period he has to present his claim for damages to the contractor not later than three months after the expiry of this period. Finally, if the damage has occurred after the expiry of the Guarantee Period the claim has to be presented to the contractor not later than three months from the date of occurrence.

(c) The insured party is, according to the insurance terms, obliged to report to his insurer as soon as he gets to know of a damage likely to lead to claims for damages. Furthermore, the insured party is obliged to pass on to his insurer claims for damages as soon as he has received them. If he fails to live up to these requirements, the insured party might find himself without an insurance cover.

(d) There is no maximum time limit stipulated regarding the time elapsing between the insurer starting the investigation in co-operation with the insured party and the conclusion of the investigation.

[1] Basic amount according to the National Insurance Act; presently some SEK 34 000.

3.5.4 TIME ALLOWED BETWEEN THE INSURER'S RECOGNITION OF A CLAIMS EVENT AND HIS PAYING OUT

No such time limit is stipulated, either by law or in the prevalent insurance cover.

3.5.5 KINDS OF DEFECTS AND DAMAGE WHICH CONSTITUTE CLAIMS EVENTS; ONUS OF PROOF

Defects: see answer given under 1.2.5.

Damage: see answer given under 1.2.5.

When the contractual party is a main contractor or a main consultant engineer/architect, there is no need for the client to worry about who specifically is the liable party. All those who are assigned to the producer as sub-contractors or sub-consultants will, from the client's point of view, be regarded as being under the hat of the main contractor/main consultant and it is therefore sufficient for the client to show that a breach of duty has been committed on the main contractor's/consultant's side.

If, on the other hand, the client has contracted with all the contractors and designers concerned individually, he will have to identify the producer that has committed the breach of duty.

3.5.6 EXCLUSIONS

The following forms of breach of duty are excluded from the prevalent professional indemnity cover for designers:

- Injury, damage or loss which the insured has undertaken to indemnify in so far as the undertaking entails to pay damages over and above the provisions of ABK 87 or current laws relating to claims for damages.
- Injury, damage or loss to the extent that a claim for damages is founded solely on a pledge or guarantee.
- Liability to pay damages for minor injuries, damage or losses which would have been excluded if the liability limit in ABK 87, para. 5 had applied to the assignment.
- Injury, damage or loss which is due to the fact that an object or part thereof has, with respect to design, materials etc. not been given an attractive appearance.
- Injury, damage or loss arising under such circumstances that it should have been clear to the insured or those of his employees in responsible positions that injury, damage or loss would occur.
- Injury, damage or loss caused by the insured through gross negligence or intentionally.

United States

John B. Miller and Mark C. Fell

1 LIABILITY

1.1 General

1.1.0 LAW OR CONTRACT OR BOTH

Under the law generally prevailing in the US, producers are potentially liable for post-construction defects and damage both as a matter of law and, separately, as a matter of contract. Liability under contract is typically determined by the text of detailed contracts between producers and their clients, which define with specificity the contractual duties of each to the other. The breach of these contract duties in turn defines the scope of liability for resulting damage caused by such breach.

Liability at law is generally governed by the common law tort of negligence. At law, the distinction between those producers engaged in the design of the building, such as architects, engineers and consultants, i.e. 'designers', and those engaged by the client to execute works in connection with the building, i.e. 'contractors', is important, since the former are required to exercise reasonable *professional* skill in the performance of their function, while the latter group is typically held to the level of skill of a reasonable contractor in performing construction work.

Thus, there is a different standard of care applicable to designers in negligence. This standard does not require or guarantee perfect results, but rather the exercise of reasonably skilled professional judgement in the performance of the designer's duties. The preferred practice in the US is to define the duties of producers who are designers in a written contract in such detail that the contract also serves to define the duties for which the designer may be held liable in negligence.

The difference between the negligence liability standard for designers and the liability standard for contractors in contract creates the possibility that involuntary errors and omissions in the design are not a basis for liability of either the designer in negligence or the contractor in contract.[1]

In addition, producers are potentially liable under the common law theory of indemnity, independent of liability in negligence. The frequency of claims for indemnity is increasing in the US, in part because of the expanded use of contract indemnity provisions.

Producers are infrequently liable to the client for post-construction defects under statutory enactments, common law principles of strict liability in tort and common law principles of intentional interference with contractual relationships.

1.1.1 TERMS AND CONCEPTS

'BUILDING'

There is no unique definition of this term in American law, although the principles of post-construction liability of producers to clients are applicable to the design, planning or construction of any man-made improvement to real property, including dwellings, commercial office space, power plants, roads, bridges and public works.

There is no general distinction between the principles of liability applicable to new buildings and to rebuilding. Fixed installations are automatically included in the definition of the term 'building'. This is so well settled that questions as to fixed installations rarely arise. However, in large manufacturing facilities, the manufacturing equipment is typically bought by the client for installation subsequent to completion of the works. In such situations, installation of the equipment is not within the contract for construction.

'PRODUCER'

The term 'producer', as used in the Questionnaire (see Appendix A), is broad enough under American law to include:

(a) the client;
(b) the architect or engineer commissioned by the client;
(c) other consultants commissioned by the client to design or supply information for the design of the building or any part of it;

---

[1] The US legal system is comprised of 50 separate state common law jurisdictions, each with its own legislature, which have separately developed the common law of contracts as well as the common law of negligence. The courts created by the federal Constitution to decide disputes under federal law comprise another 'common law' court system, which interprets and develops the federal law of contracts as it relates to the construction of buildings for the federal government. Such contracts typically incorporate the common law of negligence of the individual state in which the building is located.

(d) the contractor or contractors engaged by the client to execute works in connection with the production of the building, or any part of it, together with consultants, sub-contractors, and suppliers engaged by the architect, engineer or contractor(s) in connection with the production of the building or any part of it.

The documents which identify producers are the contracts with the client. General principles of negligence may identify producers as a matter of law.

Public control bureaux (with municipal councils or building departments of localities) cannot generally be held liable to the client under American law. Thus, public control bureaux are not producers.

The liability of sub-contractors and suppliers to the client may result from contract or from law under principles of negligence. Contract liability may arise from warranties in written form provided by the sub-contractors or the suppliers directly to the client. Liability at law may arise from the negligent performance of work by the sub-contractors or from the supply of defective materials or negligent design of equipment provided by sub-contractors or suppliers which results in personal or property damage.

'TENANT'

A tenant generally enjoys no special rights under American law unless such rights are provided specifically as a matter of contract between the client and the producers. In limited circumstances, where a tenant can show that the tenant was the intended beneficiary of contractual agreements between the client and a specific producer, the tenant may be able to enforce such contract rights as if the tenant were a party to the contract between the owner and the producer. The doctrine is known as third party beneficiary contracts and is available under American law generally, but rarely applied in the construction industry because the specific tenant is generally not known at the time of the contract between the client and the producer.

1.2   During the Guarantee (Maintenance, Defects Liability) Period

The liability of those producers engaged by the client to design the building is generally not subject to a Guarantee Period.

In general, post-construction liability during the Guarantee Period is in accordance with specific contract terms between the client and some of the producers. Only those producers engaged to execute the works (that is, contractors, sub-contractors and suppliers) have such liability during the Guarantee Period. The liability of sub-contractors is in accordance with the contract terms between the sub-contractor and the contractor. The liability of suppliers during the Guarantee Period is in accordance with the terms of the contract between the supplier and either the sub-contractor or contractor with whom the supplier has contracted, or in accordance with the terms of specific written warranties on materials supplied or equipment provided by the supplier directly to the client.

The obligations of producers during the Guarantee Period are typically *in addition to* such obligations as the producers may have after the expiration of the Guarantee Period. For example, a typical construction contract in the United States provides that during the Guarantee Period the contractor will correct defects and will perform required maintenance services *with its own personnel*. Under such a contract, the contractor would still be liable for remedy of defects after the Guarantee Period, although the contractor is no longer required to perform corrective work with its own personnel.

1.2.1 DETERMINATION OF THE COMMENCEMENT OF THE PERIOD

The commencement of the Guarantee Period is typically determined by the date of a Certificate of Substantial Completion. Substantial completion is defined as the time when the building is sufficiently complete for the client to occupy and use it for the purposes intended. Substantial completion is not equivalent to Final Completion, the latter term being defined as the time when all items, including minor corrective work, are complete.

1.2.2 SIGNATORIES TO AND NAME OF THE DOCUMENT STATING THE COMMENCEMENT OF THE PERIOD

The document which commences the Guarantee Period is known as a Certificate of Substantial Completion. Signatories are typically the client, and *some* of the producers, namely, the contractor, and the primary architect or engineer engaged by the client. Public inspectors and municipal representatives are generally *not* signatories to the Certificate of Substantial Completion, nor generally are sub-contractors and suppliers.

Individual written warranties given by producers, including sub-contractors and suppliers, are typically provided separately to the client on or shortly after the date the Certificate of Substantial Completion is provided. Most such warranties commence on the date of substantial completion.

The Certificate of Substantial Completion is not equivalent to the right of the client to occupy the building. This right is generally granted through the issuance of a separate Certificate of Occupancy by a public inspector. The Certificate of Occupancy may or may not have issued at the time of substantial completion, although the issuance of a Certificate of Occupancy is conclusive that the building is sufficiently complete for the client to occupy and use for the purposes intended, that is, 'substantially complete'. The contract between the client and the producers may, and usually does, contain a specific definition of substantial completion, which may alter the definition by mutual consent.

1.2.3 THE PERIOD'S DURATION AND TERMINATION

The duration of the Guarantee Period is generally fixed as a matter of contract between the client and those producers engaged by the client to execute the work. The usual duration of the Guarantee Period as to maintenance and repair of defective work is one year. The duration of specific warranties provided directly to the client by suppliers varies with the nature of the equipment supplied. Many such warranties extend multiple years, typically between two and five years in duration.

There are generally no formalities exchanged between the client and the producers to mark the end of the Guarantee Period.

The Guarantee Period could be extended by mutual consent between the client and individual producers by contract, but this is not typical.

1.2.4 TIME ALLOWED FOR DIFFERENT CATEGORIES OF PRODUCERS FOR MAKING GOOD, REPAIRING OR PAYING DAMAGES

There are no particular procedures set forth in law to provide the period of time allowed for making good, repairing or paying damages to clients. In isolated circumstances, where the client is a local, state, or federal government, individual statutes may provide such time periods but this is generally not true with respect to buildings built between private clients and producers. Industry form contracts produced by the American Institute of Architects (AIA), and the Engineers Joint Contract Documents Committee (EJCDC)[2] provide that producers must make good or repair or pay damages within a reasonable time after notice of such defects or maintenance is given. There are no particular procedures laid down in these model contracts, although if such repairs or payments are not timely made, the client has the right to perform the work itself and assert a claim to recover the costs from the producer who is liable therefore.

---

[2] EJCDC members represent the following American engineering societies: National Society of Professional Engineers, the American Consulting Engineers Council and the American Society of Civil Engineers.

1.2.5 KINDS OF DEFECTS AND DAMAGE COVERED BY THE GUARANTEE

Defects that have not led to damage generally do not qualify for a claim for damages under American contract law. Thus, there must be a causal relationship between a defect and resulting damage in order for a claim to qualify for damages.

Defects are broadly defined in the US, primarily by the terms of the written specifications and plans provided by the designer as to how the building is to be erected. Non-conformance with the written specifications for the work, and non-compliance with applicable public building code requirements, are generally recognized as defects causing damage for which damages may be recovered.

Since the Guarantee Period is not generally applicable to designers, defects in the design of the building, including errors and omissions by the designer, are not within the scope of protection in the Guarantee Period. The terms of the written contract between those producers who are designers and the client may alter this rule, but such changes are unusual.

The rules for determining the kind and extent of damage which qualify for damages are as follows.

CONTRACT

The fundamental concept of damages under American contract law is compensation. This rule follows from the English common law rule expressed in *Hadley v Baxendale*,[3] probably the most quoted English damages case in the US. The rule is that the injured party (the non-breaching party) should be placed in the same financial position he would have been in had the contract been performed, so far as loss can be ascertained to have flowed from the breach as a natural and probable consequence or to have been within the contemplation of the parties at the time the contract was made, and so far as such compensation in money damages can be computed by rational methods upon a reasonably firm basis of fact.

Thus, for example, the damages recoverable in a typical construction case for a defect is the cost of remedying the defect. For example, where a pump is defective and must be replaced, the amount of damages is the cost of removing and replacing the defective pump with a proper pump.

In certain circumstances, however, this rule is altered, because the defect is of such a nature that the value of the building is not diminished in proportion to the cost of remedying the defect. For example, where a building is painted with a slightly different colour than that specified, the amount of damages recoverable is not the cost to repaint the entire building

---

[3] (1854) 9 Exch 341.

but, rather, the 'diminution in value' of the building with the slightly different paint.

A breaching party is only liable for compensation, not for the cost of placing the client in a better position than the client would have been in had there been no breach. Thus, changes in design which remedy a defect, but significantly improve the building at the same time, are compensated proportionately. Absolute certainty in calculating damages is not required, but only a reasonable likelihood of such damages on a reasonably firm basis.[4]

NEGLIGENCE CLAIMS

Personal injury

Damages for personal injury are also meant to be compensatory. The general rule is that the amount of damages is that sum of money which fairly compensates the injured person for the medical expenses, loss of earning capacity, physical pain and mental suffering that have been experienced in the past resulting from the negligent acts or omissions, and are likely to occur in the future.

Property damage

The amount of damages recoverable for property damage is the cost of placing the property in the same condition as it was prior to the negligent acts, or placing the property in the condition it would have been in but for the negligent acts.

Economic loss damages

There is substantial disagreement among the American states as to recovery under a negligence theory for purely economic losses sustained by third parties *not* contracting for the erection of the building. The states are widely split on this point.

Note that the client is generally not able to recover from the producer who executed the work, where the defects in the work are the result of errors or omissions in the design of the building as prepared by the architect or engineer engaged by the client. A series of cases in the US, beginning with the Supreme Court decision in *United States v Spearin*[5] hold that the client represents to those producers erecting the work that written plans and specifications provided by the client are sufficient and fit for their intended purpose with respect to the building. Thus, claims by the owner

---

[4] Litigants in *civil* cases generally have the right to trial by jury. Thus, juries often determine the amount of damages.
[5] 248 US 132 (1918).

against those producers who erected the works will generally fail if based upon alleged defects, errors or omissions in the plans and specifications. See 1.2.6.

1.2.6 EXCULPATION OF PRODUCER

Different rules apply to different categories of producers, the most significant distinction being between those who designed the building ('designers') and those retained by the client to erect the building ('contractors').

EXCULPATION OF DESIGNERS

Designers are typically exculpated from liability to the client for a series of events generally outlined in the contract between the client and designer. The terms of such contracts generally exculpate the designers from:

(a) defects in the work of contractors in erecting the building not reasonably discoverable by the designer;
(b) the construction means and methods chosen by the contractor to erect the building which means and methods result in damage;
(c) the safety programme used by the contractor, which safety programme results in damage;
(d) geotechnical information and services provided to the designer by other consultants retained by the client;
(e) variations in the final cost of construction from that estimated by the designer;
(f) damage which arises from use by the client of the plans and specifications provided by the designer on other buildings without authorization and approval by the designer;
(g) other damage caused as a result of faulty information provided by the client to the designer on which the designer was entitled to rely, including the evaluation and discovery of hazardous substances.

Under the most recent form of standard contract issued by the AIA, the client now indemnifies both the designer and the contractor from any damage arising from the presence of PCB(s) or asbestos.

EXCULPATION OF CONTRACTORS

Under typical contracts in the US, the contractor is not liable for a series of events which may cause damage to the client. For example:

(a) the contractor is not responsible to determine the compliance of the design of the building with applicable building codes and other public requirements,

(b) the contractor typically is not responsible for damage incurred by the client for loss of use of the building.

The making of final payment by the client to the contractor exculpates the contractor from further liability to the client after construction except for:

(a) any liens recorded by sub-contractors or suppliers against the title of the land on which the building is erected; or
(b) the failure of the building to comply with the drawings and specifications for the building; or
(c) the express terms of special warranties provided to the client by any producer.

Similarly, the acceptance of final payment by the contractor from the client is generally a waiver of all claims except those identified in writing as unsettled at the time the contractor makes the application for final payment.

EXCULPATION GENERALLY

Defects which are known to the client prior to final payment are waived with the making of final payment, as described above. Hidden defects are not waived.

Producers are not generally liable after substantial completion for damage caused by acts of third parties (including vandalism, sabotage and the like) or by Acts of God unless the risk of such damage has been contractually allocated to the producers. A contractual provision to such effect would be most unusual.

Other events such as the use of experimental techniques at the client's demand or with the client's consent or the failure to maintain the building or components thereof in accordance with maintenance instructions would also excuse producers from any post-construction liability for damage resulting from the failure of such experimental techniques or the failure of components not properly maintained.

The legal rules governing exculpation of producers depend largely on the particular terms of the contract between the producer and the client.

Generally, designers are under a duty of care to act without negligence. The terms of typical contracts with designers describe the duties owed by the designer to the client which particularity.

Contractors are held to a duty of strict compliance with the terms of plans and specifications provided by the client to the contractor for the erection of the building.

The 'duty of result', that the building must be without defects, is generally not applicable in American law, although a client and an individual producer may so contract. Such contracts are the exception to ordinary practice.

1.2.7 CONSEQUENCES OF PRODUCER'S NON-COMPLIANCE WITH UNEQUIVOCAL DUTY TO MAKE GOOD, REPAIR OR PAY DAMAGES

PRODUCER CAN BUT REFUSES

In this situation, the client is free to make good, repair, or pay damages, and thereupon charge the producer for the cost thereof. If funds are still owing to the producer from the client, the client generally can offset against the balance owed the producer.

PRODUCER CANNOT (E.G. BECAUSE OF BANKRUPTCY)

If a producer is in bankruptcy, the American bankruptcy laws suspend the producer's contractual obligations until it is either (a) finally liquidated, or (b) reorganized and removed from bankruptcy. Clients in this situation must themselves make good, repair or pay damages and file a claim in the bankruptcy court against the producer for the cost thereof. Generally, only a partial recovery is possible in a bankruptcy, if any recovery is made at all.

PRODUCER CANNOT BE TRACED (E.G. COMPANY DISSOLVED)

Corporate producers are generally liable to the client only to the extent of their assets. Generally, the client may not recover the cost of making good, repair or paying damages, from shareholders or stockholders of companies that are dissolved. A client who cannot trace a producer has only a very small chance of recovering damages.

SUCCESSIVE OWNERS

Unless the specific terms of the contract between the client and the producers or the terms of transferable warranties issued by the producers to the clients provide otherwise, successive owners are generally not entitled to recover damages from producers.

OTHER PRODUCERS (SAME BUILDING)

To the extent other producers of the same building are also responsible to make good, repair, or pay damages, the client can require such producers to do so. If such producer fails to do so, the client may perform such repairs itself, and charge the producer therefor.

PROPERTY AND LIABILITY INSURERS, IF ANY

Most property and personal injury insurance in the US is made on an occurrence basis.

Thus, a claim of property or personal injury made against a producer during the period of a property or personal injury policy will generally be defended by the producer's insurance carrier. This would include errors and omissions insurance typically carried by those producers who are designers. Insurers typically determine whether the policy covers the property or personal injury which occurred and, if so, either settle the claim against the insured producer, or defend any law suits or other proceedings brought against that producer.

While insurance is not generally required by law, except perhaps for public construction in isolated jurisdictions, most clients require producers to give evidence of continuous coverage in sizable money amounts, and this is the general business practice.

1.5 From expiry of the Guarantee (Maintenance, Defects Liability) Period

The post-construction liability of producers after expiry of the Guarantee Period is the same as that during the Guarantee Period, and has been described in 1.2.

1.5.1 DETERMINATION OF THE COMMENCEMENT OF THE (POST-GUARANTEE) PERIOD DURING WHICH A PRODUCER MAY BE HELD LIABLE

See 1.2.1.

1.5.2 NAME OF THE DOCUMENT STATING THE COMMENCEMENT OF THE POST-GUARANTEE PERIOD; ITS SIGNATORIES

See 1.2.2.

1.5.3 PERIOD(S) OF LIMITATION

See 1.2.3.

1.5.4 TIME ALLOWED (DIFFERENT CATEGORIES OF PRODUCERS) FOR MAKING GOOD, REPAIRING OR PAYING DAMAGES

See 1.2.4.

1.5.5 KINDS OF DEFECTS AND DAMAGE QUALIFYING FOR DAMAGES

See 1.2.5.

1.5.6 EXCULPATION OF PRODUCER

See 1.2.6.

1.5.7 CONSEQUENCES OF PRODUCER'S NON-COMPLIANCE WITH UNEQUIVOCAL DUTY TO REPAIR

See 1.2.7.

1.6 Transfer of ownership and producers' liability

1.6.1 RULES

Whether an initial successive owner has the same rights against producers as are enjoyed by the client is almost always a matter of contract. If the original agreement between producers and the client permits the assignment of the client's rights to successive owners, or successive entities, such as financing banks, and sometimes tenants, then, provided such an assignment is made by the client in accordance with the terms of the contract, such a successive owner has the same rights as the client against producers.

If the successive owner does not acquire the rights of the client against the producers by assignment, such successive owners or tenants lose the contract rights held by the client against the producers, the chain of contractual privity having been severed. Standard contract documents between those producers who are designers and those who are contractors permit written assignment of the client's rights to certain successors, generally financing banks, and sometimes tenants.

Without such an assignment, successive owners and tenants may only bring claims against producers at law under traditional doctines of negligence, or based on the specific terms of warranties that were given by producers to the client and were subsequently transferred to successive owners or tenants.

[2–3: *Not applicable*]

The questionnaire

1 LIABILITY

1.1 General

1.1.0 Law or contract or both
1.1.1 Terms and concepts

1.2 During the Guarantee (Maintenance, Defects Liability) Period

1.2.1 Determination of the commencement of the period
1.2.2 Signatories to and name of the document stating the commencement of the period
1.2.3 The period's duration and termination
1.2.4 Time allowed for different categories of producers for making good, repairing or paying damages
1.2.5 Kinds of defects and damage covered by the guarantee
1.2.6 Exculpation of producer
1.2.7 Consequences of producer's non-compliance with unequivocal duty to make good, repair or pay damages

1.5 From expiry of the Guarantee (Maintenance, Defects Liability) Period

1.5.1 Determination of the commencement of the (post-guarantee) period during which a producer may be held liable
1.5.2 Name of the document stating the commencement of the post-guarantee period; its signatories
1.5.3 Period(s) of limitation
1.5.4 Time allowed (different categories of producers) for making good, repairing or paying damages
1.5.5 Kinds of defects and damage qualifying for damages
1.5.6 Exculpation of producer
1.5.7 Consequences of producer's non-compliance with unequivocal duty to repair

1.6 Transfer of ownership and producers' liability

1.6.1 Rules

1 LIABILITY

The topic is PRODUCERS' liability concerning the quality of the BUILDING.

1.1 General

1.1.0 LAW OR CONTRACT OR BOTH

Is PRODUCERS' POST-CONSTRUCTION liability prescribed by law or in (standard) contract(s), or both?

Please give such references which you consider useful, in your own language and, if possible, in English and French.

Note: Different categories of PRODUCERS may, within one and the same system, by governed by different rules, cf. question 1.1.1 'Producers'.

1.1.1 TERMS AND CONCEPTS

Several terms have a specific meaning in each particular system.

'BUILDING'

It may not be possible to define this term; of course, if a ready definition exists in your system (for instance in the context of insurance), please quote here.

In the case of your system distinguishing between new building and rebuilding, please indicate this here.

Please state if fixed installations are automatically included (water, sewage, ventilation, air conditioning, heating, electricity etc.; lifts, escalators etc.?). The same question applies to access roads and paths, parking decks, lampposts etc.

'PRODUCER'

(See also the List of Terms, Appendix B: of course the aim here is to identify those who, according to your system, may be liable after the taking over of the BUILDING.)

This term will recur frequently in the following questions. There may, however, be systems of such a nature that it is not or hardly possible to list exhaustively all imaginable producer categories. If your system is of this type, please give an outline of the principles of delimitation.

Examples:

- In some systems the public control bureau (or the municipal council) cannot be held liable to the CLIENT, and thus is not a 'producer' as we use the term;

- in other systems there may be liability, but only under certain circumstances, and perhaps with particular limitations; in that case please comment here (and, if necessary, wherever the term PRODUCER occurs in the following questions);
- in other systems again there is no distinction between private producers and public bureaux, and the latter are, then, 'producers'.

- In some systems it is exclusively those parties with whom the client has a contractual relationship which are 'producers';
- in others even sub-consultants and sub-contractors may be held liable to the CLIENT; they would thus fall under our heading of PRODUCER;
- there are also (at least) two systems where certain suppliers, under contract with, say, a builder or an installer – and thus without any contract with the CLIENT – are, nevertheless, under certain circumstances, liable to the CLIENT, and thus must be classified as PRODUCERS; if your system is of this type, please outline the delimitation of the supplier category liable.

'TENANT'

Please indicate here whether a tenant enjoys certain rights vis-à-vis the PRODUCERS (not vis-à-vis the CLIENT or OWNER, this latter relationship being outside our scope).

1.2   During the Guarantee (Maintenance, Defects Liability) Period

Note for non-English readers: the terms 'Maintenance Period' and 'Defects Liability Period' are typically Anglo-Saxon and would, in other systems, correspond to 'Guarantee Period'.

1.2.1   DETERMINATION OF THE COMMENCEMENT OF THE PERIOD

E.g.: Certificate of Practical Completion without / with some / irrespective of reservations; other documents with the same judicial effect in the absence of Certificate; etc.

1.2.2   SIGNATORIES TO AND NAME OF THE DOCUMENT STATING THE COMMENCEMENT OF THE PERIOD

Please give the name of the certificate or other document, in your own language and, if possible, in English and/or French.

Signatories could be e.g. the CLIENT, one of / some of / all of the PRODUCERS (cf. question 1.1.1, the concept of 'producer'), including / excluding control bureau(x), public inspector(s) etc.

1.2.3 THE PERIOD'S DURATION AND TERMINATION

Please indicate:

- Whether these are fixed in law or in contract.
- If in contract, what is the usual duration?
- Formalities, if any, for marking the end of the period (for instance issuing of a certificate).
- Could the period be extended? If so, under what circumstances?

1.2.4 TIME ALLOWED FOR DIFFERENT CATEGORIES OF PRODUCERS FOR MAKING GOOD, REPAIRING OR PAYING DAMAGES

Any particular procedures laid down (in law or model contract)?

1.2.5 KINDS OF DEFECTS AND DAMAGE COVERED BY THE GUARANTEE

DEFECTS (please refer to the List of Terms, Appendix B): It is probably rare that a DEFECT not having led to DAMAGE will qualify for DAMAGES. Have you any instance of this in your system?

DAMAGE: Are there, in your system, any fixed rules for determining the kind and the extent of DAMAGE which will qualify for DAMAGES? If so, any references would be useful.

1.2.6 EXCULPATION OF PRODUCER

If different rules apply to different categories of PRODUCERS, please clarify.

Please list, or at least outline, circumstances which, in your system, will exculpate PRODUCERS from any liability for imperfections, deterioration or malfunction (including DAMAGE) of the BUILDING.

In several systems, *for instance*, no PRODUCER is liable for

- any DEFECT (cf. the List of Terms, Appendix B) which could reasonably have been discovered and ought to have either been notified to the PRODUCER(s) and included in the Certificate of Practical Completion, or prevented the issuing of the Certificate, whether or not the DEFECT had developed into DAMAGE. If such is the rule in your system, please clarify the liability of the person inspecting and/or issuing the said Certificate (i.e. in case he ought to, but did not, discover or notify the DEFECT);

- deterioration and imperfections due to
  (a) action by third party (vandalism, sabotage etc.), or caused by an interruption of normal supplies to the premises (for instance of fuel, the cold causing deterioration of the BUILDING's foundations or piping) unless the interruption is the consequence of a DEFECT (cf. the List of Terms, Appendix B), in which case the PRODUCER is, of course, liable;
  (b) natural Act of God (or force majeure). If at all possible, please comment. For instance, does your system have a definition of what constitutes *unexpected* natural forces (gales, earth tremors (high or low), ground-water levels, temperatures, rain, snow, hail, thunderstorms etc.)?
  (c) other Act of God (or force majeure), for instance ramming by a vehicle. If so, *any* vehicle?

In some systems the PRODUCER is exculpated (not liable):

- if he has used experimental techniques, or techniques with a known risk of failure, at the CLIENT's demand or with the CLIENT's consent;
- if the BUILDING's user (the CLIENT, a SUCCESSIVE OWNER, a TENANT) has not followed instructions as to maintenance or operation of the building (including, presumably, its installations, cf. your reply to the question under 1.1.1, 'Building'), provided always that adequate instructions were supplied by the PRODUCER.

Finally, much depends on the doctrine your system applies when it comes to the definition of what the PRODUCER engages himself to accomplish: which ERRORS OR OMISSIONS constitute a BREACH OF DUTY and which – if any – do not? This, of course, is another way of asking the question: 'Which ERRORS OR OMISSIONS will implicate the PRODUCER's liability?'

In building and construction one often meets the following two doctrines: The PRODUCER is either under:

(a) a 'Duty of care' (duty not to act negligently). This is often said to imply that the PRODUCER must be 'reasonably' capable, and that he is presumed to execute his work with a 'reasonable' degree of competence; there is, according to this doctrine, BREACH OF DUTY only if the PRODUCER acted negligently, and to decide whether there is negligence – and hence BREACH OF DUTY – one queries whether the ERROR OR OMISSION committed would have been committed also by a 'normal' PRODUCER within the same profession or trade.

Or he is under:

(b) a 'Duty of result', also sometimes called a 'strict duty', which implies that the BUILDING must be without DEFECTS; according to this doctrine, if the result leaves something to be desired there is BREACH OF DUTY. To decide whether an ERROR OR OMISSION constitutes a BREACH OF DUTY or not, one therefore looks at the result, the BUILDING. Whether the PRODUCER proved to be careful and competent or not is without relevance according to this doctrine: it is the 'thing' (the BUILDING) which counts (*'res ipsa loquitur'*, 'the thing speaks for itself'). If the BUILDING suffers from a DEFECT there is BREACH OF DUTY.

To illustrate: suppose that the PRODUCER followed accepted practices, but these later proved to be wrong, and DAMAGE occurs. Is the PRODUCER in BREACH OF DUTY?

Clearly this question has a strong bearing on the liability – if any – of the authors of documents setting norms or standards, including Agrément Certificates (which may, in many systems, be said to define 'accepted practices'), and an authority's (for instance the local authority's) approving of, say, foundations, but also, perhaps as importantly, the introduction of new technologies, where 'by definition' accepted practices may be insufficiently experimented.

1.2.7   CONSEQUENCES OF PRODUCER'S NON-COMPLIANCE WITH UNEQUIVOCAL DUTY TO MAKE GOOD, REPAIR OR PAY DAMAGES

(Same question as under 1.5.7, but now during the Guarantee Period.)
Please outline the following three scenarios:

- PRODUCER can but refuses
- PRODUCER cannot (e.g. because of bankruptcy)
- PRODUCER cannot be traced (e.g. company dissolved)

with respect to the situation of

- the CLIENT (or SUCCESSIVE OWNER)
- other PRODUCERS (same BUILDING)
- property and liability (or professional indemnity) insurers, if any.

1.5   From expiry of the Guarantee (Maintenance, Defects Liability) Period

Note for non-English readers: the terms 'Maintenance Period' and 'Defects Liability Period' are typically Anglo-Saxon and would, in other systems, correspond to 'Guarantee Period'. An alternative name for this section could be 'post-Guarantee Period'.

1.5.1 DETERMINATION OF THE COMMENCEMENT OF THE (POST-GUARANTEE) PERIOD DURING WHICH A PRODUCER MAY BE HELD LIABLE

Please refer to questions 1.2.1 and 1.2.3 (where the normal guarantee or 'Defects Liability' period's commencement and duration are indicated), and you may wish to refer also to question 1.5.3 concerning limitation period(s).

In some systems, the answer may be quite simple, but please state whether the commencement may be, or always is, postponed in case DAMAGE occurred and was repaired during the guarantee period.

In other systems the answer may be very intricate, cf. question 1.5.3; if your system comprises liability in tort, for instance, the commencement may be the date:

- on which (post-guarantee) DAMAGE occurred, or
- when DAMAGE was 'reasonably discoverable', whether discovered or not ('the doctrine of discoverability'), or
- when the DAMAGE was, in fact, discovered, or
- on which the BREACH OF DUTY from which the DAMAGE originated was committed, or again
- on which a SUCCESSIVE OWNER suffered a loss (by buying the property tainted by its unknown DEFECT),

to name a few doctrines important to some systems.

If your system is such that the answer is complicated, please refer to the comment at the end of 1.2.6.

1.5.2 NAME OF THE DOCUMENT STATING THE COMMENCEMENT OF THE POST-GUARANTEE PERIOD; ITS SIGNATORIES

In some systems there is no formal certificate or other document stating the commencement, in others there is, cf. question 1.5.1. If, in your system, there is a formal document, please mention its name in your own language and if possible in English or French.

1.5.3 PERIOD(S) OF LIMITATION

(*Note*: If applicable, please indicate period(s) of limitation in tort and in contract, cf. question 1.5.1.)

Are the limitation periods the same for all categories of PRODUCERS?

1.5.4 TIME ALLOWED (DIFFERENT CATEGORIES OF PRODUCERS) FOR MAKING GOOD, REPAIRING OR PAYING DAMAGES

Particular procedures laid down (in law or model contract)?
Same rules in tort and in contract, cf. 1.5.1 and 1.5.3?

1.5.5 KINDS OF DEFECTS AND DAMAGE QUALIFYING FOR DAMAGES

DEFECTS (please refer to the List of Terms, Appendix B): It is probably rare that a DEFECT not having led to DAMAGE will qualify for DAMAGES. Have you any instance of this in your system?

DAMAGE: Are there, in your system, any fixed rules for determining the kind and the extent of DAMAGE which will qualify for DAMAGES? If so, any references would be useful.

1.5.6 EXCULPATION OF PRODUCER

This question need be treated only if the rules applied after the expiry of the Guarantee Period are different from those valid during the said period (treated under 1.2.6).

1.5.7 CONSEQUENCES OF PRODUCER'S NON-COMPLIANCE WITH UNEQUIVOCAL DUTY TO REPAIR

(Same question as under 1.2.7, but here after the Guarantee Period.)
Please outline the following three scenarios:

- PRODUCER can but refuses
- PRODUCER cannot (e.g. because of bankruptcy)
- PRODUCER cannot be traced (e.g. company dissolved)

with respect to the situation of

- the CLIENT (or SUCCESSIVE OWNER)
- other PRODUCERS (same BUILDING)
- property and liability (or professional indemnity) insurers, if any.

1.6 Transfer of ownership and producers' liability

1.6.1 RULES

Please compare a SUCCESSIVE OWNER's rights (to claim DAMAGES from a PRODUCER) to those enjoyed by the CLIENT; see also 1.1.1 ('Tenant') and the List of Terms (Appendix B: 'Succesive owner(s)').

2 PROPERTY (OR MATERIAL DAMAGE) INSURANCE

2.1 General

2.1.0 Law or contract or both
2.1.1 Terms and concepts
2.1.2 Technology and insurance terms and conditions
2.1.3 Insurer's technical control
2.1.4 Insurers' possible influence upon building technology
2.1.5 New technology
2.1.6 The insurance market, general impression

2.2 During the Guarantee (Maintenance, Defects Liability) Period

2.2.0 Existence of this kind of insurance
[2.2.1–2.2.6: *Vacant*]
2.2.7 Right of recourse against the producer

2.5 After the Guarantee (Maintenance, Defects Liability) Period

2.5.1 Normal duration of cover; cancellation
2.5.2 Levels of premiums; excess; franchise
[2.5.3: *Vacant*]
2.5.4 Time allowed for notice of claim and for payment of damages
2.5.5 Kinds of defects and damage which constitute claims events
2.5.6 Exclusions

2.6 Transfer of ownership and insurance cover

2.6.1 Rules

2 PROPERTY (OR MATERIAL DAMAGE) INSURANCE

We are interested in insurance covering the whole or part of any loss suffered by a POLICY HOLDER – normally the CLIENT or a SUCCESSIVE OWNER, sometimes a TENANT – caused by DAMAGE (or, possibly, a

DEFECT) affecting the BUILDING, as these words are defined or described in the List of Terms (Appendix B).

In case no such insurance exists in your country, please state this and proceed to section 3.

Other insurance arrangements offered to CLIENTS, OWNERS or, as the case may be, TENANTS – such as third party, fire, water, burglary, house-breaking, household – are considered to be outside our scope; you may, however, also treat those kinds of insurance if you find this useful, but only if they cover losses due to DAMAGE (in our sense, see the List of Terms, Appendix B).

Particularly for the US: if you find difficulties in describing the OCIP system (Owner Controlled Insurance Program) in accordance with the present format, please feel free to arrange your own format.

2.1 General

2.1.0 LAW OR CONTRACT OR BOTH

Please state whether property insurance (covering the BUILDING) is imposed by law, or otherwise is always/often/sometimes required, for instance by financiers or mortgagees.

If such insurance is legally mandatory, please quote the legal reference in your own language and, if possible, in English and/or French.

2.1.1 TERMS AND CONCEPTS

'BUILDING'

Same question as under 1.1.1.

'POLICY HOLDER'

(See the List of Terms, Appendix B.)

Please list typical policy holders, e.g. the CLIENT, each (?) SUCCESSIVE OWNER, a TENANT? See also question 2.6.1.

2.1.2 TECHNOLOGY AND INSURANCE TERMS AND CONDITIONS

By 'terms and conditions' we particularly mean premium(s), EXCESS, FRANCHISE(S) and exclusion(s).

The following questions are listed in order to facilitate structuring your reply, but please feel free to deviate and elaborate as you find useful.

Do terms and conditions depend:

(a) upon the latitude given to the insurer to collect technical information about the project in hand

- during the progamming stage (where the CLIENT, in his brief, states his wishes concerning the BUILDING and conveys them to his future PRODUCERS),
- during the design stage (architecture, technology etc.),
- during the building stage (including installation and equipment),
- at the issuing of the Certificate of Practical Completion.

(b) on the geographical location of the BUILDING, its height, free spans, useful loads and other data (normally) determined in the brief;

(c) on the actual technology to be used (geotechnique, spans, wind bracing, materials etc.);

(d) on the organization of the works, e.g. the levels of competence and skill of different participants, including different suppliers;

(e) on whether the insurer uses his own control bureau (see also question 2.1.3 on the matter of technical control).

2.1.3 INSURER'S TECHNICAL CONTROL

Please comment. For example, does the insurer always/often/seldom/never

- have his own organization for control?
- consult the CLIENT's control bureau?
- consult the public control bureau?

Does the manner of organizing the control – if any – influence terms and conditions as dealt with under question (e) in 2.1.2?

2.1.4 INSURERS' POSSIBLE INFLUENCE UPON BUILDING TECHNOLOGY

Do the insurers in your country:

- pool their technological experiences ('feed-back')?
- convey their technological experience, in a more or less systematic manner, to PRODUCERS (including private and public control bureaux)? To CLIENTS?

Would you say that their experience is taken into account:

- by authorities issuing codes and norms?
- by standardization organizations?
- by the organization responsible for Agrément Certificates (or similar)?
- by authors of textbooks and handbooks?

2.1.5 NEW TECHNOLOGY

Assuming that the insurer has access to all relevant information in the case of a project where new techniques are to be used, is it then

(a) theoretically, and
(b) practically

possible to insure such (partly) innovative buildings? If so, how would one negotiate terms and conditions?

2.1.6 THE INSURANCE MARKET, GENERAL IMPRESSION

This question is rather vague, but a rough idea would be useful. This is, of course, in practice linked to questions 2.1.4 and 2.1.5. For instance:

Is the competition normally considered efficacious (or is there a tendency towards monopoly or oligopoly)? Is competition national or international?

2.2 During the Guarantee (Maintenance, Defects Liability) Period

(See 1.2 if in doubt about the meaning of the words in brackets.)

2.2.0 EXISTENCE OF THIS KIND OF INSURANCE

Does property insurance covering losses due to DAMAGE during the Guarantee Period (or Maintenance or Defects Liability Period) exist in your system? If not, please state this and proceed to 2.5.

[2.2.1–2.2.6: *Vacant*]

2.2.7 RIGHT OF RECOURSE AGAINST THE PRODUCER

Presumably the insurance will be triggered only if the PRODUCER fails to make good, repair or pay DAMAGES, cf. question 1.2.7. Does the insurer normally have the right of recourse against the PRODUCER?

2.5 After the Guarantee (Maintenance, Defects Liability) Period

(See 1.2 if in doubt about the meaning of the words in brackets.)

2.5.1 NORMAL DURATION OF COVER; CANCELLATION

Is the cover always/normally annual? If so, will the premium and the possible franchise be the same for all consecutive years?

If more than one year, how many years? Is the premium paid once for the whole period? If not, what is the periodicity? Is the premium the same during the whole period? If not, is there a model formula?

Under which circumstances is each party permitted to cancel the insurance contract prior to normal expiry?

2.5.2 LEVELS OF PREMIUMS; EXCESS; FRANCHISE

Please indicate, roughly, preferably for different kinds of BUILDINGS:

- some typical premiums (per period covered, cf. question 2.5.1),
- typical levels of EXCESS,
- typical levels of FRANCHISE.

[2.5.3: *Vacant*]

2.5.4 TIME ALLOWED FOR NOTICE OF CLAIM AND FOR PAYMENT OF DAMAGES

Please outline and comment on the rules, if any, for both parties.

2.5.5 KINDS OF DEFECTS AND DAMAGE WHICH CONSTITUTE CLAIMS EVENTS

DEFECTS (please refer to the List of Terms, Appendix B): It is probably rare that a DEFECT not having led to DAMAGE will constitute a CLAIMS EVENT. Have you any instance of this in your system?

DAMAGE: Are there, in your system, any fixed rules for determining the kind and the extent of DAMAGE which will constitute a CLAIMS EVENT? If so, any references would be useful.

2.5.6 EXCLUSIONS

(You may wish to refer also to 1.2.6 and 1.5.6.)

In several systems the insurance may not cover:

- any DEFECT (cf. the List of Terms, Appendix B) which could reasonably have been discovered and ought to have either been notified to the PRODUCER(s) and included in the Certificate of Practical Completion, or prevented the issuing of the Certificate, whether or not the DEFECT had developed into DAMAGE,
- any loss due to action by third party,
- any loss due to an Act of God (or force majeure) from natural sources. If at all possible, please comment. For instance, does your system have a definition of what constitutes *unexpected* natural forces (gales, earth tremors (high or low) ground-water levels, temperatures, rain, snow, hail, thunder etc.)?
- any loss due to other Acts of God (or force majeure), for instance ramming by a vehicle. If so, *any* vehicle?
- any loss caused by the BUILDING's user not having followed instructions as to maintenance or operation of the BUILDING.

Please comment on the above suggestions and indicate other exclusions, if any.

2.6   Transfer of ownership and insurance cover

2.6.1   RULES

Presumably the insurance cover 'follows' the BUILDING and not its OWNER; are there any procedures to be complied with when transferring the insurance cover? For example, is a (new) inspection of the premises called for?

3 LIABILITY AND PROFESSIONAL INDEMNITY INSURANCE

3.1   General

3.1.0   Law or contract or both
3.1.1   Terms and concepts
3.1.2   Technology and insurance terms and conditions
3.1.3   Insurer's technical control
3.1.4   Insurers' possible influence upon building technology
3.1.5   New technologies
3.1.6   The insurance market, general impression
3.1.7   Collective insurance for professional or occupational groupings. Gradation of insurance conditions

3.2   During the Guarantee (Maintenance, Defects Liability) Period

3.2.0   Existence of liability and professional indemnity insurance
[**3.2.1–3.2.6**: *Vacant*]

3.2.7 Right of recourse against a producer

3.5 After the Guarantee (Maintenance, Defects Liability) Period

3.5.1 Normal duration of cover; cancellation
3.5.2 Level of premiums, excess and franchise
3.5.3 Time lapses between observation of damage and investigation
3.5.4 Time allowed between the insurer's recognition of a claims event and his paying out
3.5.5 Kinds of defects and damage which constitute claims events; onus of proof
3.5.6 Exclusions

3 LIABILITY AND PROFESSIONAL INDEMNITY INSURANCE

We are interested in insurance taken out by any PRODUCER to cover himself (normally only in part) against his risk of having to pay DAMAGES awarded to the CLIENT, a SUCCESSIVE OWNER, or – in certain systems – a TENANT because of a DEFECT within or DAMAGE to the BUILDING after the takeover, as these words are defined or described in the List of Terms (Appendix B), i.e. as the consequence of a BREACH OF DUTY.

Our field of interest is accordingly limited to DEFECTS and DAMAGE occurring after the CLIENT has taken over the BUILDING, cf. 2; we thus exclude other forms of liability insurance that PRODUCERS may take out (see 'POST-CONSTRUCTION' in the List of Terms, Appendix B).

Particularly for the US: If you find difficulties in describing the OCIP system (Owner Controlled Insurance Program) in accordance with the present format, please feel free to arrange your own format.

3.1 General

3.1.0 LAW OR CONTRACT OR BOTH

(You may wish to refer to 3.1.1 if different rules apply to different categories of PRODUCERS.)

Please state whether liability and/or professional indemnity insurance (covering the PRODUCER) is imposed by law, or otherwise is always/often/sometimes required, for instance by financiers or mortgagees.

If legally mandatory, please quote the name of the law in your own language and, if possible, in English and/or French.

3.1.1   TERMS AND CONCEPTS

'BUILDING'

Same question as under 1.1.1.

'INSURANCE POLICY'

Please list and elucidate at least the most important principles, such as 'project insurance' and 'running insurance'; kindly explain, for example, whether the insurance 'follows' the project (= the BUILDING), or the insured PRODUCER. See also 3.5.4.

'POLICY HOLDER', 'INSURED PARTY'

(See the List of Terms, Appendix B.)

Presumably the POLICY HOLDER is the same party as the INSURED PARTY.

3.1.2   TECHNOLOGY AND INSURANCE TERMS AND CONDITIONS

By 'terms and conditions' we particularly mean premium(s), EXCESS, FRANCHISE(S) and exclusion(s).

The following questions are listed in order to facilitate structuring your reply, but please feel free to deviate and elaborate as you find useful. You may also wish to give different replies for different categories of PRODUCERS.

Do terms and conditions depend:

(a) upon the latitude given to the insurer to collect technical information about the project in hand (for further details see 2.1.2);
(b) on the geographical location of the BUILDING, its height, free spans, useful loads or other data (normally) determined in the brief;
(c) on the actual technology to be used (geotechnique, spans, wind bracing, materials etc.);
(d) on the organization of designer's or control bureau's office, or of the works, e.g. the levels of competence and skill of different participants, including different suppliers;
(e) on whether the insurer uses his own control bureau (see also question 3.1.3 on the matter of technical control);
(f) on the claims record of the INSURED PARTY. See also 3.1.4.

3.1.3 INSURER'S TECHNICAL CONTROL

Please comment. For instance, does the insurer always/often/seldom/never

- follow design and site work through his own organization for control?
- consult the CLIENT's control bureau?
- consult the public control bureau?

Does the manner of organizing the control – if any – influence terms and conditions as dealt with under question (e) in 3.1.2?

3.1.4 INSURER'S POSSIBLE INFLUENCE UPON BUILDING TECHNOLOGY

General impressions of the kind listed under 2.1.4 would be useful here.

Do insurers categorize, classify and rate their proposers (the PRODUCERS)? If so, are categories, classification and ratings published?

3.1.5 NEW TECHNOLOGIES

Comments and observations concerning how your system 'reacts to' new technologies proposed by different categories of PRODUCERS would be most useful.

For example:

Assuming that the insurer has access to all relevant information in the case of a project where new techniques are to be used, is it then

(a) theoretically, and
(b) practically

possible to insure such (partly) innovative buildings? If so, how would one negotiate terms and conditions?

3.1.6 THE INSURANCE MARKET, GENERAL IMPRESSION

Obviously, this question is rather strongly linked to questions 2.1.4, 2.1.5, 3.1.4 and 3.1.5, but please refer to 2.1.6 for a suggested list of points to discuss.

3.1.7 COLLECTIVE INSURANCE FOR PROFESSIONAL OR OCCUPATIONAL GROUPINGS. GRADATION OF INSURANCE CONDITIONS

In several countries certain groups of professions or trades have established model professional indemnity or liability insurance contracts. Is this the case in your country?

In at least one country architects who formally engage themselves to follow continuous professional upgrading courses obtain more favourable professional indemnity insurance conditions than their colleagues who do not undertake such an engagement. Have you any experience of this or any similar arrangements in your country?

3.2 During the Guarantee (Maintenance, Defects Liability) Period

(See 1.2 if in doubt about the meaning of the words in brackets.)

3.2.0 EXISTENCE OF LIABILITY AND PROFESSIONAL INDEMNITY INSURANCE

Do liability and professional indemnity insurance covering the Guarantee (Defects Liability, Maintenance) Period exist in your system? If not, please state this and proceed to 3.5.

[3.2.1–3.2.6: *Vacant*]

3.2.7 RIGHT OF RECOURSE AGAINST A PRODUCER

If these forms of insurance do exist, will the insurer be entitled to claim reimbursement from the monies (usually 5 (?) per cent) withheld by the CLIENT until the end of the Guarantee Period?

3.5 After the Guarantee (Maintenance, Defects Liability) Period

(See 1.2 if in doubt about the meaning of the words in brackets.)

3.5.1 NORMAL DURATION OF COVER; CANCELLATION

The question of duration of cover may prove extremely difficult because of the several doctrines in force, even, sometimes, within one and the same country.

That is why it seems difficult to present a series of questions, but the following are nevertheless quoted:

- Is the cover always/normally annual or does it run over more than one year?
- Does non-payment of the premium necessarily entail cancellation of the insurance contract?
- Is it possible for each of the parties to cancel the insurance contract prior to its normal expiry?

It is not, on the other hand, necessary to describe the 'doctrine(s)' applied in your system concerning the manner in which one decides whether a CLAIMS EVENT falls or does not fall within the period covered (e.g. the date of the deed which generated the CLAIM; the date the DAMAGE occurred; the date on which the PRODUCER handed over his work; the date of the CLIENT's or a SUCCESSIVE OWNER's declaration; and so on, cf. question 1.5.1). Of course, if there is only one doctrine within your system, please describe it.

3.5.2 LEVEL OF PREMIUMS, EXCESS AND FRANCHISE

It will suffice to indicate approximative levels, but per category of PRODUCER and, if necessary (cf. question 3.1.2), according to the gradations which may follow from different degrees of technical difficulty.

Please indicate the duration of cover for each premium (which each premium will buy).

Concerning EXCESS: is it possible for the PRODUCER to reduce this to zero, or is this expressly forbidden (as in at least one country)?

Could you possibly, preferably per professional or vocational category, describe tendencies observed over the last few years within your system, in what concerns premiums, EXCESS and FRANCHISE? Are there discernible trends for the future?

3.5.3 TIME LAPSES BETWEEN OBSERVATION OF DAMAGE AND INVESTIGATION

Please consider the following points in time:

(a) the CLIENT's (or a SUCCESSIVE OWNER's or possibly a TENANT's) discovery of DAMAGE (or possibly of a DEFECT);
(b) the CLIENT's (or the SUCCESSIVE OWNER's or possibly the TENANT's) lodging a claim for DAMAGES against one or several PRODUCER(s);
(c) the PRODUCER or his insurer starting an investigation of the claim;
(d) the conclusion of the investigation.

Please indicate whether the time lapses between these – or other relevant – points in time are regulated in some way or another (custom, contract, law).

You may wish to refer also to question 1.5.3; please also see question 3.5.4 and, in case your system includes property insurance, question 2.5.4.

3.5.4 TIME ALLOWED BETWEEN THE INSURER'S RECOGNITION OF A CLAIMS EVENT AND HIS PAYING OUT

Determined by law (on insurance or other law)?

3.5.5 KINDS OF DEFECTS AND DAMAGE WHICH CONSTITUTE CLAIMS EVENTS; ONUS OF PROOF

DEFECTS (please refer to the List of Terms, Appendix B): It is probably rare that a DEFECT not having led to DAMAGE will constitue a CLAIMS EVENT. Have you any instance of this in your system?

DAMAGE: Are there, in your system, any fixed rules for determining the kind and the extent of DAMAGE which will constitute a CLAIMS EVENT? If so, any references would be useful.

Does it rest with the plaintiff (the CLIENT, a SUCCESSIVE OWNER etc.) to identify the individual PRODUCER who has committed the BREACH OF DUTY which has led to observed DAMAGE (possibly an observed DEFECT) in order to file a claim for DAMAGES against him? Or will the 'thing speak for itself' (is the '*res ipsa loquitur*' principle applied)?

3.5.6 EXCLUSIONS

May there, in your system, be forms of BREACH OF DUTY having caused DAMAGE such that the PRODUCER's liability or professional indemnity insurance would not cover him? If such forms exist, please outline the doctrine(s) concerning exclusions applied, and, if possible, list the forms of BREACH OF DUTY excluded from the insurance cover.

Cases of fraud etc. may be left aside, but you may want to treat them in order to elucidate how your liability and professional indemnity insurance works.

List of terms

This list of terms is arranged according to three headings:

- General terms   Client, Damages, Post-Construction, Producer, Successive owner(s), Tenant
- Aetiological terms   (from 'Aetiology: Assignment of a cause'): Breach of Duty, Error or Omission, Damage, Defect.
- Insurance terms   Claims Event, Excess, Franchise, Insurance, Insured Party, Policy Holder

GENERAL TERMS

CLIENT

(Fr.: *Maître de l'ouvrage*)
The entity for whom the BUILDING is produced.

    Avoid such terms as 'employer' (double meaning), 'purchaser' (not necessarily the client), 'owner' (see SUCCESSIVE OWNER(S))

DAMAGES*

(Fr.: *Indemnité*)
Awards intended to compensate for losses incurred as a result of a DEFECT or of DAMAGE.

POST-CONSTRUCTION

(Fr.: *Post-réception*)
By post-construction we mean the conditions valid during the whole period commencing by the CLIENT's taking over of the BUILDING (or any part of it; this part then becomes post-construction even if the rest of the premises do not).

---

* Do not confuse with DAMAGE (see under 'Aetiological terms', below).

PRODUCER

(Fr.: *Constructeur*)
Collective term encompassing at least the following:

- the architect commissioned★ by the CLIENT;
- any (other) consultant commissioned★ by the CLIENT to design or supply information for the design of the BUILDING or any part of it;
- anybody whom the CLIENT engaged★ to execute works in connection with the production of the BUILDING or any part of it.

National considerations will include or exclude

- any technical control officers comissioned★ by the CLIENT;
- any public authority acting as a consultant or control officer;
- any party not engaged★ or commissioned★ by the CLIENT who participates in design of, works on or technical control of the BUILDING or any part of it (e.g. the UK with its principle of liability in tort; France (after 1979) where certain suppliers are liable to the CLIENT).

Avoid 'engineer', 'builder', 'contractor', 'installer' as these terms are exclusive.

Note that if the CLIENT actively takes part in or intervenes in the design of the BUILDING, the execution of the works or technical control he becomes a 'producer' in what concerns this part of his activity.

SUCCESSIVE OWNER(S)

(Fr.: *Propriétaire(s) successif(s)*)
Anyone having acquired the BUILDING from the CLIENT in a legal manner; anyone having done so from a SUCCESSIVE OWNER.

TENANT

(Fr.: *Locataire*)
Entity or person holding the BUILDING under lease.

★ The words 'engaged' and 'commissioned' need not be explained.

AETIOLOGICAL TERMS

BREACH OF DUTY

(Fr.: *Faute*)
Any ERROR OR OMISSION which gives rise to a DEFECT *and** which
entails LIABILITY.

DAMAGE

(Fr.: *Désordre*)
The material manifestation of a DEFECT.

This definition excludes damage due to action by third parties (sabotage,
vandalism etc.), normal wear and tear and Act of God.**

Concerning losses due to erroneous utilization or maintenance there is
damage in our sense only if instructions for use were lacking or inadequate.[†]

(Do not confuse with 'Damages', see General terms.)

DEFECT

(Fr.: *Défaut*)

(a) The fact that the BUILDING, as actually built, does not conform to
recognized rules (*secundum artem* or *lege artis* or *selon les règles de l'art*),
to qualitative specifications in contract or in normative documents.
(b) The fact that instructions for maintenance and operation are inadequate
or lacking.

ERROR OR OMISSION

(Fr.: *Erreur ou omission*)

(a) Any departure from correct[†] building.
(b) Any absence of adequate[†] instructions as to maintenance and operation
of the BUILDING.

---

* Please refer to question 1.2.6.
** Also sometimes called 'force majeure'.
[†] Thus it is presumed that PRODUCERS know what consitutes correct building and adequate
instructions for operation and maintenance. This may not be the case in practice, so ERRORS
or OMISSIONS may be voluntary or involuntary.

INSURANCE TERMS

CLAIM(S) EVENT

(Fr.: *Sinistre*)
The event triggering the claim against the insurer.

EXCESS

(Fr.: *Franchise*)
The part of the DAMAGES left to be paid by the POLICY HOLDER.

FRANCHISE

(Fr.: *Minimum d'avaire*)
A specified amount under which no sum will be recoverable under the policy.

INSURED PARTY

(Fr.: *Le souscripteur*)
The party paying the premium(s) to the insurer (also sometimes called 'the proposer').

POLICY HOLDER

(Fr.: *Bénéficiaire*)
The party to whom the insurer pays in case of a CLAIMS EVENT. (Not necessarily the same as the INSURED PARTY.)

Index